T4-AUB-474

INTRODUCTION TO MULTICHIP MODULES

INTRODUCTION TO MULTICHIP MODULES

NAVEED SHERWANI
Intel Corporation

QIONG YU
Cadence Design Systems

SANDEEP BADIDA
Advanced Micro Devices

A Wiley-Interscience Publication
JOHN WILEY & SONS, INC.
New York / Chichester / Brisbane / Toronto / Singapore

This text is printed on acid-free paper.

Library of Congress Cataloging in Publication Data:

Sherwani, N. A. (Naveed A.)
Introduction to multichip modules / Naveed Sherwani, Qiong Yu,
Sandeep Badida.
p. cm.
"A Wiley-Interscience publication."
Includes bibliographical references (p.).
ISBN 0-471-11438-3 (alk. paper)
1. Multichip modules (Microelectronics) I. Yu, Qiong.
II. Sandeep, Badida, 1965- . III. Title.
TK7870. 15.S54 1995
621.381' 046-dc20 95-1483

Printed in the United States of America

10 9 8 7 6 5 4 3 2 1

Naveed dedicates this book
to his wife Sabahat

Qiong dedicates this book
to his wife Lihong

Sandeep dedicates this book
to his wife, sister, and parents

CONTENTS

PREFACE

The increasing complexity and density of semiconductor devices is the key driving force behind the development of more advanced very large scale integration (VLSI) packaging and interconnection approaches. The revolutionary growth in integrated circuit (IC) technology, which has gone through the stages of small scale integration, medium scale integration, and VLSI, compels the evolution of electronic packaging. The electronic packaging of chips, in turn, has evolved from printed circuit boards (PCBs) with the through-hole technology of the 1960s and the surface-mount technology the 1980s, to the current multichip module (MCM) technology.

The traditional approach of single chip packages on a PCB has intrinsic limitations in terms of silicon density, system size, and contribution to propagation delay. If a PCB were completely covered with chip carriers the board would have at most a 6 percent efficiency of holding silicon. In other words, 94 percent or more of the board area would be wasted space, unavailable to active silicon and contributing to increased propagation delays.

A higher performance packaging and interconnection approach is necessary to achieve the performance improvements promised by VLSI technologies. The need for high performance system packaging is to be met by the developments taking place in MCMs. MCMs act as a space transformer between the advancing ICs and PCB technology.

MCMs provide a very high level of system integration, with hundreds of bare IC chips that can be placed very close to each other on a substrate. As a result, systems based on MCMs can achieve much denser circuits and much shorter interconnect distances among the bare chips than the systems in which chips are packaged in the traditional single chip module and placed on PCBs. The quantum leap from the single chip modules on PCBs to bare chips on MCMs has resulted in the dramatic improvement in the performance and reliability of the electronic systems as well as the weight and size of the systems. The performance gain is due to the reduction in signal delay as the interconnect distances among the chips is far shorter than the distances between chips on a PCB. The reliability improves because there are fewer packaging levels. The size and weight of the systems are reduced, since the packages of the individual dies are eliminated. MCMs are therefore considered critical for the development of all computing systems, especially high performance systems.

As the bare chips are placed very close to each other, many new problems and challenges arise in the design of MCMs that do not exist in the design of traditional electronic packaging. The heat generated by chips, owing to their close proximity, is concentrated in a small area and has to be removed efficiently to guarantee the reliable operation of the system. In addition, many new electrical phenomena occur as the operating frequency of the system increases. As a result, studies of traditional electronic packaging do not apply to the understanding of MCM technology.

This book is an introduction to this rapidly growing technology. The book not only covers the technical material in a systematic and organized manner but also gives information on the industrial developments. The book covers the classification, design considerations, and CAD tools of MCMs. Our book is written not only for the electrical engineers or the electrical engineering students who want to design the MCMs or MCM based systems but also for the computer scientists and the computer science students who want to develop the CAD tools for MCMs. It will serve as an introductory textbook for students as well as a reference book in the area of MCMs.

The book is organized in ten chapters. The chapters cover all the important aspects of MCMs.

Chapter 1 introduces electronic packaging and its evolution. Traditional packaging technologies, including through-hole and surface-mount technologies, are presented. Then the basic concepts and the major components of MCMs are introduced. The current status of MCMs and their applications are also discussed.

Chapter 2 gives details of the methods and materials used in MCMs. The parameters that characterize an MCM are then discussed. The impact of the different methods and materials on the parameters of MCMs are given. The traditional classification of MCM technologies and the characteristics of each type of MCM technology are presented. Finally, the limitations of the traditional classification method are discussed.

The design of MCMs is a long and complicated process. The overall design flow, in terms of system design, logic design, and physical design, is discussed in Chapter 3. As the MCMs become more complicated, there is an increasing demand for CAD tools to handle highly complex MCM problems. The design automation problems in MCMs are also discussed in Chapter 3.

In Chapter 4 the electrical design of MCMs is discussed. The electrical transmission line model for the MCMs is developed, and the calculation method of signal delay is given. Electrical noise is discussed, and the methods of its control are presented.

MCMs generate a large amount of heat, which must be efficiently removed for normal operations. In Chapter 5 the thermal characteristics and the thermal management of the MCMs are presented. Several industry examples of thermal design are also given.

The substrate testing and yield problems of the MCMs are discussed in Chapter 6. Different die testing, substrate testing, and module testing methods are presented and compared. The design of testability is also discussed.

The reliability of electronic systems is critical in many MCM applications. In Chapter 7 we discuss the reliability issues in MCMs and the method of improving the reliability of an electronic system.

Chapter 8 begins with the overall physical design cycle. Then the problems in each phase of the physical design cycle are formally defined. The basic algorithms for these problems are discussed. The existing algorithms for the problems are then presented.

Although MCM CAD is relatively new, several vendors have started marketing their MCM CAD tools. The majority of these tools are enhanced versions of PCB and/or IC based tools. Chapter 9 gives the details of several commercial MCM products. Complete MCM design systems from major vendors are presented and compared.

Custom built MCMs are expensive and are therefore restricted to applications for which high costs can be justified. In addition, the turn-around time is unacceptably long for many applications. One approach to reduce the cost of MCMs and improve their turn-around time is the programmable MCM which is discussed in Chapter 10. Progammable MCMs are generic substrates that are manufactured in volume. The user can mount the chips and program the substrate to complete the interconnection required. These MCMs have quite unique place and route problems. All issues related to programmable MCMs, including architecture, CAD tools, and applications, are discussed.

This book is an introductory text on a very important technology. MCMs are the subject of many special topics courses in many universities. In addition, MCMs have attracted the attention of many different segments of industry, including thermal management, CAD, electrical simulation, and interconnect analysis groups, and many engineers working on a specific aspect of MCMs who may not have global knowledge of all aspects of the subjects. This text therefore presents material on all aspects of MCMs in a cohesive and comprehensive manner for a wide audience:

1. **Researchers:** This book is written for researchers in early design, physical design, and thermal/electrical design who want to familarize themselves with MCM technology. They will find a comprehensive coverage of MCMs, along with references to more advanced material. The problems at the end of each chapter also serve as possible research problems. MCMs offer many challenges in physical, thermal, and electrical design.

2. **Graduate students:** This book can be used as a text for an advanced topics class, for a research course in MCMs or electronic packaging, or a CAD tool for MCMs. It can be used by both computer science and electrical engineering students. It is self-contained and only assumes some basic concepts in algorithms and electric circuits. All the required material to understand a topic is presented whenever possible. Problems are included at the end of each chapter to help students review the material. Some are research oriented and may be used as research projects.
3. **Designers of MCM based systems:** This is the first attempt to discuss MCMs in a systematic way. The classification of MCMs is discussed. The design considerations of MCMs from electrical, thermal, and testing perspectives are presented. Designers who work on one particular aspect of MCMs can get a good global perspective by reading this text.
4. **CAD engineers:** This book can also serve as a quick reference for a CAD engineer involved in the design and development of CAD tools for MCMs. The problem formulation for the physical design phase and all the existing algorithms for the physical design phase are discussed in detail.

ERRORS AND OMISSIONS

Although we have tried our best, this text may still contain some errors. If you find any errors or have any suggestions, we would appreciate receiving your comments and suggestions. You may send them to the authors via e-mail at sherwani @ichips.intel.com, jyu@cadence.com, or sandeep.badida@amd.com.

An effort has been made to include all pertinent references to papers and books. If we have inadvertently omitted a reference, please feel free to remind us of the omission.

NAVEED A. SHERWANI
QIONG YU
SANDEEP BADIDA

Portland, OR
Chelmsford, MA
Sunyvale, CA
July 1995

ACKNOWLEDGMENTS

Our book is the result of the efforts of many people. We would like to take this opportunity to thank everyone for their contributions. First, we thank all members of the *nitegroup*. In particular we thank Srinivasa Rao Danda, Rameshwar Donakanti, Praveen Jayamohan, and Sreekrishna Madhwapathy. Special thanks are due to Aman Gopal Sureka, who helped us in making the figures and editing the text. His enthusiasm stimulated everyone in the team. He provided the critical central link to manage the project, when the three authors were in three corners of the country. Thanks are also due to Surendra Burman, who helped us start this book and provided us with valuable advice.

Thanks are extended to Western Michigan University and, in particular, Donald Nelson and Douglas Ferraro, who, despite the cost, made the necessary facilities available to complete this book. The National Science Foundation deserves thanks for supporting the VLSI laboratory and the research at Western Michigan University.

We thank secretaries Phyllis Wolf and Sue Moorian at the Computer Science Department, Western Michigan University, for being very helpful during all stages of this project.

We thank the staff at Intel Corp., Cadence Design Systems, and Advanced Micro Devices for their support. In particular, we thank Jim Kawakami, Waqar Shah, Doug Van Emerik, and Jim Schultz at Advanced Micro Devices for their valuable suggestions.

We thank John Wiley & Sons for their continuous support of our work. Thanks are especially due to our editor, George Telecki, for his support.

Finally, we thank our families for their support during this project. Sabahat, deserves our thanks for reviewing many chapters. Lihong and Anupama proofread many parts of the book and helped in editing it. More importantly, their continuous encouragement and understanding helped us complete the book.

NAVEED A. SHERWANI
QIONG YU
SANDEEP BADIDA

CHAPTER 1

EVOLUTION OF PACKAGING TECHNOLOGY

The electronics industry has grown tremendously over the past four decades. The major contribution to this rapid development has been the revolutionary growth in the integrated circuit (IC) technology. To keep pace with the advances in IC technology, electronics packaging and interconnect technologies have evolved from the through-hole technology of the 1960s to the surface-mount technology of the 1980s.

Surface-mount technology eliminated the large through-holes in the packages, resulting in increased silicon density and packaging efficiency. However, the traditional approach of using single chip packages on printed circuit boards (PCBs) has intrinsic limitations in terms of silicon density, system size, and contribution to propagation delay, even if surface-mount technology is used. If a PCB were completely covered with chip carriers, the board would only have at most a 6% efficiency of holding silicon. In other words, 94% or more of the board area would be wasted space, unavailable to active silicon and contributing to increased propagation delays.

With the emergence of VLSI technology, even the surface-mount packaging technology cannot satisfy the increasing demand for higher performance and packaging efficiency. A higher performance packaging and interconnection approach is necessary to achieve the performance improvements made possible by VLSI technologies. The need for high performance system packaging will be met by the developments taking place in the multichip modules (MCMs). MCMs provide a very high level of system integration, with hundreds of bare IC chips that are placed very close to each other on a single high density substrate. As a result, MCM based systems can achieve much denser circuits and much shorter interconnect distances among the bare chips. The development in packaging technology is illustrated in Figure 1.1.

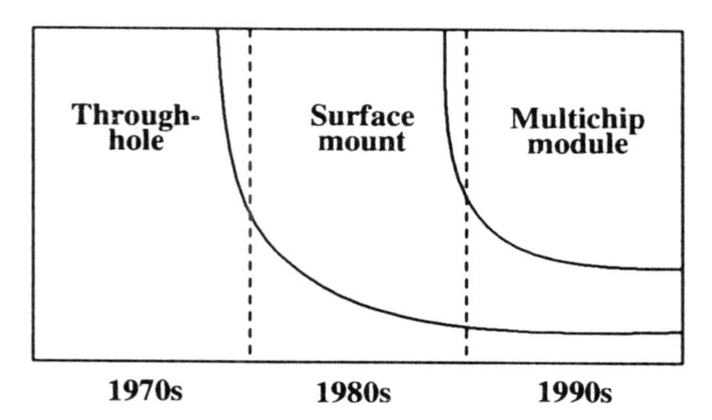

FIG. 1.1. Evolution of electronic packaging.

The quantum leap from packaging single chip modules onto PCBs to packaging bare chips onto MCMs has led to a dramatic improvement in the performance and reliability of an electronic system. In addition, MCMs lead to a reduction in the weight and size of the system. The performance gain is due to the reduction in signal delay as the interconnect distances between the bare chips is far shorter than the distances between the chips on a PCB. The reliability improves because there are fewer package levels. The size and weight are reduced since the packages of the individual dies are eliminated. In addition, MCMs can also remove a large amount of heat and provide a large number of I/Os. Due to these features, MCMs have significantly enhanced the performance and reliability of the endproducts. MCMs are therefore considered critical for the development of all computing systems, especially high performance systems.

Although, MCMs have been in use for some years, especially in high-end mainframes and military products, only recently have they become cost effective to be used widely. MCM technology is becoming the dominant packaging technology of the 1990s. It is one of the fastest growing segments of all the electronic packaging technologies available today. Compared with large application specific integrated circuits (ASICs), MCMs provide more flexibility to mix technologies within a single packaged unit. MCMs can be assembled to contain combinations of digital logic, analog logic, RAM, bipolar, and CMOS dies. This integration is not possible or cost effective on a single chip or wafer.

In this chapter, the basic concepts of electronic packaging are introduced. So the reader can appreciate the benefits of MCM technology and to understand the pattern of the development in packaging technology, the evolution of electronic packaging is presented. Finally, the multichip modules and their applications are discussed.

1.1 SYSTEM, PACKAGING, AND PERFORMANCE

In this section, we discuss first the structure of an electronic system and then the relationship between the performance of the system and its packaging and IC technology. The evolution of the packaging technology is also covered.

1.1.1 Hierarchy in Computer Systems

An IC chip consists of a number of electronic components built by layering several different materials in a well-defined fashion on a silicon base called a wafer. The IC chip has gone through several revolutionary phases. From a single transistor, it went through a series of historical transitions that have been labeled small scale integration (SSI), medium scale integration (MSI), large scale integration (LSI), and very large scale integration (VLSI). An SSI chip usually integrates tens of transistors, an MSI chip integrates hundreds of transistors, an LSI chip integrates thousands of transistors, and a VLSI chip contains tens of thousands of transistors or more. Current fabrication processes can build a VLSI chip containing ten million or more transistors.

Though IC technology has advanced rapidly, a complex electronic system consisting of tens of millions of transistors cannot be fabricated on a single chip. Therefore, several chips have to be used in the system. In many cases, a large system is partitioned into several subcircuits each of which fits into an existing chip. This reduces the cost of the system as well as the design time. In either case, the system consists of several chips. Those chips need to be mounted and interconnected with each other to achieve the required functionality of the system. Conventionally, a PCB is used for mounting and interconnecting the chips. Like chips, the PCBs have the size limitation as well. Several PCBs may be required to build the system. Therefore, complex electronic systems such as supercomputers are built hierarchically. At the bottom of the hierarchy are IC chips that are mounted and interconnected on PCBs. Finally, the PCBs are inserted on a motherboard and interconnected through the system bus.

A chip is packaged into a carrier, called the chip package. The packaged chip, which includes both the bare chip and its package, is referred to as a chip in this book. A bare chip is referred to as a die. A chip package provides the following functions:

1. I/O connections for signals, power, and ground (see Fig. 1.2a)
2. Remove the heat generated by the circuit (see Fig. 1.2b)
3. Support and protect the die from hostile environments (see Fig. 1.2c)

An electronic system can be viewed as consisting of many levels of components, as it is built hierarchically. The bare chips are referred to as the *first level components*. The physical medium that provides the necessary support for the first level components, including the signal paths, power and ground paths, heat

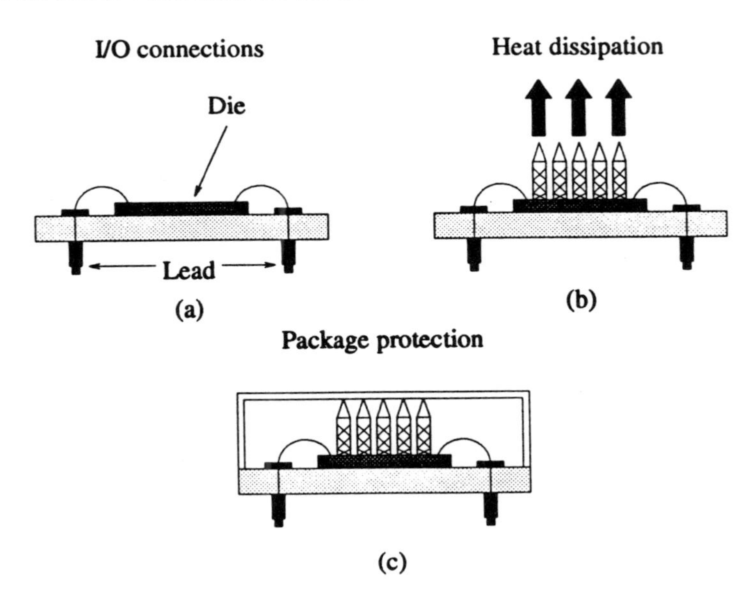

FIG. 1.2. Functions of electronic package: **(a)** I/O connections, **(b)** heat dissipation, **(c)** package protection.

removal facilities, and protection from environment, is called *first level package*. A first level component together with its first level package is called a second *level component*. Therefore, a packaged chip is a second level component in the system. Recursively, we define the second *level package* as the physical medium that provides the necessary support for the second level components. PCBs are examples of second level packages. The second level components and packages, in turn, form the *third level components*, and the physical medium such as the motherboard with the system bus and all the supporting facilities is called the *third level package* for a three-level hierarchical system. Note that the components and packages for more levels can be defined recursively for more complicated systems. Figure 1.3 illustrates the three-level hierarchy of electronic packages. The technology of packages and the study of packages is referred to as *electronic packaging*.

1.1.2 Effects of Packaging and On-Chip Delay on the Performance

The driving force in the electronics industry is the desire to improve the system performance. The performance of a processing system is measured in terms of millions of instructions executed per second (MIPS). In the last two decades, there has been a remarkable increase in performance. In 1972, Intel's 8008 microprocessor was capable of only 0.03 MIPS. The performance was raised to 0.9 MIPS in the 80286 processor by 1982. In 1992, Digital released its Alpha

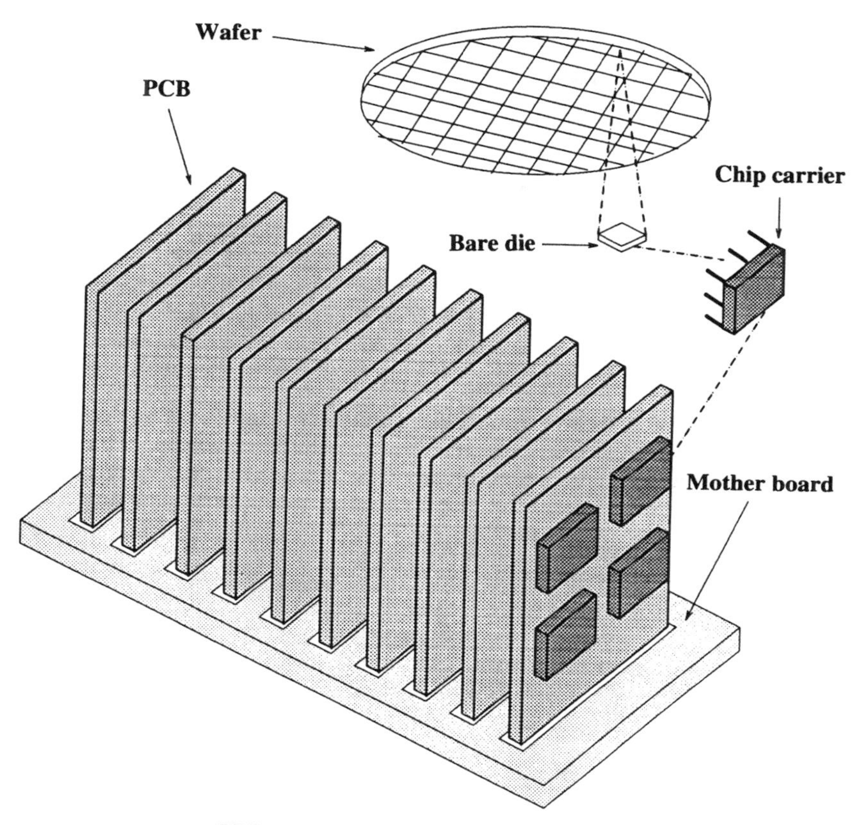

FIG. 1.3. Electronic package hierarchy.

microprocessor, which is capable of over 400 MIPS. In the years to come, it is expected that the MIPS rating will give way to BIPS, that is, billions of instructions per second.

The performance of an electronic system is mainly determined by the number of instructions executed in each cycle and the clock frequency (see Fig. 1.4). As shown in Figure 1.4, the number of instructions executed per cycle depends on the system design, whereas the clock frequency depends on the IC and electronic packaging technologies. It should be further noted that, while the performance is a fundamental parameter, it is affected by the architecture and instruction set. Thus a reduced instruction set computer (RISC) system cannot be compared easily with a complex instruction set computer (CISC) system. Accordingly, performance comparisons are only valid within a given system design.

The performance of an electronic system can be improved in three ways: by increasing the clock frequency, by increasing the number of instructions executed in one clock cycle, and by increasing the efficiency of the operations that are

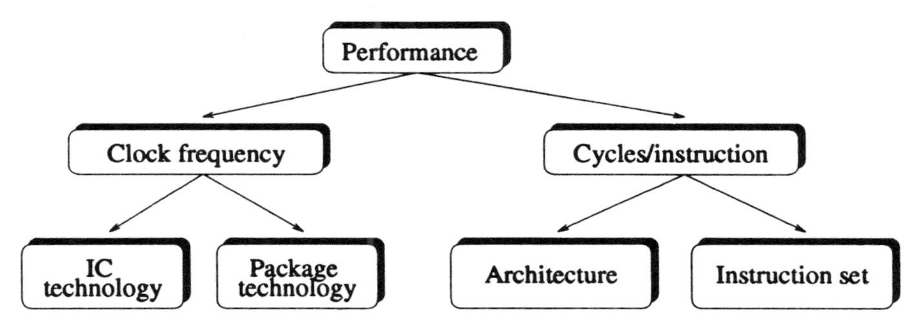

FIG. 1.4. Different factors affecting the system performance.

performed by each instruction. The number of instructions executed in one clock cycle and the efficiency of the operations that are performed by each instruction depends on the system design and architecture, whereas the clock frequency depends on the IC and electronic packaging technologies. Progress in each of these areas has been made. The emphasis on parallel processing and RISC architecture addresses the issue of the number of instructions occurring in one clock cycle and more efficiency of the operations within a single instruction. The evolution of IC and electronic packaging technologies has resulted in higher clock frequencies. Figure 1.4 shows the factors affecting the performance of an electronic system.

The clock frequency can be increased by reducing the signal delay of the critical signals. Due to the hierarchical structure of an electronic system, the signal delay in the system comprises not only the signal delay within each component but also the signal delay in the package. The signal delay within each chip is called the *on-chip delay*. Similarly the signal delay within each board is called the *on-board delay*. The signal delay in a package is called the *package delay*. The package delay could be much larger than the signal delay within a component. For example, the signal carried by a net whose terminals are on different PCBs has to travel from one board to another through the system bus. The system bus is very slow since it is shared by several boards. Two boards that need to communicate must first acquire the bus and then communicate. Other boards must wait until these boards have completed the communication. As a result, the signal delay caused by critical signals traveling between PCBs (off-board delay) plays a major role in determining the system performance since it is much longer than the on-board delay. The signal delay caused by signals traveling between chip carriers (off-chip delay) is longer than the on-chip delay because the distance between two adjacent transistors on a chip is a few microns, while the distance between two adjacent chip carriers is centimeters. In addition, the signal traveling between the chip carriers has to go through the connections between chip carriers and PCBs, which have higher capacitance and inductance than the wiring on the chips and thus contributes to longer delays. Figure 1.5 shows the different kinds of delay in a computer system. In Figure 1.5b, the off-board delay

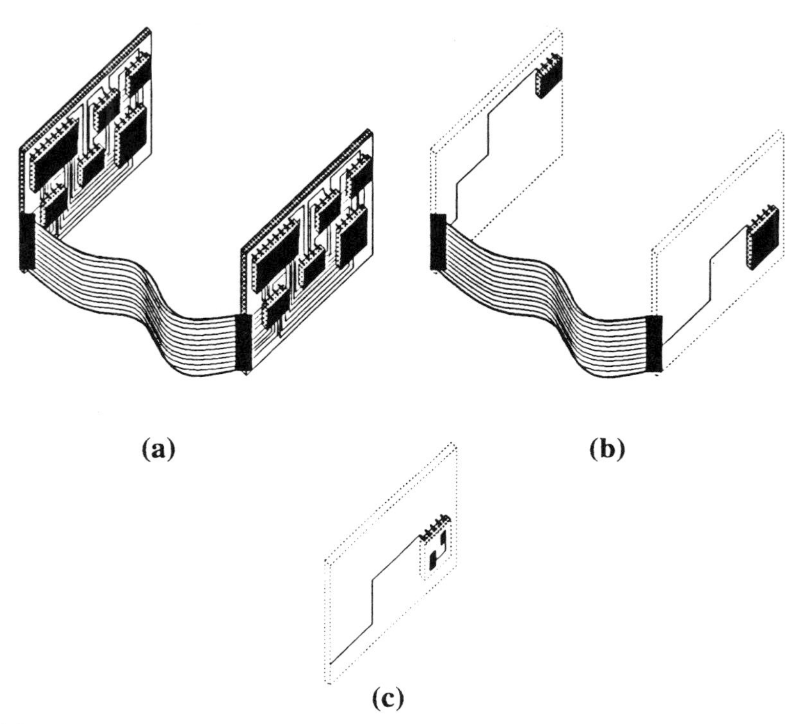

FIG. 1.5. Comparing delays in a computer system: **(a)** off-board delay, **(b)** off-board delay vs. on-board delay, **(c)** off-chip delay vs. on-chip delay.

is compared with the on-board delay, while the off-chip delay is compared with the on-chip delay in Figure 1.5c. To reduce the signal delay, it is necessary to reduce both the on-chip delay and the package delay. The effect of package delay can be reduced by either increasing the number of components within a package level so that the number of packaging levels is reduced or reducing the package delay. In fact, giant strides have been made in both directions. For example, the IC densities on a chip and the chip size have both been increased in the previous decades so that the number of transistors on a chip has increased to about 9 million. On the other hand, the electronic packaging has gone through a significant growth so that the signal delay between different package levels is reduced. Generally, the reduction in the time for signals to travel between the package levels provides the gradual improvement in system performance while a reduction of the package levels by just one can provide a quantum leap in the performance improvement of an electronic system (see Fig. 1.6).

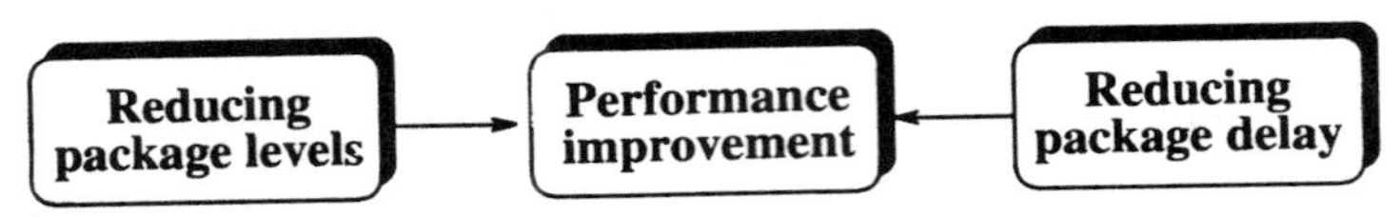

FIG. 1.6. Reducing the package levels or the package delay can improve system performance.

1.1.3 IC Development Drives Packaging Development

The ability to integrate more transistors on a tiny piece of silicon has grown tremendously in the past two decades. This growth can be appreciated by examining Figure 1.7. Correspondingly, so has the clock frequency. The key to the improvement in the clock frequency is the shrinking feature size, i.e., the minimum gate or line width on a device. Feature sizes have been reduced from 4 μm or larger for MSI devices to 2 μm or larger for LSI devices. Currently, the feature size for VLSI or VHSIC (very high speed IC) is 0.7 μm. It is predicted that the feature size for the next generation, ULSI (Ultra Large Scale Integration) or VHSIC-II, will be between 0.3 and 0.1 μm. As the feature size is reduced, the number of transistors on a chip increases dramatically from several dozens on an SSI chip to several millions on a VLSI chip. As the gate size is reduced, the parasitic effects are reduced, resulting in a faster switch time. In addition, the distance between the devices is reduced such that the signal travel time on the chip is shortened. Overall the on-chip delay has been significantly reduced. On the other hand, the number of I/Os (leads) on a chip increases as more transistors are integrated into the chip. The relationship between the number of transistors on a chip and the number of leads on a chip is estimated by an equation known as Rent's rule. Rent's rule states that the number of I/O leads needed in a chip is proportional to the number of transistors in the chip. It can be expressed as

$$C = \left(\frac{N}{K}\right)^n$$

where C is the average number of transistors on the chip; N is the number of leads on the chip; K is a constant that depends on the ability to share signal lines (a typical value of K for high performance applications is 2.5); and n is a constant in the range of 1.5 to 3.0. As shown in Figure 1.8 [137], the number of transistors in a microprocessor closely follows Rent's rule.

As the number of transistors has increased over the years, the clock frequency has increased as well. The clock frequency depends on the on-chip delay and the package delay. Both the on-chip and package delays are affected directly by the interconnection distance. On a chip the interconnection distance is the aluminium or silicon nitride distance. On a chip package it is the wire bound length or the TAB lead length from the chip to the solder joint. On a PCB the interconnection distance is the copper laminate trace distance. In all cases, the

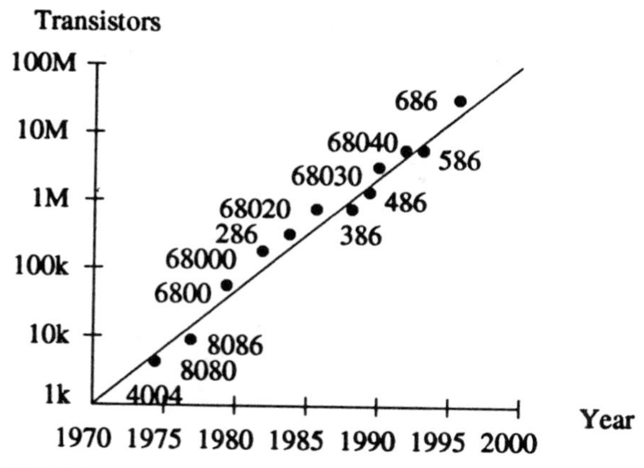

FIG. 1.7. Growth in the number of transistors on a chip.

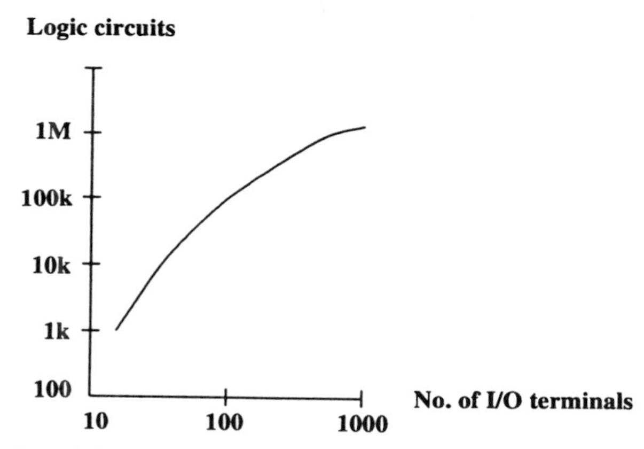

FIG. 1.8. Number of I/O terminals required increases as the number of logic circuits on a chip increases.

delay is reduced when the interconnecting distance decreases. On the other hand, the contribution of the delay of each type depends on the interconnection distance, which is shown in Figure 1.9. Figure 1.9 shows the signal delay as a percentage of the total delay versus the interconnection distance. For example, if the total interconnection distance is 30 mm, then the on-chip delay accounts for approximately 20% of the total signal delay, whereas the chip-package delay contributes 30% and on-board delay is about 50%. As a result, the package delay becomes more significant than the on-chip delay when the interconnection distance decreases due to the technology advances.

As the IC technology advances, the on-chip delay is reduced. The package delay becomes the dominating factor in determining the system performance. At

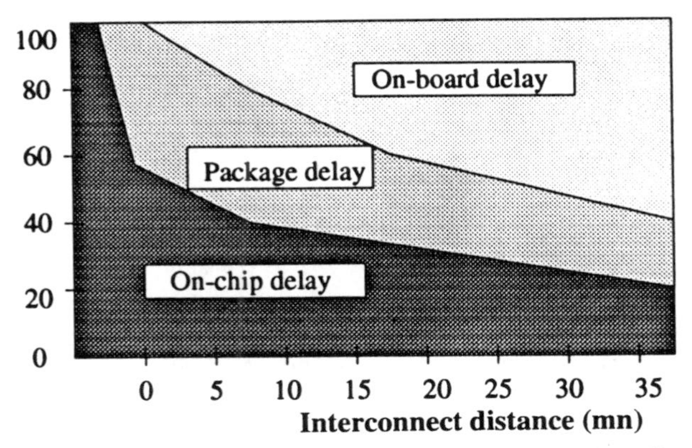

FIG. 1.9. Signal delay as a function of the interconnection distance.

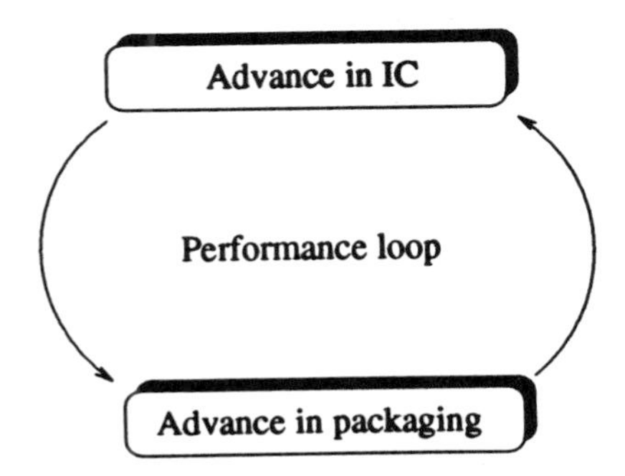

FIG. 1.10. IC advances drive the evolution of packaging.

the same time, the number of leads on the package increases as estimated by Rent's rule. The advance in IC technology drives the evolution of electronic packaging. This is due to the demand on more leads on a package as the number of transistors of a chip increases. In addition, the reduction in on-chip delay alone cannot increase system performance, and the packaging factor plays more of a role in determining the system performance when the on-chip delay is reduced to a level such that the package delay becomes more significant than the on-chip delay. Each time, as the level of integration and the transistor switching speeds increase, the chip interconnection becomes the limiting factor in realizing high speed system performance. As a result, the packaging technology is driven to evolve. But before long the IC technology advances so that the packaging factor becomes a bottleneck again. This loop continues so that the overall performance is improved (see Fig. 1.10). Figure 1.11 shows the evolution in the clock cycle time of IBM mainframes and the contributions of

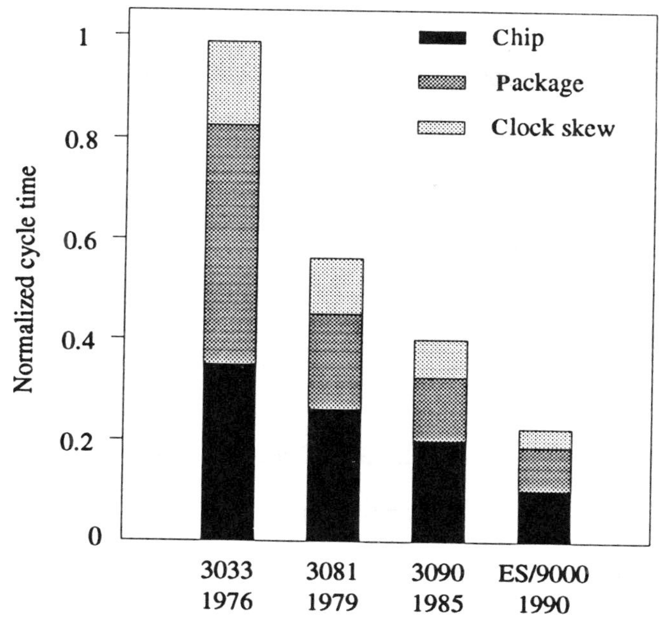

FIG. 1.11. Evolution of mainframe cycle time.

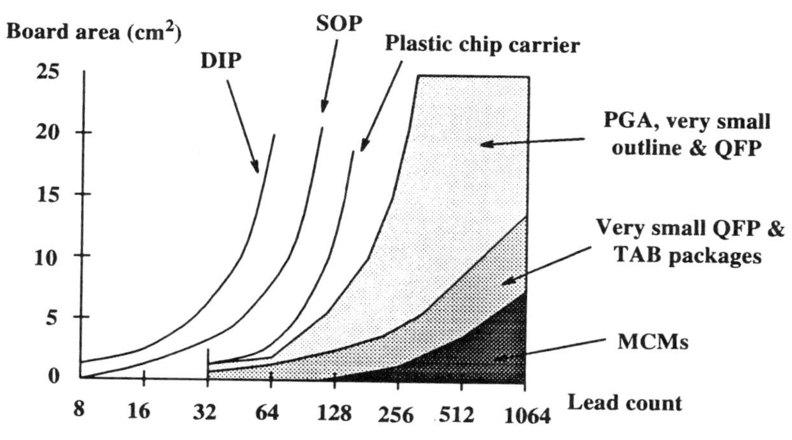

FIG. 1.12. Lead count for different packages

the on-chip delay and package delay in the clock cycle time. It can be seen in Figure 1.11 that both the on-chip and package delays are reduced to increase the clock frequency in each generation of IBM mainframe. Clearly, the system performance gain is obtained by advancing both IC and packaging technologies.

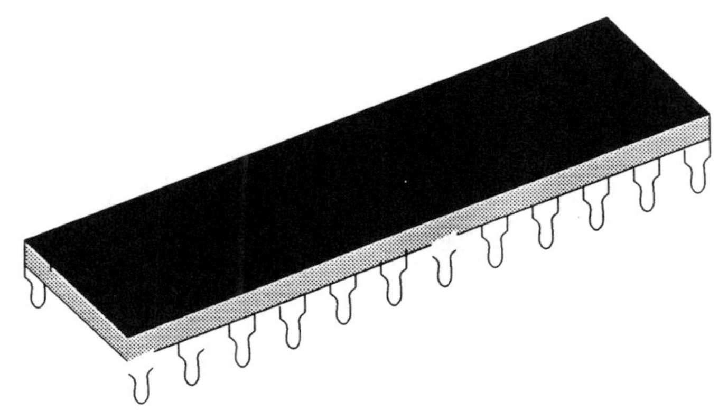

FIG. 1.13. Through-hole DIP.

1.2 TRADITIONAL PACKAGING TECHNOLOGIES

From the 1960s to the mid-1980s, through-hole has been the prevailing packaging technology. Then Surface-mount became the dominant package, offering significant size savings but mixed cost savings. In this section, we discuss several different traditional packaging technologies. We also discuss the limitations of each technology.

1.2.1 Through-Hole Technology

The first semiconductor packages were metal cans, resembling vacuum tubes that were assembled into the PCB substrate just like vacuum tubes, using through-holes. Chip carriers mounted by inserting small legs into plated through-holes in the PCB which are mass soldered, have evolved as ceramic or plastic carriers. Through-hole technology (THT) generally uses packages with leads on 2.5 mm centres. These leads are inserted and soldered into plated holes that have been drilled in the PCB. The soldering is done by transporting the board and package leads over a pot containing molten solder. THT was the prevailing assembly technology from the 1960s through the mid-1980s.

The through-hole chip carriers can be classified as periphery carriers and area carriers. A dual in-line package (DIP) with leads on 100 mil centres on two sides of a rectangular package, is a typical kind of periphery carrier (see Fig. 1.13). Pin grid arrays (PGAs), whose pins are distributed on a grid, are the most widely used area carrier (see Fig. 1.14). The leads of either kind of package are bent down and are soldered into a plated hole that goes through the PCB (see Fig. 1.15). The DIP footprint (required board mounting area) for MSI devices with up to 20 leads is 400 by 100 mil. The footprint was increased to 1,000 (1.0 inch) by 3,000 (3.0 inches) mil for the largest Dips—the 64 pin LSI devices—resulting in a seven-fold increase. PGA packages with up to 400 pins on 100 mil grids are also in production.

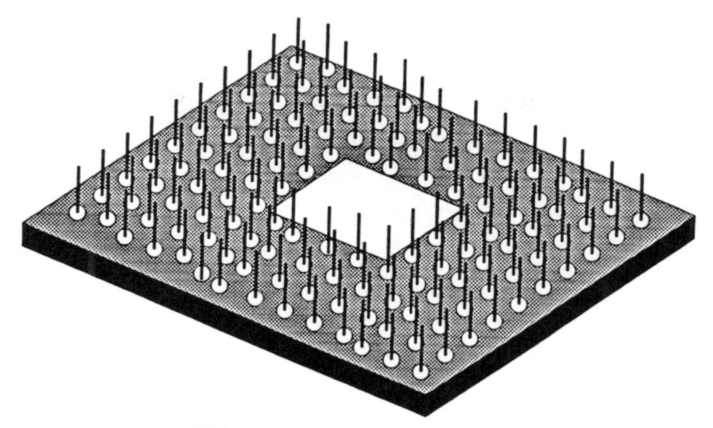

FIG. 1.14. Through-hole PGA.

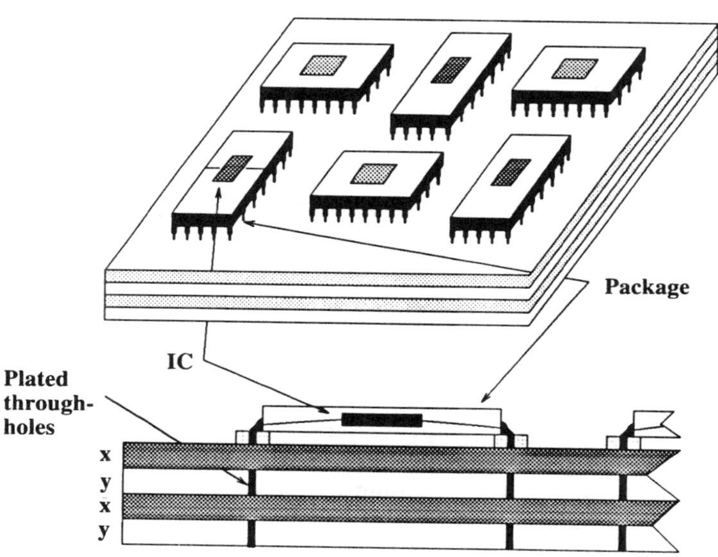

FIG. 1.15. A PCB assembly using through-hole technology.

1.2.2 Limitations of Through-Hole Technology

The increase in density is primarily paced by IC technology. The increasing scale of integration at the chip level increased the number of external connections that the chip package must provide. However, the number of external connections that the chip package can provide is severely limited by the large through-holes in the PCBS. The second major limitation of THT is the package size. The DIP is about 1,500% larger than a bare die. As a result, these packages lead to larger system size degrading performance. The higher performance and higher I/O requirements of the advancing ICs forced a fundamental change in assembly technology. The packaging evolved from wave-soldered, through-hole assembly to a higher density surface-mount technology.

1.2.3 Surface-Mount Technology

Beginning in the late 1970s surface-mounting of some components, such as passive, low-lead-count IC and several through-hole components, became popular. The low availability surface-mount components and their higher cost kept this technology from overtaking through-hole until the mid-1980s. Surface-mount IC packages have leads on centers ranging from 0.8 to 1.3 mm.

Surface-mount assembly technology consists of applying a solder-bearing paste to the board, placing the package leads into the paste, and forming the solder joints by reflowing the solder in an oven. Chip carriers no longer have wire-like leads that are soldered into holes in PCBs with minimum spaces of 100 mil. Surface-mount assembly achieves higher interconnection densities by reducing the separation between leads to 25 mil and by eliminating space-consuming drill holes. The first revolutionary change in the packaging approach was the trend toward surface-mounted packages, with the introduction of chip carriers having peripheral leads on all four sides. These chip carriers are designed for butt solder termination to solder pads on the surface of the PCB. A chip carrier with a 25 mil pitch and about 300 pins has a smaller footprint than a through-hole package and puts less demand on the density of the PCB. With surface-mount technology the holes are only there for communication from the surface into the circuit planes and thus the size of the holes can be decreased. The leads are attached to the surface rather than inside the holes (see Fig. 1.16). If, for example, the holes could be as small as 0.20 mm (0.008 inch), with a decrease in the spacing between lines and holes to 0.10 mm (0.004 inch), a design could be supported with three lines in 1.25 mm (0.050 inch) or six lines in 2.5 mm (0.100 inch). Thus the density increase could be about 50% in proceeding from pin-in-hole technology to surface-mount technology (six lines vs. four lines per channel of 2.5 mm [100 mil]). An extension of SMT is the fine-pitch surface-mount technology (FPT), which has a smaller distance between leads.

Most of the products using IC chips – computers, peripherals, telecommunications – can be reduced in size using SMT. Although there are many other factors that play important roles in determining the size, SMT boards still contribute significantly to size reduction. Because no through-holes penetrate the board, components may be placed close together. This has allowed a reduction in component size and hence a reduction in the circuit board size.

1.2.4 Advantages of Surface-Mount Technology

SMT offers many benefits over conventional through-hole technology.

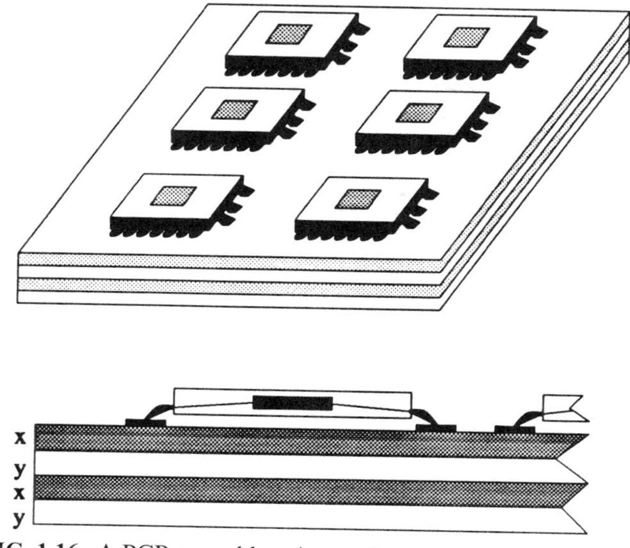

FIG. 1.16. A PCB assembly using surface-mount technology.

1.2.4.1 Size The most obvious benefits of SMT packaging are the smaller package size, the higher number of leads, and the resulting increased packaging efficiency. SMT packages offer a 12-fold savings over the conventional THT devices, with the exception of the PGA package, which offers a two- to fivefold savings. The relative area comparison is illustrated in Figure 1.12. Surface-mount package lead capacity currently ranges from 40 to 1,256 on lead centers ranging from 0.65 to 0.15 mm.

Leaded and leadless chip carriers are available in quad packages with leads on 50, 40, or 25 mil centers. This reduces package footprints, decreases chip-to-chip distances, and permits higher pin count ICs. A 64 pin leadless chip carrier requires only a 0.5 by 0.5 inch footprint with a 25 mil pitch. This represents a 12-fold density improvement and a fourfold reduction in interconnection distances over DIP assemblies. Again, the chip carriers can be classified as periphery or area ones.

Smaller package area resulting in denser circuits is only one size saving benefit. Surface packaging also offers size advantages to the IC chip itself. For many high volume ICs, reducing the size of the IC chip translates into a competitive advantage. Smaller chips generate more die per silicon wafer and higher sales dollars per wafer. Table 1.1 shows the impact that smaller IC chip size has on the number of good chips and sales dollars per wafer.

1.2.4.2 Cost A smaller package also results in smaller ICs, which increases the yield per wafer and thus reduces the cost of the ICs. Table 1.2 shows the estimated cost for manufacturing a small computer board. This example shows

TABLE 1.1. Smaller Size ICs Result in Cost Savings

Pad Size/ Pad Pitch (in/in)	Chip Size (in × in)	Good Chips Per Wafer	Wafer Cost ($)	Chip Cost ($)	Savings (%)
0.01/0.006	0.23 × 0.23	509	500	0.98	–
0.008/0.005	0.18 × 0.18	725	500	0.69	30
0.006/0.004	0.15 × 0.15	1104	500	0.45	54

TABLE 1.2. Estimated Cost for Manufacturing a Small Computer Board

		Total Costs ($)	
Qty	Elements	SMT	THT
1	80386SX	103.00	198.00
2	27128	9.00	7.00
2	27C512-250	8.90	8.20
9	74S74	1.80	1.80
10	74LS00	1.00	0.90
10	1N4148	0.40	0.40
40	0.1mfCAPS	2.44	2.52
20	1%RES	0.34	0.22
total cost		126.88	219.04

that a significant area reduction is possible by using SMT. In this example, there is a savings in the board area. The board area is 19 square inches for the SMT and 36 square inches for the THT. This represents a 47% area savings and a 16% board cost savings.

1.2.4.3 Performance The reduction in the peripheral area of the surface-mount package and the distance between the surface-mount packages increase the speed and signal integrity compared with the larger through-hole packaging. Figure 1.17 demonstrates the impact on a signal as it traverses the diagonal length of a package. In this example, a wire is bonded to the two faethest leads of the packages shown in Figure 1.18, and then a step function voltage signal is applied to one lead and the output is monitored on the other end. The output wave form and the time required to settle to within 99% of the final amplitude is recorded.

Table 1.3 illustrates the advantages of surface-mounting over the THT.

1.2.5 Need for New Packaging Technology

For the more complex VLSI devices, with 120–196 I/Os, even the surface-mounted approach becomes inefficient and begins to limit the system perfor-

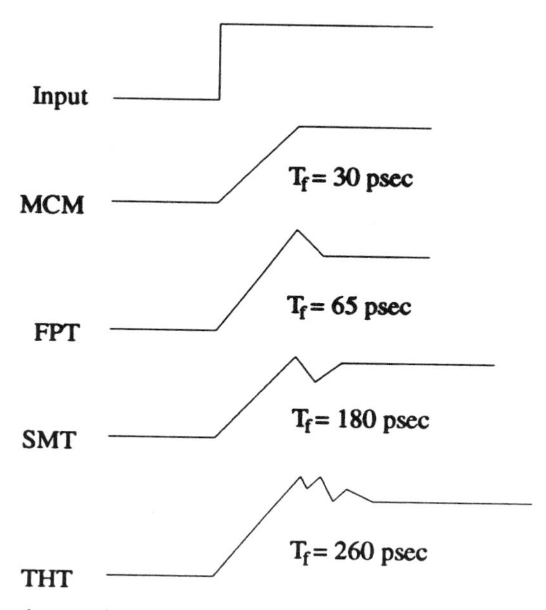

FIG. 1.17. The shorter the interconnect length, the cleaner the signal propagation.

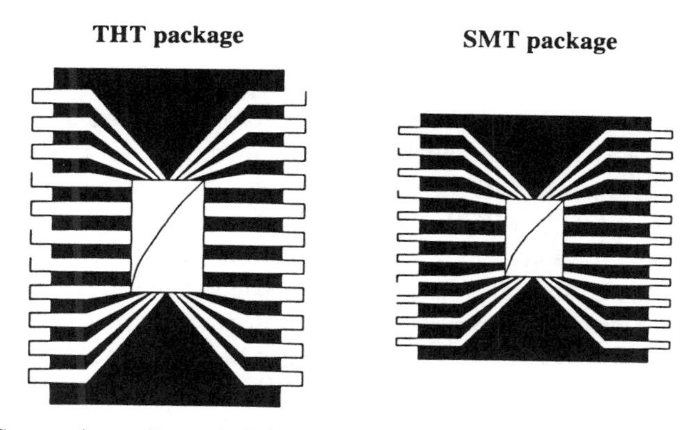

FIG. 1.18. Comparison of a typical through-hole package and a surface-mount package.

mance. A 132 pin device in a 25 mil pitch carrier requires a 1,000–1,500 mil square footprint. This represents a four- to six-fold density loss and a two times increase in interconnect distances compared with a 64 pin device. The interconnect density for current packaging technology is at least one order of magnitude lower than the interconnect density at the chip level. In the last 10 years, integrated circuits have advanced rapidly in terms of the number of I/O pads, the switching edge rates, and in-power dissipation. But the impact on system performance has not

TABLE 1.3. Comparisons Among FPT, SMT, and THT

	THT DIP 40L	SMT PCC 44L	FPT TapePak 40L
Lead thickness (mils)	10.0	10.0	2.8
Lead C-C (mils)	100.0	50.0	20.0
Pkg. length (mils)	2050.0	650.0	350.0
Pkg. width (mils)	600.0	650.0	350.0
Pkg. thickness	175.0	190.0	72.0
Volume ratio	24.4	9.1	1.0
Lead length (inches)	0.99	0.20	0.12
Resistance (mOhm)	125.0	98.0	3.6
Inductance (nH)	22.0	4.6	2.1
Capacitance (pF)	0.68	0.12	0.04

been fully utilized due to the lack in progress in the packaging and interconnect technologies.

The traditional approach of single chip packages on a PCB have intrinsic limitations in terms of silicon density, system size, and the contribution to the propagation delay. The IC packaging efficiency, is given in as follows.

$$\varepsilon = \frac{S}{P} = \left(\frac{c}{p}\right)^2$$

where S is the active silicon area, P is the total package area, c is the chip pitch, and p is the package pitch. For example, the typical inner lead bond pitch on VLSI chips is 6 mil. The finest pitch for a leaded chip carrier is 25 mil. The ratio of the area of the silicon inside the package to the package area given by the above equation is about 6%. If a PCB were completely covered with chip carriers, the board would only have at most a 6% efficiency of holding silicon. In other words, 94% or more of the board area would be wasted space, unavailable to active silicon and contributing to increased propagation delays.

With the introduction of VLSI chips, even SMTs became inefficient for high performance system design. Chip packaging becomes the bottleneck to further improvement in system performance. Extending semiconductor techniques is one approach to filling the interconnection density gap. However, with the continued rapid growth in IC technology, the gradual change in the package delay even by introducing the FPT cannot meet the increasing demand. A quantum leap requiring a reduction in package levels is necessary to meet the increasing demands of IC technology.

From a packaging point of view, fabrication of an entire system on a single wafer should be ideal. The idea of integrating a complete system on a wafer is referred to as wafer scale integration (WSI). WSI can provide the smallest size for the complete system and the shortest interconnect distance for the best possible performance. However, it is plagued by several limitations:

1. **Yield:** The yield of a large wafer is very low, necessitating redundancy schemes. This is perhaps the largest hurdle to the widespread use of WSI.
2. **Mix of technology:** A system consists of different components, such as microprocessor and memory. Different fabrication processes have been optimized to produce cost effective solutions for each individual component. For example, a process optimized for microprocessors may not be optimal for memories. WSI requires that all components be produced on the same process, yielding a suboptimal and expensive solution.

Due to these limitations, WSI cannot be considered a viable packaging technology for high performance systems. The packaging technology that is required for high performance systems should include the following features:

1. Better performance, smaller size, and less weight than existing technologies
2. Allow the mixing of chips produced on different fabrication lines
3. Reduce the number of levels of packaging
4. Provide good thermal management
5. Reduce total system cost

Such requirements can be met by multichip modules.

1.3 WHAT ARE MULTICHIP MODULES?

The multichip module is a new packaging approach in which multiple bare chips are mounted and interconnected on a substrate. Since the substrate, or the chip carrier, has substantially finer conductor lines, smaller dielectric thickness, and denser via grid than the board, they are not subject to the conventional PCB design rules and assembly restrictions. Correspondingly, a chip carrier that has only one chip in it is called a *single chip module* (SCM). Figure 1.19 shows the diagram of a typical multichip module.

The substrate is used for supporting the dies as well as the media for the interconnections among the dies. The interconnections among the dies are made using metal wires that are laid out on layers called *signal planes* in the substrate. Typically, different layers are used for making horizontal and vertical connections. Additional layers are provided for power and ground connections. The bare dies are attached to the substrate by using a method called *chip bonding*. The chip bonding technique determines how the electrical connections are made between the dies and the substrate. The electrical connections between the dies and the substrate are called *chip I/Os*. Like an SCM, an MCM is packaged to protect it from the hostile environment. In addition, some cooling methods must be used to remove the heat generated by the dies. Electrical connections between the MCM and the outside world such as a PCB are needed to transfer the signals on and off the MCM. Such electrical connections are called *module I/Os*.

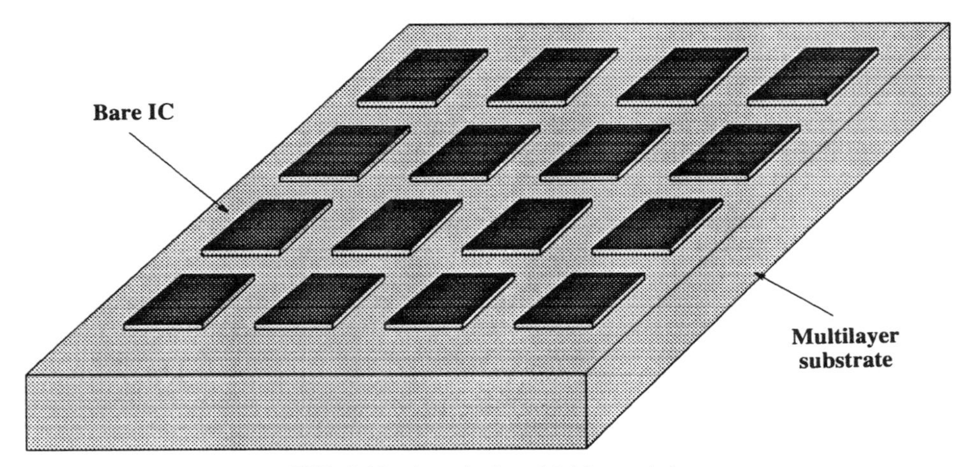

FIG. 1.19. A typical multichip module.

Far from being passive carriers for microelectronic devices, the packages in today's high performance computers pose at least as many engineering challenges as the semiconductor chips that they interconnect, power, cool, and protect. While the semiconductor circuit performance as measured in picoseconds continues to be improved upon relentlessly, the computer performance is expected to be in nanoseconds for the rest of this century. The factor of 1,000 is attributable to the packaging. Thus the packaging, which interconnects all the chips to form a particular function, such as a central processor, is likely to set the limits on how far computers can evolve. Multichip packaging, which minimizes these limits drastically, is expected to be the basis of all advanced computers in the future. In addition, since this technology allows chips to be placed closer and to occupy less space, it has the added advantages of being used in portable consumer electronics and medical, aerospace, automotive, and telecommunication applications.

Multichip modules have been developing rapidly. As a result, different types of MCMs have been developed. In the following section, we discuss the different types of MCMs.

1.3.1 Substrate Technologies

MCMs are generally categorized on the basis of the substrate technology. While some MCMs are based on older PCB technology, others use state-of-the-art VLSI technology. In this section, we briefly review MCM substrate technologies. The details of the substrate and interconnection technologies are discussed in Chapter 2. If a substrate can be programmed by the user, it is referred to as a programmable MCM (PMCM). We will not discuss PMCMs here; they are treated in detail in Chapter 10.

MCMs are usually categorized by substrate type and have been defined by the Institute for Interconnecting and Packaging Electronic Circuits (IPC):

1. **MCM-L:** Substrates based on laminated, multilayer PCB technology
2. **MCM-C:** Substrates based on co-fired ceramic or glass-ceramic technology
3. **MCM-D:** Interconnection pattern formed by the deposition of dielectrics and conductors, on a base substrate, typically by thin film process

MCM-L is the oldest available MCM technology. MCM-L is essentially an advanced PCB on which bare IC chips are mounted using chip-on-board (COB) technology. The well established PCB infrastructure can be used to produce MCM-L modules at low cost, making them an attractive electronic packaging alternative for many low-end MCM applications with low interconnect densities. MCM-L becomes less cost effective at higher densities when many additional layers are required. For cost effectiveness, MCM technology must increase the functionality of each layer instead of adding more layers. MCM-L is considered a suitable technology for applications that require a low risk packaging approach and most of the steps have already been automated.

MCM-C (ceramic) refers to MCMs with substrates fabricated with co-fired ceramic or glass-ceramic techniques. These have been used for many years, and MCM-C has been the primary packaging choice in many advanced applications requiring both performance and reliability. Ceramic substrates, having excellent thermal conductivity and low thermal expansion, have also been used to serve as the package. While interconnect densities are in the range of 200–400 cm/cm^2, this is still not enough for high-end applications.

MCM-D (deposited) technology is closest to the IC technology. It consists of substrates that have alternating deposited layers of high density thin film metals and low dielectric materials such as poly or silicon dioxide. MCM-D technology is an extension of conventional IC technology and was developed specifically for high performance applications demanding superior electrical performance and high interconnect density. This technology, being relatively new, has neither a cost effective manufacturing infrastructure nor a high volume application that serves as a driving force. Table 1.4 compares the MCM families in terms of line widths, line density, line separation, turn-around time, and the number of years for which these technologies have been available.

Variations of technologies mentioned above have also been considered. For example, a variation of MCM-L using plastic packages, involving molding compounds and lead frames, the so-called multichip plastic quad packs (MCM-P), containing four chips, is also available. There is another variation based on deposition of thin films (MCM-D) on MCM-C. The wiring in this type of multichip is shared in co-fired ceramic thick film and polymer–metal thin film, offering the best of both technologies in cost and function. It is referred to as MCM-CD or MCM-DC. The details of the substrate and interconnect technologies are presented in Chapter 2.

TABLE 1.4. Multichip Module Classifications

Characteristics	MCM-L	MCM-C	MCM-D
Line density (cm/cm^2)	250–400	200–400	>400
Line width/separation (μm)	750/2,250	125/125–375	10/10–30
Turn around time	9–13 weeks	1 month	10–25 days
Years of availability	50	>10	>5

1.3.2 Chip Bonding Techniques

Irrespective of the type of the MCM technology used, bare chips are to be attached to the substrates. Bare chips are attached to the MCM substrates in three ways:, wire bonding, tape automated bonding (TAB), and *flip-chip* bonding.

In wire bonding (shown in Fig. 1.20a), the rear side of a chip (the nondevice side) is attached to the substrate, and the electrical connections are made by attaching very small wires from the I/O pads on the device side of the chip to the appropriate point on the substrate. The wires are attached to the chip by thermal compression. TAB is a relatively new method of attaching chips to a substrate. It uses a thin polymer tape containing metallic circuitry. The connection pattern is simply etched on a polymer tape. As shown in Figure 1.20b, the actual path is simply a set of connections from inner leads to outer leads. The inner leads are positioned on the I/O pads of the chips, while the outer leads are positioned on the connection points on the substrate. The tape is placed on top of the chip and the substrate and pressed. The metallic material on the tape is deposited on the chip and the substrate to make the desired connections. Flip-chip bonding uses small solder balls on the I/O pads of the chip to both physically attach the chip and make required electrical connections (see Fig. 1.20c). This is also called *face down bonding or controlled-collapse chip connections* (C4).

1.3.3 Development in MCMs

COB and hybrid circuits are the earliest forms of MCMs. Subsystems based on ICs wire bonded directly to PCBs, have been used extensively since the early 1970s. The packaging efficiencies of these early systems were quite low. Systems built with this process were said to be based on COB technology. In conventional COB packaging, the bare dies were wire bonded directly to a substrate (PCB or flexible tape) and protected with a glob top encapsulant. This approach was mainly used in low cost, low density applications such as watches and calculators. This approach has also been used for years in applications requiring maximum space conservation. Intel, for example, used COB technology to create an IBM XT-compatible personal computer on board slightly larger than a credit card. The microprocessor was then mounted directly on the board [58].

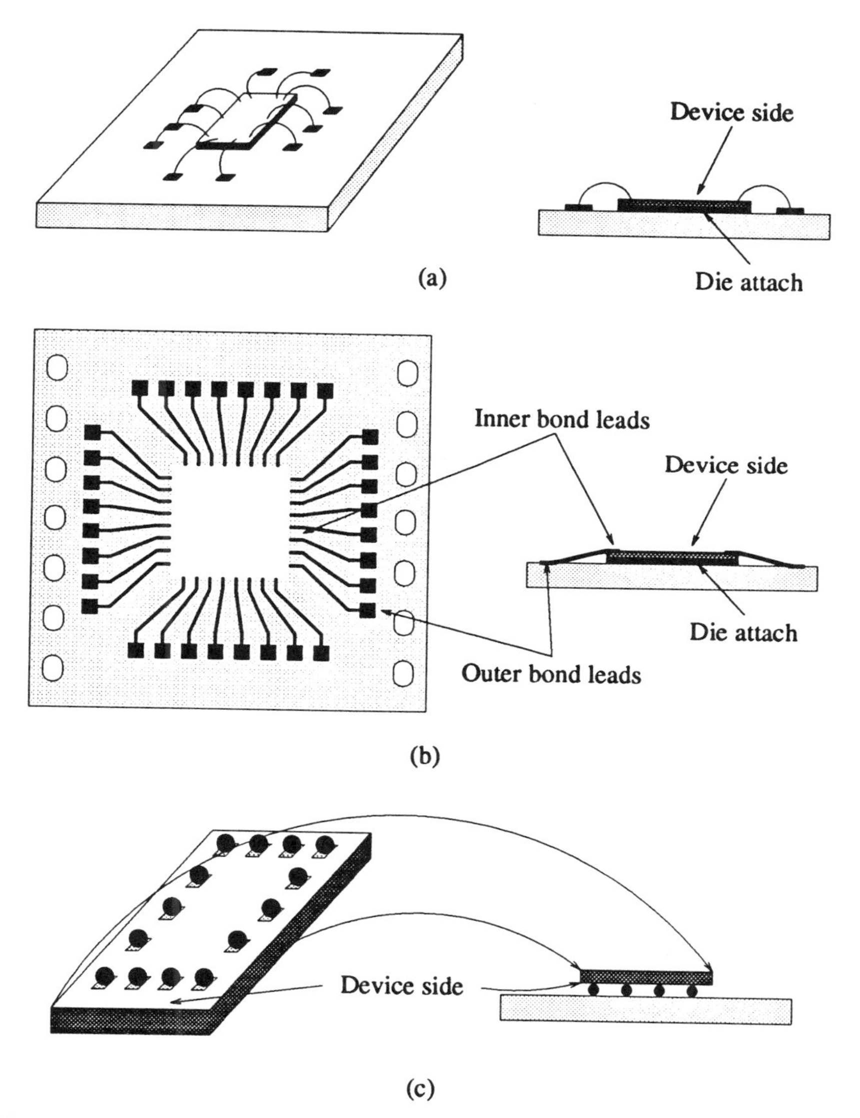

FIG. 1.20. Die attachment techniques: **(a)** wire bonding, **(b)** tape automated bonding, **(c)** flip-chip bonding.

MCM-Cs have been used in the electronics industry for over 20 years. Ceramic hybrid circuits have been used in automobiles to cope with harsh environmental conditions. Plastic packages provide insufficient protection in this hot hydrocarbon-filled environment, and hermetic hybrid packages provide good protection at a reasonable cost. Thick film technology was originally developed to interface

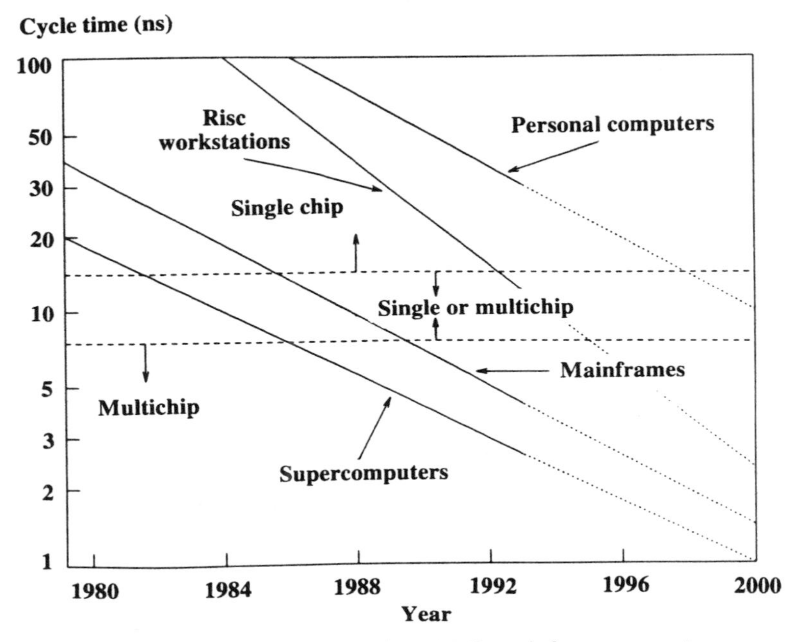

FIG. 1.21. Multichip module in IBM's mainframe computer.

and interconnect bare ICs on a hybrid circuit. MCM-C is a sophisticated extension of a simple hybrid circuit in which bare chips mounted on a common substrate are formed by using screen printing and co-firing processes. The thick film multilayer (TFM) hybrid circuits may be the oldest type of MCM-C. In summary, high density MCMs owe much of their heritage to three diverse packaging technologies: hybrid microcircuits (assembly and packaging), integrated circuit semiconductors (micrometer-dimension photolithography), and PCBs (multilayer constructions). All three technologies blossomed in the 1960s and have grown tremendously over the past decades. IBM pioneered the use of MCMs in its mainframe computers like IBM 3081 (see Fig. 1.21). The pioneering work on using a silicon substrate to build high density MCMs also emerged in the 1960s at PhilcoFord [118]. Further refinements in thick film ceramic constructions were made in the 1970s such as the IBM designs [24, 39]. Significant developmental work was carried out during the 1980s on MCMs using organic dielectrics (such as polyimide between multilayer thin film conductor layers) in pioneering development projects at Honeywell [100, 106, 107, 190], General Electric [32, 148–150], AT&T [11, 12, 92, 131], and Rockwell [64, 82–84].

The past few years have seen enormous activity in multichip packaging. The key factors that led to the development of MCMs are

1. Development of algorithms and CAD tools specifically designed for MCMs [23, 105, 112, 140]

2. Development of multilayer thin film MCM substrate insulator materials with lower dielectric constants, lower moisture absorption, and lower processing cost photosensitive formulations [29, 144, 174]
3. Introduction of "transparent" MCM products (two to eight dies in a package identical to SCMs) [25, 95, 188]
4. Development of multiple vendors offering fabrication of MCM substrates or complete merchant assembly and testing of MCMs [99, 117, 169, 192, 205]
5. Development of multiple vendors offering packages specially tailored for MCMs [138, 198]
6. Increasing availability of known-good dies (required for high yield assembly) in formats suitable for assembly in MCMs [86]
7. Substantial government investments in technologies and infrastructures needed to mature a domestic MCM industry
8. Substantial growth of MCM support technologies and vendors for substrate testing, sockets, connectors, manufacturing equipment and so forth [57, 62, 76, 109, 158]

1.3.4 Advantages of MCMs Over Traditional Packaging Approaches

MCMs completely eliminate chip packages and offer a cost effective solution to achieve high system performance. In this section, we discuss some of the advantages of MCMs over traditional packaging technologies.

The first benefit of the MCM approach over the traditional packaging approaches is the increased IC packaging efficiency ε. The comparison of IC packaging efficiency among all the packaging approaches is shown in Figure 1.22. High density MCM substrates allow closer chip-to-chip spacings and their improved high-frequency performance. Higher density packaging obviously also allows significant miniaturization. For example, a PCB of 16×16 cm can be replaced by a 4.4×4.4 cm MCM, resulting in area reduction by 13 times. Another advantage of MCMs lies in the area of inherent reliability improvement. An entire level of packaging is eliminated in chip-to-chip connections, as shown in Figure 1.23. This results in a dramatic reduction in the number of fatigue-prone solder joints, hermetic seals, and, in general, the total number of parts prone to failure. Improved reliability comes from the use of high thermal conductivity packaging materials, which improve heat removal paths and allow the MCMs to run with lower junction temperatures. Finally, the miniaturization is accompanied by significant weight reduction, which translates directly into lower G-load forces during shock or vibration. A summary of the advantages of MCM packaging compared with other packaging technologies is given in Table 1.5.

The performance trend of computers from personal computers to supercomputers is illustrated in Figure 1.24, showing the cycle time over the last couple

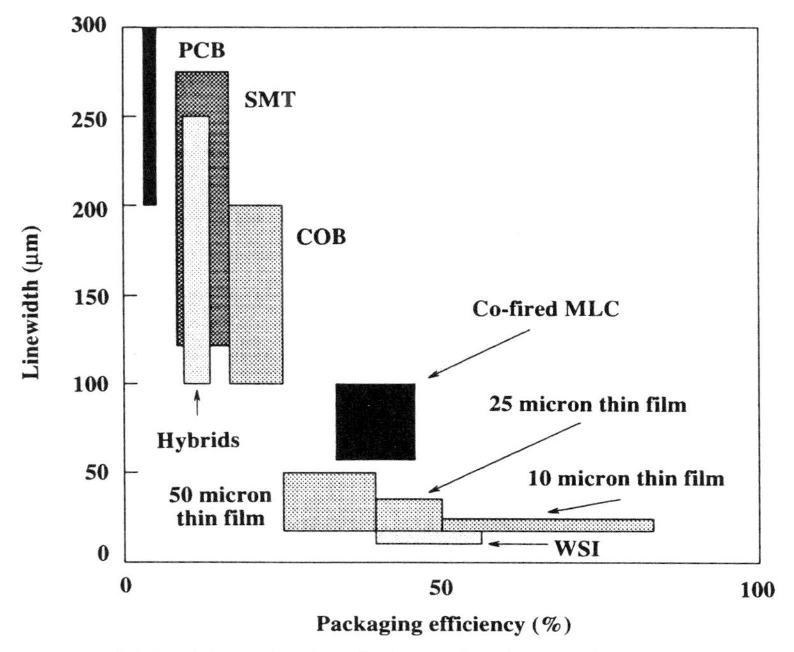

FIG. 1.22. Packaging efficiency of various packages.

of decades and projected to the year 2000. Since the contributions of packaging and semiconductors have been approximately 50% each, a system, with a cycle time below 10 ns (5 ns semiconductor and 5 ns packaging) can only be achieved by using MCMs. Currently all mainframe and supercomputer systems operating at these performances use MCMs. The multichip package in the latest IBM ES9000 system provides only about 2.5 ns delay out of about 8–9 ns cycle time. This outstanding multichip performance (20% cycle time) is due to the 121 chip packaging efficiency provided by a substrate only 127.5 mm in size. The multichip substrate is based on 63 layers of glass-ceramic of dielectric constant 5.0, providing a wiring density of 844 cm/cm^2, removing 27 W per chip and provding 648 interconnections to each of the 121 chips [202]. It also provides 2,772 signal and power pins at the bottom of the substrate.

MCM will not make PCB obsolete, but it will act as a space transformer between the PCB technology and the advancing IC technology. The ability of MCMs to integrate a mix of technologies will certainly give it an edge over wafer scale integration which is yet to become a feasible technology. Further improvement in performance is possible through the use of superconductor and optical interconnections.

In sum, the advantages of MCMs include smaller size, less weight, better performance, and improved reliability. Due to these advantages, MCMs are finding many applications, as discussed in the following section.

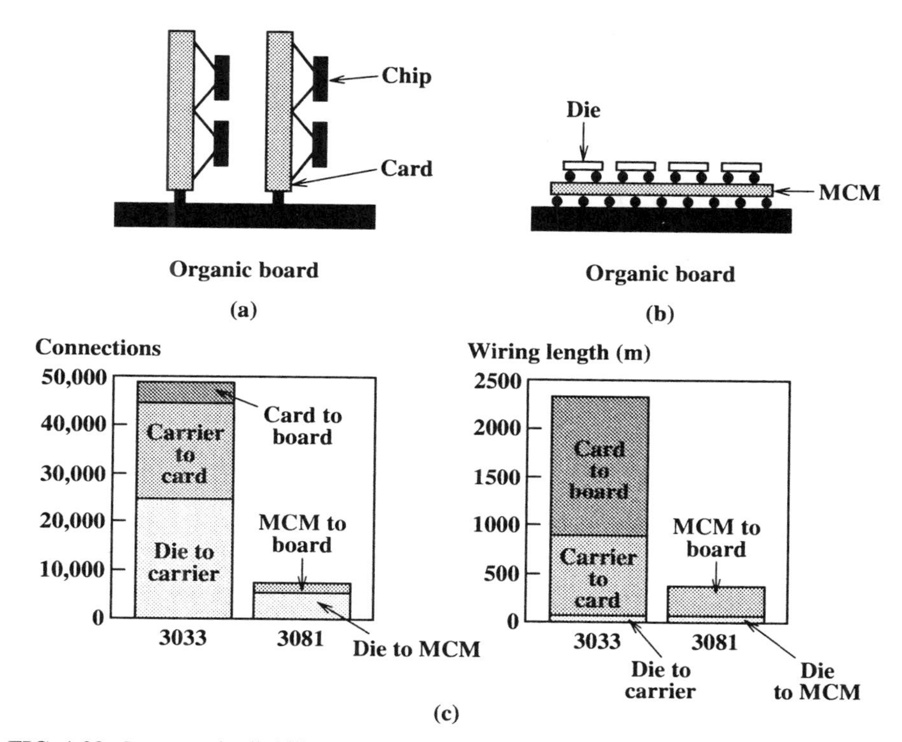

FIG. 1.23. Improved reliability accompanies reduced numbers of interconnects in multichip modules.

TABLE 1.5. Advantages of Multichip Packaging

	Compared with		
PCB	Through-hole	Surface Mount	Fine Pitch
Size reduction	10×	4×	2×
Weight reduction	6×	3×	1.5×
Signal delay reduction	3×	2×	1.5×
Junction temperature reduction	20 °C	20 °C	20 °C
Reliability improvement	4×	5×	4×
Building cost	Competitive	Competitive	Competitive
Life cycle cost	Better	Better	Better

1.4 APPLICATIONS OF MCMs

The advanced multichip packaging technology has significant advantages in electrical performance, thermal performance, size, weight, cost, and reliability. Therefore, the application spectrum of MCMs includes high performance

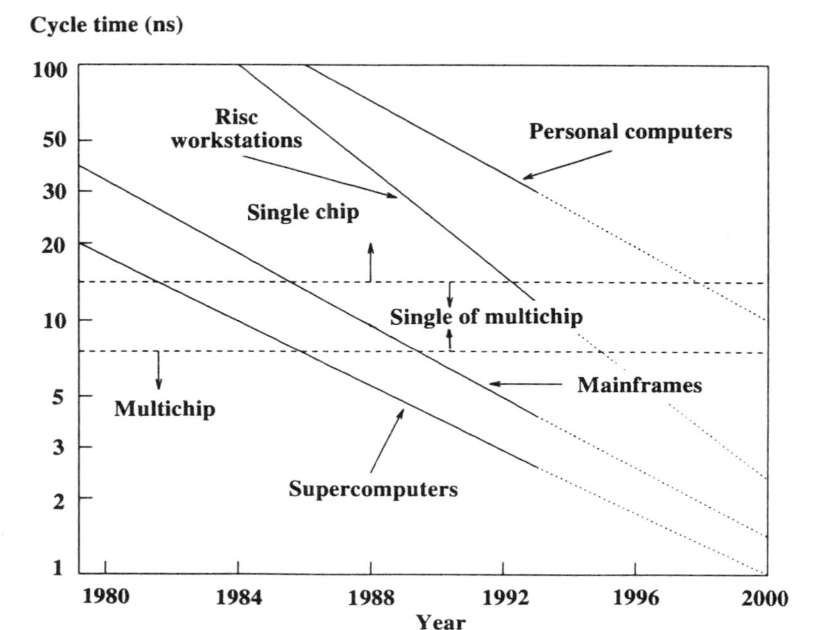

FIG. 1.24. Performance trend in computers.

computers, military, automotive, telecommunication, and space applications, avionics, and low cost consumer electronics.

1.4.1 Computers

The most obvious benefit of MCM technology is the performance gain. In fact, all the current supercomputers and mainframe computers with cycle times below 10 ns use the MCM packaging technology [201]. The alumina-Mo multichip technology was the first widely used MCM technology in computers. It has been enhanced significantly in wiring density, power, cooling, chip level connections, and size since it was introduced in IBM System 3080 in the early 1989s. The latest and most advanced computer using alumina-Mo is IBM System ES9000, which has 63 layers of ceramic and two layers of surface thinfilm redistribution. The size of the MCM is 127.5 ×127.5 mm and can contain up to 121 chips, 80 700-chip I/Os, and 2,772 module I/Os. The cooling capability of IBM System ES9000 is 600 W.

The alumina-Mo multichip technology was no longer suitable when the cycle time was reduced to 10 ns. This led to the development of a new technology with glass-ceramic/copper with a thin film multichip. The new technology achieved unprecedented interconnection wiring and power densities. This was accomplished by buried wiring in ceramic brought to the surface in thin film. The resulting MCM contributes only about 30% of the total system delay.

As MCMs become more cost effective, more and more workstations and personal computers can be built with MCMs. A thin film on silicon or ceramic MCM is capable of packaging one million CMOS circuits in a space about 12 times smaller than the PCB currently being used. A workstation constructed with MCM operates at twice the clock rate as one without the MCMs, reaching up to 100 MHz. The MCM is made of thin film on silicon, 54 mm in size, and packages 9 chips containing floating point and fixed point processors, instruction cache, storage control, four data cache chips, and a microchannel interface. Five layers of ultradense thin file polyimide and aluminum are deposited on the silicon substrate. The resolution of 13 μm line widths separated by 12 μm spaces is achieved, providing a wiring density of 750 cm/cm^2. The module has 684 peripheral I/Os, 512 of which are signals and the remaining ones are for power and ground. The I/O connections are made with TAB film that acts as a flexible, surface-mount-compatible interconnection. Four independent flex connections with a tight 0.63 mm I/O grid on the module to a looser pitch on the board are made. Gang solder bonding operation is used to connect the module to the board.

1.4.2 Avionics/Automotive

The MCM technology eliminates a whole package level with the accompanying elimination of the thermal resistances and lead and solder joint problems and thus improves the reliability of the product (see Table 1.6). Avionics equipment usually operates in a hostile environment. This equipment must operate reliably over extremes of temperature, shock, humidity, and pressure. In addition, high performance is required for the avionics applications. Thus MCMs are ideal for avionics and auto-motive applications.

Both military and civilian aircrafts carry a variety of electronics equipment. This equipment is required to perform communications, navigation, flight management, sensing, and data recording. In addition, military aircrafts use specialized electronics to manage radars, avoid events, and operate countermeasures. Missiles must perform target selection or target tracking in real time. Most of this equipment may operate from dc to gigahertz ranges. In addition, avionics experience very harsh environmental conditions. They sustain wide extrema of temperature, vibration, shock, humidity, and atmospheric pressure. In addition, they may dissipate a large amount of heat. Automotive systems tend to be significantly simpler as their failure is not catastrophic. One major difference between automotive and avionics is rapid cycling of event. While both are designed to operate in environmental extremes of very humid tropics to frigid polar regions to salt spray conditions [81], avionics may experience all these conditions several times in a week or even a day. Aircrafts fly at low pressure high altitudes and land and take off in normal pressure conditions; therefore avionics are stressed by this pressure cycling. In space applications, heat removal is the key problem, since radiative heat loss is the only option in the absence of cooling air. Another concern is the exposure to cosmic radiation.

TABLE 1.6. Reliability Improvements Provided by MCMs

Reduced parts/site count
- 95% of solder joints eliminated
- 84% of hermetic I/O eliminated
- 67% of lineal lid seal eliminated
- 75% of connectors eliminated

Thermal and G-lad stresses significantly reduced
- Matched-expansion materials
- 70% of mass eliminated
- Compliant design interfaces
- High heat removal
- Packaging materials

Reduced diffusion, grain growth, and intermetallic growth mechanisms
- Monometallic
- Design interconnects
- Typically 20°C coolder IC junctions

Reduced corrosion/moisture susceptibility
- Greater percentage parts and interconnects in protective package
- Rugged high-rel co-fired I/O (no glass-to-metal seals)
- More room for built-in self-testing, redundancy, fault-tolerant circuitry

One overriding concern in all avionics is reliability. Avionics, whether it is in a single engine private plane or a space shuttle, must perform reliably in the wide range of conditions discussed above.

Multichip technology has been very successful in meeting the challenges posed by the harsh environmental conditions and reliability concerns of avionics. The key advantages of MCMs in avionics include small size (4 × smaller than PCB), less weight (3 × lighter than of PCM), and increased reliability (5 × higher than PCB). The reliability of MCM is better than other technologies due to having fewer parts and because 95% of the solder joints can be eliminated. In addition, thermal stresses can be reduced by high heat packaging materials and the matching of thermal expansion coefficients of different materials, while G-load stresses can be reduced by the weight reduction. MCM based avionics are better protected from the environment since the greater percentage of the parts are enclosed in a protective package.

Several companies have built (and are building) MCM based avionics. Examples of avionics include, 1 × 2 inch MCMs containing 12 ICs and four chip capacitors, which replaces a 4 square inch PCB [81]. Yet another example is a co-fired aluminum nitride MCM containing 32 ICs and 64 chip capacitors in a 2.2 square inch package. These examples show the weight and size reductions possible with MCMs. For enhancement of reliability, redundancy is used. Hagge [81] describes such an application of MCMs. Four processor memory elements (PMEs) are integrated by a redundancy manager unit (RMU), which acts as a voter. PMEs are 2.2 square inch aluminum-polyimide multilayer thinfilm MCMs containing 53 ICs and 40 discrete devices, while an RMU is an MCM containing 60 ICs and 8 passive devices. This system provides highly reliable and redundant flight information to aircrafts. Its reliability enhancement through redundancy would be impractical without the high degree of miniature and high module reliability of MCMs.

1.4.3 Military

There are significant size and weight reductions in MCMs. Size and weight are crucial factors in many military applications. High performance and the ability to operate under extremely harsh conditions are also very important in military equipment. Thus the MCM is a good choice for military applications.

The military requires high performance computing for the navigational needs of troops, tanks, and aircraft. High speed computers aim guns, select targets, and provide simulation for command and control. As demonstrated in recent engagements, the involvement of computers is likely to increase manifold as military weapons become more hi-tech.

MCMs offer several benefits to the military. These include superior performance, small size, better environmental protection and reliability, and better heat dissipation. For many military applications, the weight advantage offered by MCMs is of little consequence, and its higher cost can be easily justified.

MCMs developed for military applications must satisfy a variety of specifications. While the specifications for single chip packages are relatively simple, the specifications for MCMs (or mutlichip hybrids) are complex, since MCMs are manufactured with a variety of parts, materials, and processes. Military hybrid qualifications concentrate on the processes rather than the individual hybrid by establishing requirements for incoming inspection, tests for wafer lots, package lots, lids, and bond wires, materials for substrate printing, die attachment, wire bonding, and reworking. This and other standards are controlled and applied through Rome laboratory and the Defense Electronics Supply Center (DESC).

MCMs developed for military use must be able to satisfy very strict reliability, environmental, and performance specifications. For example, MCMs developed for aiming of tank guns must be able to perform reliably in severe polar weather as well as desert conditions. MCMs developed for military specifications are tested by temperature cycling, severe G-forces, and thermal shock. They are also tested for moisture resistance and lead integrity, among other criteria.

While MCMS are being developed for military applications, there are several factors that impede this development. For example, many competing MCM technologies exist that can be used for development of MCMs for military applications. However, it is not clear how these technologies compare. In addition, standards do not cover the entire range of this fast growing field, and most importantly there are no clear military qualified sources for all of the different types of MCMs. Until many of these issues are satisfactorily addressed, the implementation of MCMs in the military will be slow.

1.5 IMPACT OF MCMs

Multichip packaging is expected to be the pervasive technology in a number of industries, including computer, telecommunication, aerospace, automotive, and consumer electronics. The initial driving force has been and continues to be the high performance computers involving a hundred or more chips on a single multichip substrate for improving the cycle time by drastically reducing the package delay. With the bipolar and CMOS integrations that are expected in the mid to late 1990s, MCMs will be needed to package a smaller number of chips on smaller sized substrates and to provide functions that cannot be fulfilled by individual chips. In view of the complex technologies involved, such as sophisticated materials, processes, tools, design systems, test burn-in and repair, as well as the capital-intensive nature of the manufacturing technology, MCMs in the early 1990s continue to be used for those computer systems that can afford the high initial cost. The knowledge gained from these systems and the resultant cost reduction will provide the thrust necessary to expand to other medium cost systems and eventually to low cost systems. Simplified and inexpensive designs, design tools, materials, processes, and manufacturing tools, therefore, are the future needs of this packaging technology.

1.6 SUMMARY

The conventional packaging approaches such as through-hole, surface-mount, COB, and hybrid technology are beginning to limit the system performance. These approaches cannot utilize the full potential of the dense circuitry packaged on an IC chip today. As a result, the new packaging techniques of MCMs have attracted the attention of electronics packaging designers. MCMs have become one of the most popular approaches for packaging, especially for high performance systems. It must, however, be noted that system packaging is a complex design process involving various trade-offs. While MCMs are not suitable for all applications, they are the only choice if small size, high performance, less weight, and high reliability are critical. The recent deluge of several consortia, conferences, and symposia and the entry of many MCM CAD and fabrication vendors into this dynamic industry is testimony of these activities in MCMs.

BIBLIOGRAPHIC NOTES

A handbook by Tummala and Rymaszewski [203] covers the fundamental concepts of microlectronic packaging. An introductory discussion on through-hole and surface-mount technologies is given by Bakoglu [6]. Discussions of many problems with MCM package selection, laminated MCMs, thick film and ceramic technologies, thin film technologies, selection criteria for MCM dielectrics, chip-to-substrate connection technologies, MCM-to-PCB connection technologies, electrical design of MCMs, thermal design of MCMs, electrical testing of MCMs, and case studies are given by Doane and Franzon [58]. A collection of papers on analysis of packaging issues, co-fired ceramic technology, co-fired glass-ceramic technology, thin film technology, thermal analysis of multichip modules, electrical analysis of multichip modules, and testing of multichip modules are presented by Johnson et al. [108]. Issues in the design of MCMs, especially the trade-off analysis, are discussed by Sandborn and Moreno [178].

The Institute for Interconnecting and Packaging Electronic Circuits (IPC), the International Society for Hybrid Microelectronics (ISHM), and the International Electronics Packaging Society (IEPS) are three representative societies that deal with the professional development of the people involved, technical aspects of the MCMs, and other electronic packaging fields. These committees hold conferences, publish journals, develop standards, and support research in MCMs and other electronic packaging field.

In recent years, research on MCM technology has been very active, and several hundred papers on MCMs have been published. MCM is an interdisciplinary area. The papers on MCMs are published in many conferences and journals. The major conference on MCMs is the International Conference on MCMs, organized by ISHM and IEEE Multichip Module Conference. The other conferences on MCMs and other packaging technologies are International Symposium on Microelectronics organized by ISHM, International Conference on Electronic Packaging organized by IEPS, National Electronic Packaging and Production Conference (NEPCON), and Electronic Components and Technology Conference.

CHAPTER 2

CLASSIFICATION OF MULTICHIP MODULES

The MCM technology has been advancing rapidly in recent years. As a result, significantly different fabrication processes, materials, and chip attachment methods are now used in MCMs. Some fabrication processes resemble the traditional PCB technology, whereas others use more advanced and sophisticated VLSI technologies. As a result, technologies with varying properties exist. These properties include linepitch, number of layers, number of dies, performance, and cost. For example, the line pitch of an MCM derived from the traditional PCB technology is about 150 μm, whereas the line pitch of an MCM based on IC technology can be as small as 10 μm. Similarly, there are restricted number of layers available for VLSI technology based MCMs whereas there are virtually unlimited number of layers available for PCB technology based MCMs. The MCMs for consumer electronics may contain as few as two dies, whereas the MCMs for mainframes may consist of hundreds of dies. As a result, it is virtually impossible to identify the properties of a general MCM and discuss problems of the design, fabrication, testing, and maintenance of a general MCM. The classification will allow us to concentrate on the properties of each type. Based on the properties of a particular type of MCM, one can develop a CAD tool or perform the electrical analysis that suits that particular type. For example, if only a limited number of layers is allowed in a particular type of MCM, then the objective of the CAD tool may be to complete the routing in the given number of layers. On the other hand, if large number of layers are available, the minimization of the number of layers may not be an important issue for the CAD tool. Therefore, it is critical to know the type of MCM while designing or using CAD tools. In the electrical design process, one needs to know the actual line pitch, which is determined by the type of MCM. For example, if the line pitch is small (such as 10 μm), then the crosstalk is an important issue. On the other hand, if the line pitch is large (such as 150 μm), then the crosstalk may not be a major concern. Furthermore, the classification of the MCM technologies helps in understanding the differences among them. The classification helps a systems designer to choose a specific MCM technology to meet the system performance and cost goals. The classification of MCMs also helps in identifying the problems associated with each type.

From the above discussion, it is clear that a good classification scheme would indeed be helpful in all stages of design and development of MCMs. However, the basis of such a classification scheme is a complex issue. A good classification method should be able to distinguish among MCMs based on all the important parameters such as the fabrication methods, chip bonding methods, materials, thermal management methods, and packaging methods.

The traditional classification method is based solely on the fabrication processes of MCM substrates. As a result, MCMs are classified into MCM-C, MCM-D and MCM-L. The traditional classification method is too simple to provide a comprehensive view of MCMs.

In this chapter, the classification of MCMs is discussed. We first discuss the important classification parameters. Then different materials and methods used in MCMs and their impact on these parameters of MCMs are presented. The traditional classification method is discussed next. We conclude with a discussion of limitations of existing classification method.

2.1 CLASSIFICATION PARAMETERS

It is important to identify all the criteria of the MCM technologies to derive a classification method. Basic criteria for an MCM based system are wiring capacity, performance, and cost. In fact, the entire computer industry is driven by the goal to acheive the highest performance in the most cost effective manner. The wiring capacity of an MCM is the total length of wiring that can be laid out in the MCM. As more funtions are integrated onto the MCM to achieve higher performance, the wiring capacity must be increased accordingly. Other important criteria for MCMs include thermal capacity and ease of repair. As the number of dies on an MCM increases, the heat generated increases as well. The maximum number of dies on an MCM is restricted by the ability of the MCM to remove the heat generated by the dies. This heat removal capacity is referred to as the *thermal capacity* of the MCM. Another issue is the ease of repair. It is not cost effective to discard an entire MCM due to a single or even a few faulty dies. Therefore, it should be easy to replace faulty dies; that is, MCMs should be easy to repair. In the following, we discuss these criteria in detail.

2.1.1 Performance

The performance of a system is measured in terms of the clock frequency used by that system. The clock frequency depends on the signal delay. Detailed analyses of performance and signal delay are given in Chapter 3. Here we present a brief overview.

As a significant part of the total signal delay is due to the delay in the MCM, it is necessary to reduce the delay in the MCM to increase the system performance. The delay in the MCM is estimated by

$$T_d = 0.033L\sqrt{\varepsilon_r}$$

where T_d is the signal delay in an MCM (in nanoseconds), L is the chip pitch (in centimeters), and ε_r is the dielectric constant of the substrate material. The *chip pitch* is the distance between two chips.

To achieve the best performance in an MCM system, signal delay in MCMs should be kept at a minimum. As the signal delay is proportional to the chip pitch and the square root of dielectric constant, both the chip pitch and the dielectric constant of the substrate material should be minimized to achieve the highest performance. Note that using a material of lower dielectric constant may not always reduce the signal delay. For comparison, consider the following: Glass-ceramic with a dielectric constant 5.0 when used as the substrate material in an MCM achieves a chip pitch of 10 mm. On the other hand, MCMs using polyamide with a dielectric constant of 3.5 acheives a minimum chip pitch of 20 mm. However, the signal delay in the first and the second are 760 and 1,320 ps, respectively. Clearly this is a complex trade off, and, in addition to the dielectric constant, several factors have to be taken into account to choose the substrate material to achieve the best performance.

2.1.2 Wiring Capacity

Another imporatant criterion of an MCM is the wiring capacity. The number of package levels can be reduced by increasing the number of circuits in MCMs, resulting in the increased performance as explained in Chapter 1. Furthermore, reduction in the number of package levels improves reliability and reduces the weight and the space occupied by the system. However, the maximum number of circuits that can be contained in an MCM depends on its wiring capacity. As a result, the wiring capacity of an MCM is an important design criteria.

The wiring capacity of an MCM is affected by several factors, one of which is the *wiring density* of the MCM. The *wiring density* of an MCM, also called the *interconnect density*, is defined as the length of signal lines per unit area per layer in the MCM. The wiring density of an MCM is also defined as the length of wiring lines per unit area in the MCM in some literature. We restrict ourselves to the first definition. The wiring capacity (W) of an MCM can be computed by

$$W = w \times S \times L$$

where w is the wiring density, S is the routing area available on each layer, and L is the number of layers (called *signal layers*) for routing the signals.

Given the number of signal layers and the area available for routing on each layer, the wiring capacity can be increased by increasing the wiring density. The wiring density of an MCM is determined mainly by the minimum line width, line separation, and the via size. The characteristic length of the minimum line width, line separation, and the via size are specified by the *design rules.* The design rules are primarily determined by fabrication technology. For example, the minimum separation design rule in the laminating process is about 150 μm, whereas the minimum separation design rule in the thin film process can be made as small as 10 μm. The design rules are also determined by electrical effects. For example, two signal lines have to be separated by a certain minimum distance in order to avoid excessive crosstalk noise. The details of the electrical effects are discussed in Chapter 3. By improving any of the design rules, such as minimizing the line separation, improves the wiring density. The increased wiring density allows more chips on the MCM. At the same time, the chip pitch is also reduced, which may result in increasing the system performance.

On the other hand, given the wiring density and the area available for routing on each layer, the wiring capacity can be drastically improved by increasing the number of layers. The maximum number of layers that can be achieved in an MCM depends on the technology used in fabricating the MCM. For example, the maximum number of layers in a thin film process is 8, whereas the maximum number of layers in a thick film process is 63. However, there is a drawback in using more layers. As more vias are used for each signal to travel between layers, the signal delay is increased, resulting in performance degradation.

Given the wiring capacity, we can estimate the maximum number of chips that can be mounted on the MCM substrate as

$$N \leq \frac{0.44W}{PI}$$

where N is the number of chips, W is the total wiring length, P is the chip pitch, and I is the number of I/Os per chip. I is determined by the number of circuits per chip as is predicted by Rent's rule, as discussed in Chapter 1. This equation is obtained by assuming that there are 3 interconnections for every 4 terminals, the average wiring length for an interconnection is 1.5 × chip pitch, and the routing efficiency is 50%.

It can be seen from the equation that more chips can be mounted on an MCM substrate as the wiring capacity of the MCM increases. This leads to the reduction of package levels. As a result, the system performance increases, the size and weight of the final product are reduced, the reliability of the system is improved, and, finally, the cost of final product is reduced.

2.1.3 Thermal Capability

In addition to the performance and wiring capacity, thermal capacity is an important criterion. Each die generates a certain amount of heat, and, if this heat

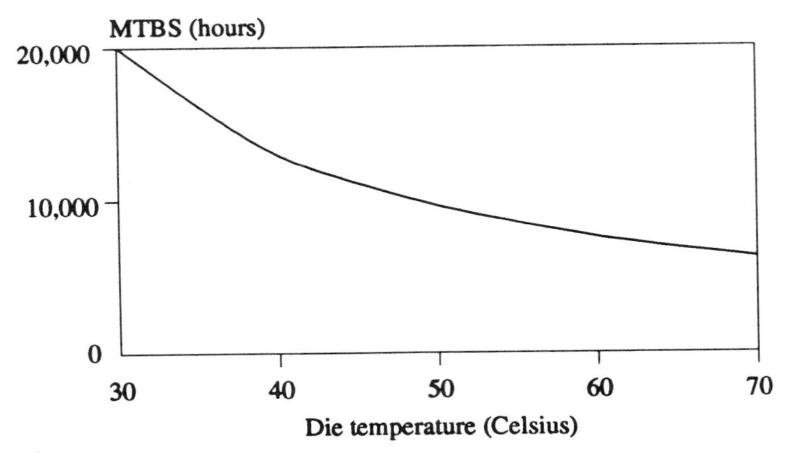

FIG. 2.1. Relationship between MTBS and temperature.

is not properly removed, it may lead to a high die temperature. The temperature of a die should be maintained below a specified temperature for proper functioning of the die with high reliability. In addition, the lifetime of an IC die depends on its operating temperature. Figure 2.1 shows the effect of temperature on the mean time between service (MTBS) for dies used in mainframes. It is clear that the heat generated must be removed efficiently so that the temperature is maintained below the specified temperature. The amount of heat that can be removed in unit time in an MCM is referred to as the *thermal capacity* of the MCM. The maximum number of dies mounted on an MCM depends not only on the wiring capacity but also on its thermal capacity. Therefore, the thermal capacity on an MCM is an important criterion.

The limitation imposed by the thermal capacity of the MCM is given by

$$N \leq \frac{T}{H}$$

where N is the number of dies in the MCM, T is the thermal capacity of the MCM, and H is the amount of heat generated by a die per unit time.

2.1.4 Ease of Repair and Rework

As more and more chips are integrated into an MCM, repairing and reworking become important tasks since it may not be cost effective to discard an entire MCM that has a few defects. As a result, the ease of repairing and reworking is an essential ingredient to MCM technology's cost effectiveness. *Repairing* refers to the repair of shorts between lines, opens in the lines, opens at the line-to-via connections, and via misalignment. Lines and vias are the most common items to repair. *Reworking* is to remove bad chips and replace them with good ones. Thorough electrical testing is used to detect any defect and to monitor the process. Testing is done to detect line shorts and opens; measure electrical parameters such as propagation delay, impedance, capacitance, and resistance of signal lines; measure noise levels such as crosstalk, reflection, and skin effects; and monitor processes such as vial alignment, thickness, and uniformity.

2.1.5 Cost

Cost is one of the most important factors in any electronic product. The objective of the design for most electronic systems is to achieve the desired performance at the minimum cost. It must be noted that the cost to be considered should be the cost of the entire system when the endproduct of the design is a complete system. This is due to the fact that even though the cost of individual high density MCMs is high, the total cost of the system using high density MCMs may still be lower than one using individually packaged chips or PCBs because of the reduced package levels in the system.

The total cost of an MCM based system depends on the design, production, and maintenance costs. The *design cost* is the cost incurred during the design process, which includes the cost incurred in training and designing, cost of CAD tools, and costs of fabrication and testing of prototypes. The cost incurred in the fabrication or manufacturing is called the *production cost*, and it consists of the cost of equipment and materials, testing, loss due to defective products such as cost of rework and repair, and, finally, the manpower required in all the production stages. After products are delivered to customers, there is after-sale service, including repairing or replacing defective parts. The cost of this service is the *maintenance cost*.

To reduce the total cost, all three of its components need to be reduced. Let us now consider the design cost. One way of reducing the design cost is to automate the design process. As more stages of design process are automated, less manpower is required, thus reducing the design cost. Design automation tools have been used extensively in VLSI and PCB technologies. As a result, significant progress has been made in both technologies. Design automation tools for MCMs are currently under development. However, few tools have been designed specially for MCMs. Many tools are simply borrowed from VLSI or PCB domains and are not exactly suitable for MCMs. As a result, the lack of good MCM design automation tools remains one of the major obstacles for the further development of MCM technology.

Let us now consider the production cost, which can be classified into two types. One is the *recurrent cost*, which recurs for every part made. The other cost is the *nonrecurrent cost*, which is a cost incurred only once in the entire production process. Manpower, materials, and energy used in the production process are recurrent, whereas purchasing equipment is nonrecurrent. High volume production helps to reduce the nonrecurrent cost since it will be distributed among more parts. Increasing the yield of MCMs, on the other hand, can reduce the loss due to defective products. The cost of repair and rework can also be reduced if MCMs are properly designed such that they are easy to repair and rework. Other methods of reducing production costs include using less expensive machinery and materials.

Another cost to be considered is the maintenance cost. The maintenance cost depends on the reliability of the system. The reliability improves by reduction or elimination of possible failure mechanisms by employing a better thermal management system to ensure that the MCM operates below a specified temperature. The improved reliability, in turn, will reduce the maintenance cost. The maintenance cost can also be reduced if the field repair and replacement of failed parts can be done at low cost. This should be considered during the design phase, and one should design MCMs that are easy to repair or replace in the field. Another way is to reduce the size of the field replaceable unit. If the field replaceable unit is a die, then the failed die can be replaced without affecting other parts of the MCM or the system. On the other hand, if the field replaceable unit is an MCM, then the entire MCM has to be replaced if there is any failed chip in the MCM.

Another important factor to be considred is time-to-market. To reduce the time-to-market, one should use a well-established technology or infrastructure. One of the ways to reduce the time-to-market is to use automation tools extensively at each stage of the design.

2.2 BASIS OF CLASSIFICATION

MCM technology has been growing rapidly in recent years. As a result, MCMs with significantly different properties have been developed. To classify MCMs, we need to study the factors or parameters that affect their properties. Such factors include the substrate fabrication method, materials, and chip bonding techniques (see Fig. 2.2). The substrate fabrication method used in an MCM affects its performance, wiring capacity, and cost. The materials used, including dielectric material and conductive material, influence the performance, wiring capacity, thermal capacity, and cost of the MCM, whereas the chip bonding method affects the performance, wiring capacity, ease of repair and rework, and cost. In this section, we discuss different substrate fabrication methods, materials, and chip bonding techniques, as well as thermal management and MCM packaging methods.

2.2.1 Substrate Fabrication Methods

The substrate fabrication method is one of the most important factors in determining the properties of an MCM. It is also the basis of the traditional classification of MCMs. There are three popular substrate fabrication methods, namely, the laminating process, co-firing or thick film process, and depositing or thin film process.

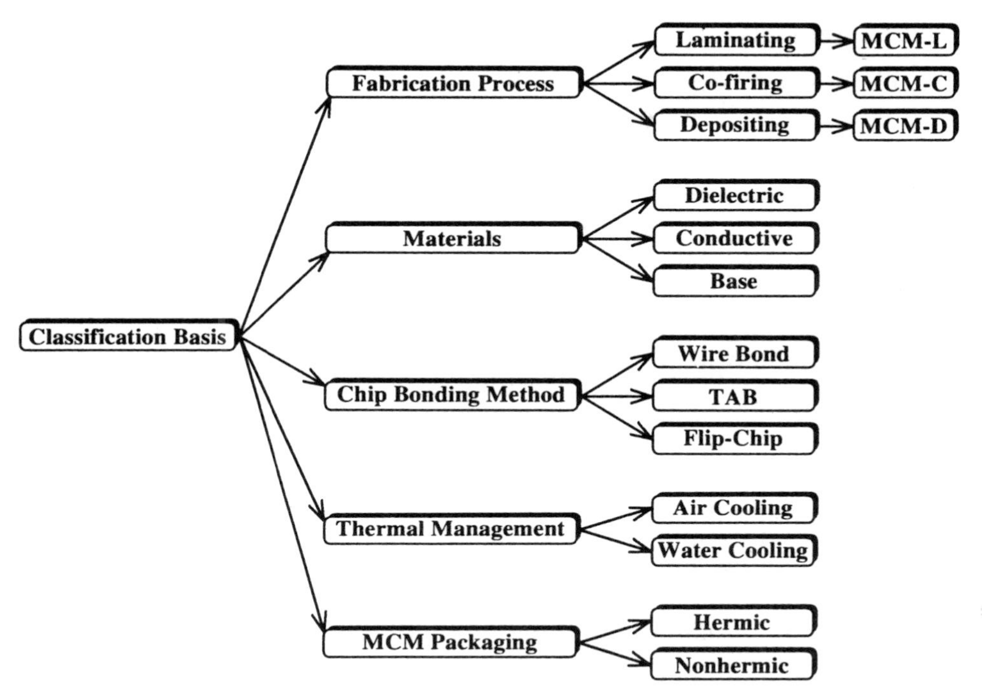

FIG. 2.2. Basis of classification.

In the fabrication process of MCM-Ls, the sheets on which signal paths are laid out are prepared individually and then laminated together to form the substrate. This process is called *lamination.*

The fabrication method of MCM-Cs is called *co-firing*. In this process, signal paths are formed by screen printing conductive material onto green tapes, and the substrate is formed by co-firing the green tapes together. Since the layers formed in the co-firing process are relatively thick, it is also called a *thick film process.* A layer that is formed by screen printing a conductive, resistive, or dielectric ink or paste onto a substrate is called the *thick film*. Thick films are typically 0.001 inch thick. The MCMs whose conductor, resistor, and dielectric layers are thick films are called *thick film MCMs.*

The conductive and dielectric layers in the substrate, formed by depositing the dielectric material and conductive material alternately, are relatively thin compared with the one generated by co-firing.

This method is used to fabricate MCM-Ds and is called a *depositing process* or a *thin film process.* A layer formed by sputtering or evaporating a conductive, resistive, or dielectric material onto a substrate in a vacuum is called a *thin film* and is typically less than 4×10^{-6} inch thick. The MCMs formed by a thin film process are called *thin film MCMs.* More details of these fabrication processes are presented in Section 2.3.

2.2.2 Materials

In this section, we discuss the materials used for each substrate technology and the properties of these materials. The MCM substrate consists of alternating layers of dielectric material and conductive material. In the case of MCM-D, the alternating layers of dielectric material and conductive material are deposited on a support substrate. The material of the support substrate is called the *base material.* Different dielectric and conductive materials, as well as base materials (in case the support substrate is present), are required for different substrate technologies. Table 2.1 shows the materials used in different types of MCM technologies. Bare dies are usually made of silicon, though GaAs is also used occasionally. Since there are few choices for die materials, die materials are not discussed here. In the following, we will discuss important properties of materials used in MCMs.

The important properties for dielectric material are listed as follows:

1. **Low Dielectric Constant:** The material should have a low dielectric constant to reduce the signal delay. The *dielectric constant* of a material is the ratio of the capacitance, using the material as the dielectric to the capacitance resulting when the material is replaced by air. The signal delay is affected by the dielectric constant of the material used in the MCM. The signal delay in an MCM can be reduced by using the dielectric material of a smaller dielectric constant.

2. **Good CTE Match:** The CTE (*coefficient of thermal expansion*), which is a measure of the rate of expansion of a material, is defined as the change in dimension of the material per unit dimension per 1°C rise in temperature. Thus different materials have different rates of expansion. Since an MCM contains different materials adjacent to each other, the CTE of the materials should match each other so as to prevent weakening or breakage of connections between them. For example, the CTE of the dielectric material should be close to the CTE of the

TABLE 2.1. Materials Used in Different Substrate Technologies

Materials	MCM-L	MCM-C	MCM-D
Base	N/A	N/A	Al_2O_3, AIN, BeO, Si, Cu, glass-ceramic
Conductive	Cu	Au, Ag, PdAg, Cu, W, Mo	Au, Al, Cu
Dielectric	Epoxy-glass Polyamide BT-epoxy Cyanate ester Sycar Laminate sheets	Glass-ceramic tape Ceramic tape (Al_2O_3)	SiO_2, polyamide benzocyclobutene

conductor, base substrate, and silicon die material in order to prolong the fatigue lifetime of the material. In other words, the larger the difference between the CTE's of the materials, the more likely is the failure. Furthermore, silicon dies are attached to the MCM substrate. As a result, the CTE of the dielectric material should also closely match the CTE of silicon. In case that support substrate is present, the CTEs of the dielectric material and the base material should also match each other. Consider an MCM in which silicon dies are solder bonded to the MCM substrate: The fatigue life of the solder connecting the substrate and the chip as a function of the CTE of the substrate dielectric material is shown in Figure 2.3. Note that the CTE of the silicon, which is the material of the dies, is about 6. It can be seen from Figure 2.3 that the fatigue life is reduced sharply when the CTE value of the dielectric material moves away from the CTE value of silicon. For example, the fatigue life of the solder connecting the substrate and the chip is over 10,000 if the CTE of the dielectric material is around 6. The fatigue life of the solder is less than 1,000 if the CTE of the dielectric material is 8.

3. **Good Thermal Conductivity:** MCMs contain a large number of dies, each dissipating a certain amount of heat. As a result, MCMs may generate a very large amount of heat. The heat should be removed efficiently so that the MCMs can operate properly. As a result, materials used in MCMs should be able to remove the heat efficiently. The efficiency of heat removal in solids is measured by thermal conductivity. *Thermal conductivity* of a material is the physical constant for the quantity of heat transferred through a unit cube of the material in a unit time when the temperature difference on the two faces of the cube is 1°C.

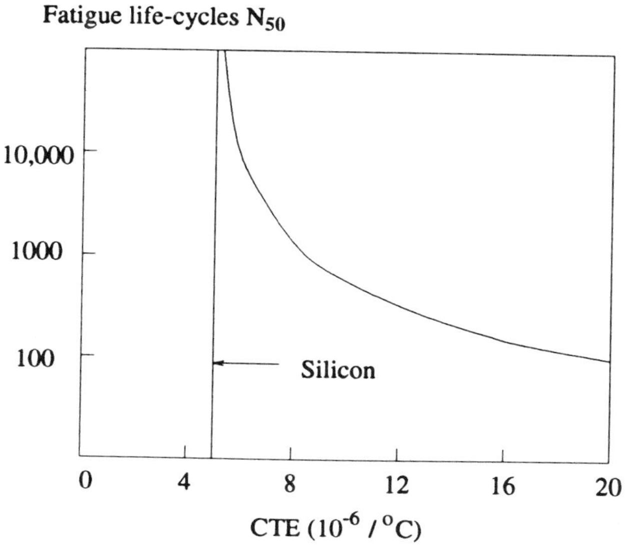

FIG. 2.3. Relationship between fatigue life and thermal expansion coefficient.

4. **Good Adhesion:** MCM substrates are formed by laying conductive material on top of dielectric material. As a result, the dielectric material should have good adhesion to the conductive material in order to avoid breaking the MCM substrate. In MCM-D technology, the dielectric material and conductive material are deposited on the support substrate. As a result, the dielectric material should also have good adhesion to the base material.

5. **Ease for Planarization:** After each layer of conductive material is formed on a layer of dielectric material, the resulting surface is highly nonplanar due to conductive patterns and vias. Thus, the surface needs to be planarized so that subsequent layers can be added with acceptable yield. Therefore, the maximum number of layers of an MCM depends on the quality of planarization. As a result, the dielectric material used should ease the process of planarization.

6. **Low Water Absorption:** All the materials absorb water. The water absorbed can change the electrical, thermal, and mechanical properties of MCMs. Therefore, the materials used in an MCM should have low water absorption so as to prevent damage caused by moisture.

7. **Thermal Stability During Repairing:** It is expensive to discard the entire MCM if it contains a defective die. As a result, the defective die should be replaced. Thus, when heat is applied to the substrate to remove the defective die, the material should be able to endure the heat.

8. **Good Mechanical Properties:** Many electronic systems are exposed to excessive mechanical forces. The material used in an MCM, therefore, should be able to resist mechanical forces. Elastic modulus is a measure of the ability of a material to resist forces.

In addition to the properties listed above, we briefly outline a few others. Conductive material is used to conduct electric currents. Therefore, the conductive material used in MCMs should have a low resistance to reduce the delay. As the power of each chip and the number of chips in an MCM increase, it becomes more difficult to supply stable, noise-free power to the chips. The power distribution in an MCM must be carefuly designed. One way to improve the power supply is to choose the proper conductive material for the MCM. The desired properties of a substrate conductive material include improved power distribution, lower noise, and higher performance. In MCM-C technology, the substrate is fired at a very high temperature (1,500°C). As a result, the conductive material used in MCM-C should remain stable and not decompose at high temperatures.

In MCM-D technology, support substrate is required. In this case, the base material should have high elastic modulus to be able to resist excessive forces applied to the substrate. Moreover, the CTE of the base material should match the CTEs of the conductive and dielectric materials used. It should have good adhesion to the dielectric substrate and to the conductor material. During the repairing process, the base material should not be damaged by the heat applied to

TABLE 2.2. Properties of Materials Used in MCMs

Material	CTE	Thermal Conductivity (w/m °C)	Elastic Modulus (GPa)
Beryllia	9.0	200.0	30–70
Silicon	5.6	40.0	12–15
Aluminum	23.6	237.0	28.0
Copper	17.7	400.0	110.0
Solder 60/40	25.0	5.1	2.0
Glass	12.0	1.1	65.0
Epoxy-glass (FR4)	15.0	0.31	15.0
Teflon (PFA)	184.0	0.055	2.8
Beryllium-copper alloy	16.0	180.0	122.0

TABLE 2.3. Properties of Dielectric Materials Used in MCMs

Material	Dielectric Constant (1 MHz)	CTE	Fabrication Temperature (°C)	Thermal Conductivity (w/m °C)
Alumina	9.4	6.6	1,600	20
Aluminum nitride	8.8	3.5	1,600	200
Mullite + glass	5.9	3.5	1,200	6
Glass-ceramic	5.0	3.0	950	7
Composite SiO_2 + glass	3.9	1.9	900	4
Polyamide				
PMDA-DDA	3.5	25–40	400	0.2
BPDA-DDA	3.0	2–6	400	0.2
Polyphenyl quinoxaline	2.7	35	450	0.2

the substrate. In addition, the base material should have good thermal conductivity to remove the heat efficiently. Tables 2.2 and 2.3 compare the properties of the materials used in MCMs.

2.2.3 Chip Bonding Methods

Each of the chip bonding methods can be used in any substrate technology. The bonding method used in an MCM has a significant impact on the properties, such as performance and cost, of the MCM. In fact, bare chips reduce board space requirements by a large factor [42]. As a result, the chip bonding method should be an important factor in classifying MCMs. Wire bonding,TAB, and flip-chip are the three most popular chip bonding methods used. In this section, we discuss these techniques.

Since the chip bonding method is used to make electric connections between dies and the MCM substrate, it determines the chip I/Os. The *chip I/Os* are the power, ground, and signal paths between a die and on MCM substrate. The technique used for attaching the chips to an MCM affects the chip I/O density and heat management in the MCM.

Among the chip bonding methods, flip-chip can achieve the maximum number of chip I/Os and the highest performance, while wire bonding is the most widely available and affordable method. The availability of flip-chip is, however, rather poor. It is relatively easy to pretest a TAB. The pretesting is moderate for flip-chip, whereas it is difficult to pretest a wire bonded chip. Both repair and rework are easy in flip-chip, while it is difficult either to repair or to rework in wire bonding and TAB. TAB and flip-chip can be either area I/O type or perimeter I/O type. However, the wire bonding can only be perimeter I/O type in which all the I/O leads of a die are located at the perimeter of the die. Both wire bonding and TAB provide excellent thermal paths in which heat can be efficiently removed, whereas there are no efficient thermal paths for flip-chip. A more detailed comparison of the three chip bonding methods is given in Table 2.4.

The method of heat removal depends on the chip bonding method. Figure 2.4 shows the different ways chips are attached to the substrate and the direction of heat dissipatation. When the active side of the chip is up, as shown in Figure 2.4 b, c, the heat is removed through the substrate, requiring high thermal

TABLE 2.4. Properties of Different Chip Bonding Techniques

	Wire Bond	TAB	Flip-Chip
Size (diagonal)	Die + 110–220 (mil)	Die + 110–220 (mil)	Die
Thickness	Die + 30 (mil)	Die + 3 (mil)	Die +3 (mil)
Handling method	Waffle pack	33 mm frame	Waffle pack
Testability method	Special probing	Socket	Sacrificial carrier
Maximum I/O	300	500	> 1,000
Lead resistance (Ω)	0.07	0.01	0.003
Lead inductance (nH)	1.0	0.7	0.05
Adjacent lead capacitance	0.02	0.006	0.001
Metal	Au, Al-Si	Cooper on polymer	Solder
Process	Thermo-ultrasonic	Thermo-compression	Melt/reflow
Chip metallurgy	Al, Ti/W/Au	Ti/W/Au	Cr/Cu/Au
Substrate metallurgy	Ni/Au	Pb-Sn/Au	Ni/Au
I/O density (cm^2)	400	400	1,600
Failure rate (% /1,000 hours)	10^{-5}	N/A	10^{-8}

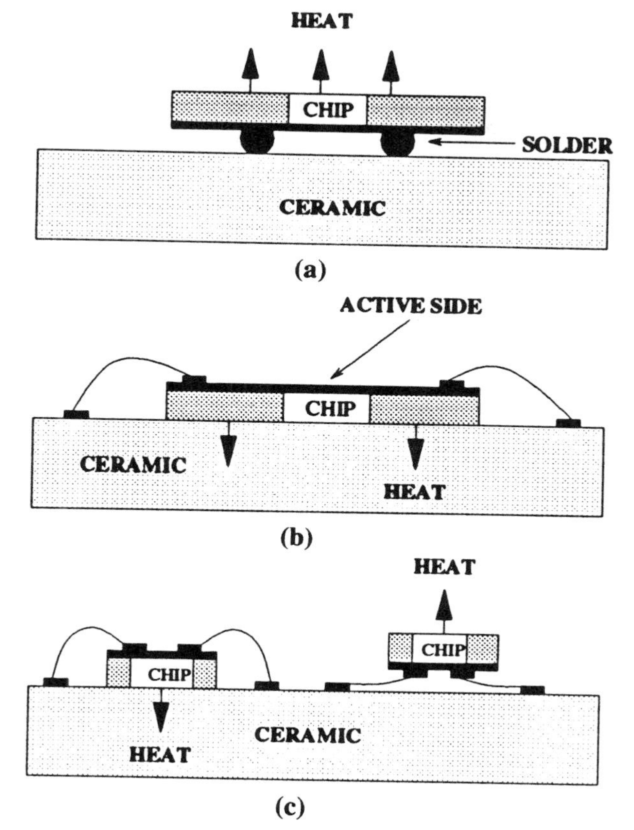

FIG. 2.4. Heat removal depends on the chip bonding method.

conductivity of the substrate materials. Sometimes thermal vias are used in the substrate to help remove the heat. However, the addition of thermal vias reduces the wiring capacity of the substrate. In this case, the heat conductivity of the substrate material is a critical factor in choosing a particular material. When the active side of the chip is down, as shown in Figure 2.4 a, c, the heat is conducted through a high conductance silicon chip to an aluminum, or other low cost, high conductance heat sink.

2.2.4 Thermal Management Methods

The thermal management method used in an MCM affects its thermal capacity and cost. As a result, the thermal management methods is an important factor in classifying MCMs.

MCMs can contain more than 100 dies. As a result, a large amount of heat is generated in a small concentrated area, creating a very challenging thermal management problem. For example, a single chip may generate up to 30 W of heat, and an MCM with more than 100 chips generates up to 3 kW of heat. In the near future, it is expected that the heat generated by each chip will reach 100 W and the heat per unit area will reach 30–100 W/cm^2. The heat must be removed efficiently so that the system can operate reliably.

The method used to move the heat from the MCM to the environment is called the *external heat removal method.* Currently air cooling and water cooling are the most frequently used methods. Air cooling is simpler and costs less than water cooling. However, water cooling is more effective at removing the heat. As a result, MCMs using water cooling methods have higher thermal capacities than the ones using air cooling. Most low-end workstations use air cooling, whereas high-end mainframes and supercomputers use water cooling. Usually, air cooling is sufficient for MCMs generating 10 W/cm^2 or less. For MCMs generating more than 10 W/cm^2 heat, water cooling should be used. For MCMs that generate 30 W/Cm^2 or more, even the water cooling method is not sufficient. A more advanced boiling and liquid gas cooling method has to be used. The cooling effects of the different methods are shown in Figure 2.5.

The *internal heat removal method* removes the heat from the interior of an MCM to the surface of the MCM. Using thermal vias or materials with high thermal conductivity is an internal heat removal method. As stated earlier, *thermal vias* are through-holes filled with a material of high thermal conductivity so that heat can be removed through the holes. However, thermal vias reduce the wiring density of the MCM. Therefore, it is not desirable to use thermal vias in a high performance system. On the other hand, the material with high thermal

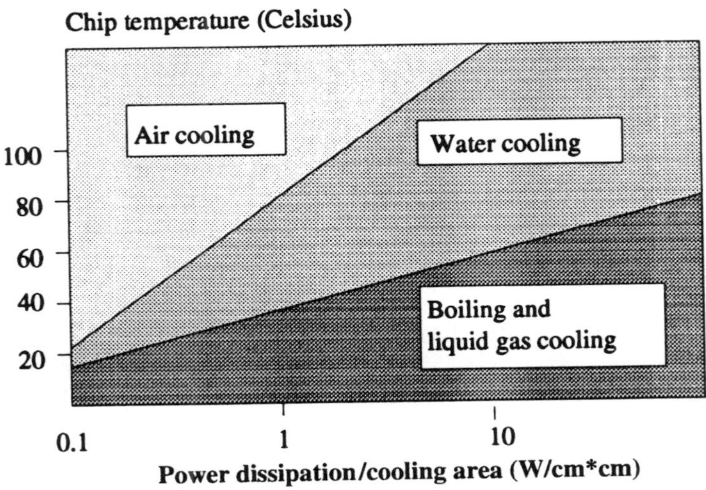

FIG. 2.5. Cooling effects of different methods.

conductivity may have high dielectric constant. As a result, using that material may increase the signal delay, thus reducing the system performance. Therefore, different factors have to be weighed against each other to choose a particular method. Often trade offs have to be made in choosing a proper thermal management method.

2.2.5 Packaging of MCMs

After an MCM has been fabricated, it is packaged to prevent transmission of moisture and air and other hostile effects from the outside environment. The packaging of an MCM affects its reliability and cost. As a result, it is an important factor in classifying MCMs.

The packages of MCMs can be either hermetic or nonhermetic. A *hermetic package* is permanently sealed by fusion or soldering. The different types of MCM packages are shown in Figure 2.6. Cost and reliability are the main concerns for the packaging of MCMs. As discussed in Section 2.1.5, the maintenance cost of MCM based systems depends on the size of field replaceable units and the ease of field repair or replacement. The field replaceable unit is a board if MCMs are soldered onto the board whereas the field replaceable unit is an MCM if the MCM is socketed onto the board. If an MCM is sealed in epoxy, it must be scrapped, whereas the MCM may be repairable if it has a resealable lid.

The package of an MCM also affects the module I/Os of the MCM. The *module I/Os* of an MCM are used to communicate with the rest of the system. There are different module I/O structures in the MCMs. The approaches used for module I/Os can be characterized as area versus periphery terminals, through-hole versus surface-mount interconnections, leaded versus leadless, and soldered versus separable connections. The module I/O structure affects the number of module I/Os, module test methods, and maintenance approaches. The number of module I/O has been increased steadily over the years; it has reached 11,500 in Fujitsu's latest mainframe [197].

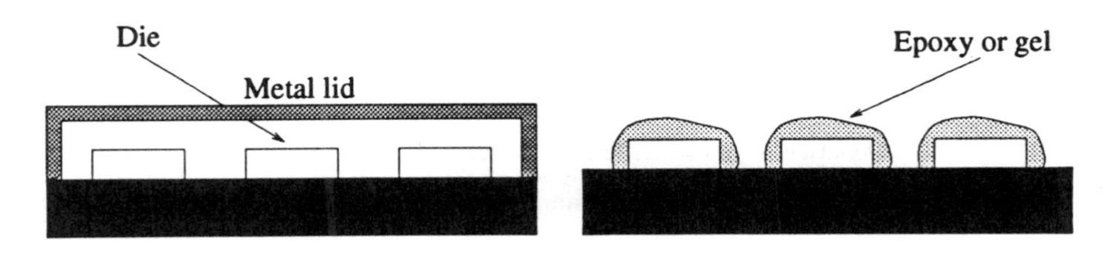

FIG. 2.6. MCM packaging options.

2.3 TRADITIONAL CLASSIFICATION METHOD

We have discussed the factors in classifying MCMs. The MCMs are traditionally classified as MCM-L, MCM-C, and MCM-D based on the fabrication method. In this section, we discuss the fabrication process, characteristics, and applications of each type of MCMs. Then the different types of MCMs are compared.

2.3.1 Laminated MCMs

Laminated MCM (MCM-L) substrates are based on laminated, multilayer PCB technology. MCM-L can be considered as fine-line PCB technology. The manufacturing process and materials used in PCBs have been used in MCM-Ls. COB, the earliest form of MCM-L, has been used since 1960s, not in high performance computer systems but in mass produced, low cost consumer applications. MCM-Ls are widely used because of the availability and low cost of the technology.

2.3.1.1 Fabrication Process The dielectric material that is most frequently used in MCM-L is FR4 epoxy-glass due to its low cost and wide availability. The usual choice of conductor is copper foil. There are other materials used for the substrate. There are trade-offs between the performance and costs in these materials. A comparison of these materials is given in Table 2.5.

As stated earlier, an MCM-L is fabricated in a way similar to a PCB. The steps of the standard fabrication process of MCM-L are listed in Figure 2.7. First a dielectric sheet covered with conductive material is prepared. Then pattern etching is carried out on the prepared dielectric sheet to form a layer with signal paths on it. After several layers have been prepared, these layers are bonded together with heat and pressure to form a sublaminate. The process of bonding layers together is called *lamination.* After a sublaminate has been formed, through-holes are drilled and and plated to form buried vias. The plated through-holes that go through a few adjacent layers but not the whole substrate are referred to as *buried vias* or *blind vias.* This process is repeated until all the sublaminates are formed. These sublaminates are laminated together to form a laminate that is the substrate. After the laminate is formed, through-holes are drilled and plated to form the *through-vias*. If the plated through-holes connecting two adjacent layers go through the whole substrate, the presence of the hole in the layers to which it is not connected will reduce the routing capacity in these

TABLE 2.5. Comparisons Among Different Dielectric Materials in MCM-L

Material	Dielectric Constant	CTE	Relative Cost
FR-4	4.8	16	1
BT epoxy	4.2	14	1.8
Polyamide	4.5	13	3.5
Gyanate ester	4.5	15	4.5

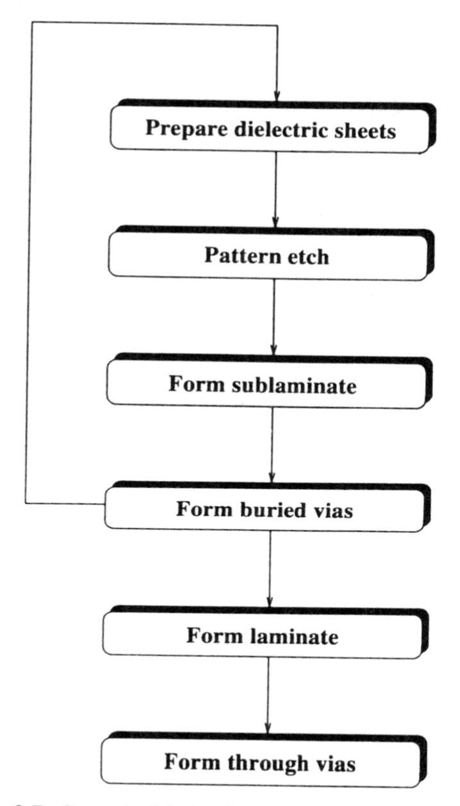

FIG. 2.7. Steps in fabrication process of MCM-Ls.

layers. In addition, if the plated through-hole goes through a reference layer to which it is not connected, it will reduce the efficiency of the reference plane to carry current uniformly. It also marginally increases the inductance of the reference plane. Therefore, buried vias should be used whenever possible.

The fabrication process is illustrated in Figure 2.8. Figure 2.8a shows the pattern etching of a single layer. Figure 2.8b shows the formation of a sublaminate by three adjacent layers. Figure 2.8c shows the through-hole drilled in the sublaminate, while Figure 2.8d shows the plated through-hole. Finally, Figure 2.8e shows the laminated MCM.

Wire bonding is the most commonly used method for attaching chips to the substrate. However, both TAB and flip-chip have been successfully used in commercial products. In the future, TAB is predicted to be the dominate bonding technique used in MCM-Ls because of its ability to pretest chips and the ease of outer lead bonding (OLB) formation by soldering.

MCM-L technology used mainly for high volume, low cost applications, repair, and rework is usually avoided. The MCM-L modules are encapsulated by using a "glob-top" method.

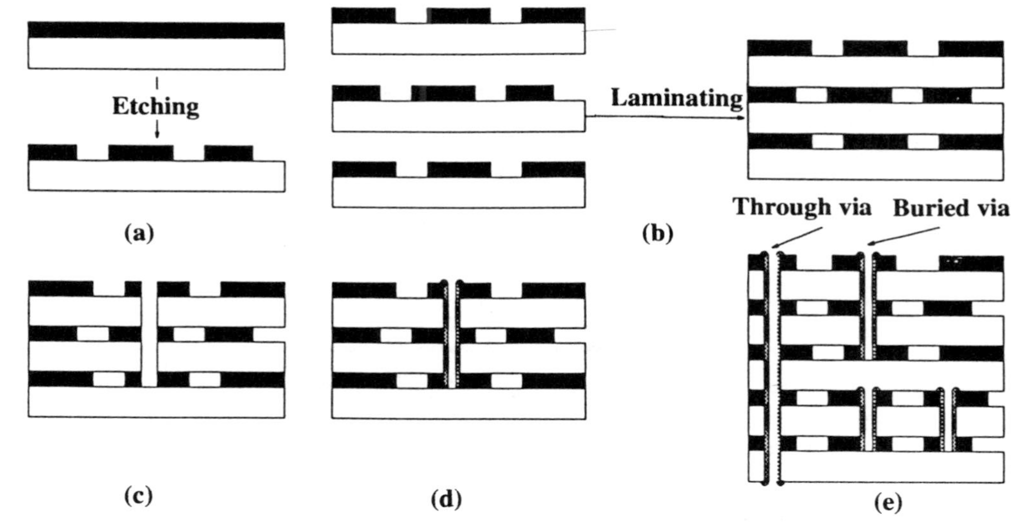

FIG. 2.8. Illustration of the fabrication process of MCM-L.

2.3.1.2 Characteristics MCM-Ls have many advantages, including low cost, parallel process, and well established infrastructure. As a result, MCM-Ls find applications in many different areas.

The most obvious advantage of MCM-L technology is its low cost due to its maturity. Table 2.6 compares the cost of different substrate technologies. It can be seen that the cost of MCM-L technology is significantly lower than the other MCM technologies.

MCM-L substrates are fabricated in large panels, and then each substrate is cut from the panel, thus allowing high volume substrates to be fabricated in a cost effective manner. A parallel process technique is used in MCM-Ls. Each substrate layer is fabricated, inspected, and yielded individually prior to stacking them into a single multilayer substrate. A defect in one layer (sublaminate) can be required or reworked in that layer (sublaminate) without affecting the other layers (sublaminate) under this parallel process. Therefore, yield losses are minimized.

TABLE 2.6. Cost Ratios Among Different Substrate Technologies

MCM Type	Cost Ratio
MCM-L	1
MCM-C	1.5–2.0
MCM-D (thin film on alumina)	2–3
MCM-D (silicon substrate)	7–10

Another important advantage of MCM-L technology is its well established infrastructure. Many manufacturers of MCM-L follow the same standards and techniques of the PCB industry, which are well established. This allows lower cost, higher volumes, and shorter design cycles for MCM-L products. MCM-Ls have other advantages such as the ability to have assemblies with components mounted on both sides of the substrate and simple module I/O structures.

In spite of the many advantages of MCM-Ls, there are several disadvantages that prevent MCM-Ls from being used in some applications. Among these drawbacks are low performance, low wiring density, poor thermal conductivity of the substrate, significant mismatch of CTE between the substrate and the silicon chips, moisture sensitivity of materials, and, usually, high crosstalk noise. Thermal vias and embedded metal plates can be used as heat spreaders. However, this heat transfer method degrades the wiring density of the substrate.

To solve or alleviate these problems, new materials such as advanced polyamides, aramids, fluoropolymers, and composites are currently being developed. In addition, new processing techniques using thinner dielectric layers and smaller vias are being developed.

In summary, the advantages and disadvantages of MCM-Ls are listed as follows:

1. Advantages of MCM-L technology
   a. Low cost
   b. Parallel fabrication process
   c. Ease of repairing or reworking individual layers
   d. Well established infrastructure
   e. Assemblies with components on both sides
2. Disadvantages of MCM-L technology
   a. Low performance
   b. Low wiring density
   c. Poor thermal conductivity of substrate
   d. Significant CTE mismatch between substrate and die materials
   e. Moisture sensitivity of materials
   f. High crosstalk noise

2.3.1.3 Applications The low cost and well established infrastructure of MCM-L technology allows it to be used in all market segments. The primary applications of MCM-Ls include communications, computer, and automotive. Another active area for MCM-L applications is in test equipment, measurement equipment, and other portable equipment. Though MCM-L is not suitable for high performance applications, it is a cost effective packaging technology for most high volume and portable electronic systems.

2.3.2 Co-fired MCMs

The co-fired multilayer ceramic MCMs (MCM-C) have substrates based on co-fired ceramic or glass-ceramic technology. The MCM-C technology was originally developed to manufacture chip capacitors, and it was then used for single chip packaging. In fact, MCM-C technology has been around for more than 25 years. Lately, its capability to make high density packages has been realized. The major breakthough in the increased wiring density was achieved by IBM, which uses alumina ceramic for their mainframes.

2.3.2.1 Substrate Technology Alumina is the most mature and cost effective material and is the most widely used dielectric material in co-fired multilayer MCM-Cs. The next choice is low temperature co-fired ceramic (LTCC). The newest family member of MCM-C dielectric materials is aluminum nitride (A/N). Alumina is widely used due to its low cost, high wiring density, and high strength. Alumina is a high temperature co-fired ceramic (HTCC). Its firing temperature is above 1,500 °C. Other advantages of alumina include compatibility with high temperature braze to low-expansion ferrous alloys and good thermal conductivity. The thermal conductivity of alumina is approximately 10 times better than that of LTCC and 100 times better than that of organic materials.

Though different materials may be used, the fabrication processes of all MCM-C substrates are similar. The fabrication sequence of MCM-C consists of formation of green tape, via drilling or punching, screen printing of conductive material, and, finally, stacking and co-firing all the layers to form a complete interconnect structure.

First, a liquid slurry is formed from ceramic particles and organic binders and then cast into a solid sheet called *green tape* (see Fig. 2.9a). Then via holes are drilled or punched on the green tape sheets (Fig. 2.9b). A conductive ink is applied to each green tape sheet to form the conductive pattern by using a screen

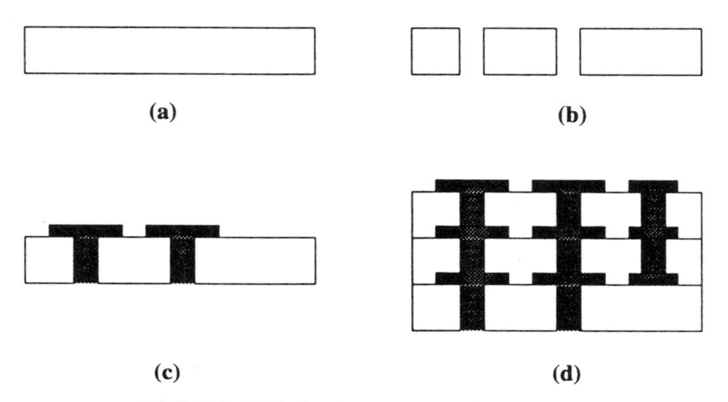

FIG. 2.9. Fabrication process of MCM-C.

printing process while a vacuum is applied to the other side of the punched green sheet to pull the ink into the holes to coat the side walls of the holes or to fill the holes completely (see Fig. 2.9c). *Screen printing* refers to a thick film process in which a paste or ink is squeezed through open areas of a screen and transferred to the surface of a green tape sheet to form certain geometric patterns such as signal paths. The hole is filled completely with conductive ink and is called a *solid via*. Solid vias have the advantage of being able to be stacked directly on top of each other. Finally, all the green tapes with formed conductive patterns on them are stacked on top of each other and co-fired to form a complete structure of interconnections (see Fig. 2.9d). The green tapes of HTCC materials are co-fired at temperatures above 1,500 °C, while the green tapes of LTCC materials are co-fired at temperatures about 800 °C.

2.3.2.2 Advantages and Disadvantages The major advantages of MCM-C technology are flexibility, high wiring capacity, better electrical conductivity, better thermal conductivity, flexible packaging, superior strength and rigidity, and parallel process. In addition, MCM-Cs allow dies to be attached on both sides.

MCM-C is one of the most flexible packaging systems. There is virtually no restriction on the number of layers used. For example, 63 ceramic layers are used in the MCM of the IBM ES/9000 processor [48]. The green tape process allows flexibility in the layer thickness. Therefore, impedance control circuits can easily be fabricated. Both through-vias and buried vias can be processed without increasing the cost, resulting in high wiring density. In addition, MCM-C technology allows metal components on its assembly, the combination of package and module, and the combinations of MCM technologies. For example, thin film conductors on polymers can be used together with MCM-C interconnect modules to provide a grid transformation to flip-chips.

High total wiring capacity is another advantage of MCM-C technology. MCM-C offers the same total wiring capacity as MCM-D by using additional layers at a much lower cost.

The larger cross-sectional area of conductors permitted by screen printing results in better electrical conductivity. The electrical conductivity for MCM-C is typically two to four times greater than for MCM-D.

Improved thermal conductivity is another significant advantage of MCM-Cs. The thermal conductivity of co-fired ceramic is approximately 10 to 1,000 times better than the counterparts in MCM-D or MCM-L.

Like MCM-Ls, MCM-Cs also allow the chips mounted on both sides of the substrate, resulting in a shorter signal path and thus improved signal delay. With the unlimited number of layers that can be used in an MCM-C, diagonal routing is possible on some of the ceramic layers. Diagonal routing can reduce the length of conductor paths by as much as 40%, resulting in significant performance improvement.

MCM-C technology provides a hermetic cavity, integral pins, seal rings, and heat sinks. MCM-C offers the best high temperature stability and reliability among all MCM technologies. The materials used in MCM-C are impervious to moisture and require minimum additional cost if hermetic housing is added. The package and module can be combined in one structure in MCM-C. Therefore, one level of electrical and thermal interconnection is eliminated. By eliminating the metal seal ring, the substrate is ready for coating if no-circuit changes are required. Thus, MCM-C technology offers system design flexibility.

MCM-C has better strength and rigidity than other MCM technologies. MCM-C substrates are resistant to elevated temperature assembly and operating conditions. The metallization adhesion on high temperature MCM-C is one order of magnitude greater than what is available with MCM-L or MCM-D. With MCM-C, there is no issue of glass transition temperatures, nor is there a highly stressed condition due to mixing materials of significantly different thermal expansion coefficients. Both assembly and repair are simplified by this superior metallization adhesion and dimensional stability at elevated temperatures. In addition, the fabrication process of MCM-C is a parallel in nature, allowing each layer to be fabricated, tested, repaired, and reworked individually before cofiring.

Though MCM-Cs have many advantages, there are some shortcomings that need to be addressed. Its disadvantages include lower wiring density than MCM-D, shrinkage of the substrate during co-firing process, high dielectric constant of the substrate material, and CTE mismatch between substrate material and die material.

One of the major disadvantages of MCM-C technology is the lower wiring density than MCM-D technology. There are continuous efforts toward finer line pitch and smaller vias so that fewer layers are required to achieve the MCM-D level of wiring. Another disadvantage is the shrinkage of the substrate caused by the firing process. This makes it difficult to attach the chips by a TAB or flipchip method, as the location of the contact points on the substrate cannot be accurately predicted. There is an on-going effort to improve the predictability of size variation due to firing shrinkage for co-fired ceramic, while new technologies of zero shrinkage approaches are being developed. Yet another limitation of the traditional MCM-C is high dielectric constant. The typical dielectric constant of traditional MCM-Cs is in the range of 7 to 10, contributing to longer signal delays. To reduce the signal delay, new low dielectric constant ceramics have been developed. The low dielectric constant materials include several glass-ceramic materials with dielectric constants as low as 3.9. However, the thermal conductivity of these materials is very poor compared with alumina ceramics. Therefore, they can only be used in very high performance systems.

Another problem has been the differential thermal expansion of silicon die, ceramic substrate, and system level package. As a result, the use of alumina as a ceramic is not considered suitable for high speed applications. Recently, low dielectric ceramics have been developed. These low dielectric ceramics can serve as the substrate, or as the dielectric surrounding the wires in the wire coating applied to the substrate, or as both. A lower dielectric constant decreases the dielectric thickness, reduces crosstalk, and allows smaller spacing.

The low dielectric coefficient is not the only desirable property. Some ceramics with low dielectric and other desirable properties conduct heat rather poorly. For such ceramics, heat must be removed from the die itself, or the low dielectric ceramic can be used simply as the thin film insulator around wiring coated on a strong thermal conductive ceramic substrate.

From the above discussion, it can be seen that the major advantages and disadvantages of MCM-C technology are as follows:

1. Advantages of MCM-C technology
   a. High wiring capacity
   b. Better electrical conductivity
   c. Better thermal conductivity
   d. Assemblies with components on both sides
   e. Flexible packaging
   f. Superior strength and rigidity
   g. Parallel process
2. Disadvantages of MCM-C technology
   a. Lower wiring density than MCM-D
   b. Shrinkage of substrate during cofiring
   c. High dielectric constant of substrate material
   d. CTE mismatch between substrate and die materials

2.3.2.3 Applications Multichip interconnections in co-fired ceramic consisting of 35 layers have been used for many years. IBM was the first to introduce this technology mainly because, as a vertically integrated company, it could obtain bare chips and could design chip sets that could be mounted in the regular arrays needed for the co-fired ceramic technology. The IBM thermal conduction modules (TCM) have been around for over 10 years. With up to 35 layers of metallization on the doctor-bladed green ceramic tape, assemblies could be laminated and co-fired to produce the desired dense interconnect assemblies.

In the past, the system cost using MCMCs was comparable to other alternatives, but the cost and the complexity of MCM-C technology was only possible for mainframe companies like IBM, NEC, and Mitsubishi. The perception was that this type of product was only suitable for high performance mainframes and not really acceptable for cost-performance applications or even military computers. However, the cost of MCM-Cs is being reduced due to advanced production techniques, and it is expected that these MCMs will be used in many other applications as well.

2.3.3 Deposited MCMs

The deposited MCMs (MCM-D) are the MCMs whose interconnection patterns are formed by deposited dielectrics and conductors on a base substrate, typically by thin film process. Therefore, MCM-Ds are also called *thin film MCMs.* This type of MCM, introduced in the 1980s, is the youngest of the three MCM technologies. MCM-D implementation uses a naked die on a silicon or aluminum base substrate that supports a multilayer, high density metal-polyamide interconnection structure. The MCM structure provides very high wiring density, supports a high frequency signal and clock distribution, and also provides minimal interconnection parasitics.

In the following, we first discuss the MCM-D fabrication technology. Then the materials used in MCM-D are presented followed by the bonding methods. Finally, the characteristics and applications of MCM-Ds are discussed.

2.3.3.1 Fabrication Technology IC technology and IC equipment are used to fabricate thin film interconections on an MCM-D. Such IC technology can produce substrates with the finest line pitch and highest wiring density of the other MCM alternatives. The MCM-D technology is the process of depositing thin film metal conductors and un-reinforced polymer dielectrics on a base (support) substrate. The supporting substrate can be silicon, ceramic, metal, or PCB laminate. A cross section of an MCM-D is illustrated in Figure 2.10.

Dielectric and conductive materials are deposited on the base substrate alternately. Masks are used to create specific patterns of conductive material in a sequential manner and create a complex pattern of several layers of conductive path ways. The complete fabrication process, which is a repetition of the basic three-step process (shown in Fig. 2.11), can involve many steps.

1. **Create:** This step creates material on or in the surface of the silicon wafer with a variety of methods. Deposition and thermal growth are used to create materials on the wafer, while ion implantation and diffusion are used to create material (actually, they alter the characteristics of the existing material) in the wafer.

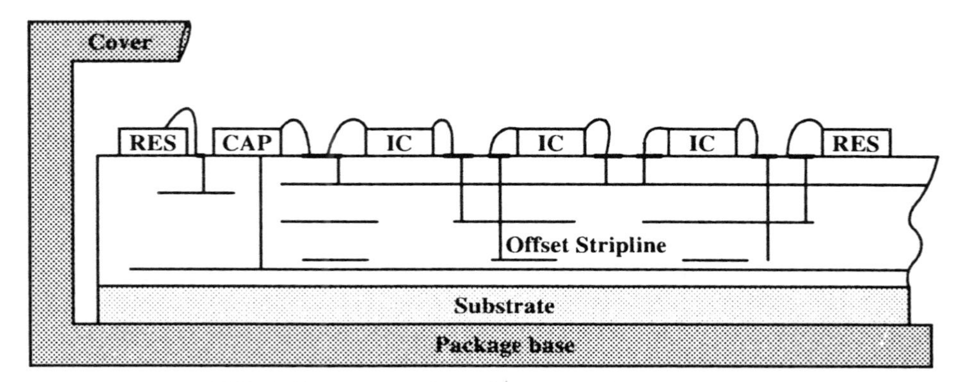

FIG. 2.10. Cross section of a prototypical MCM-D.

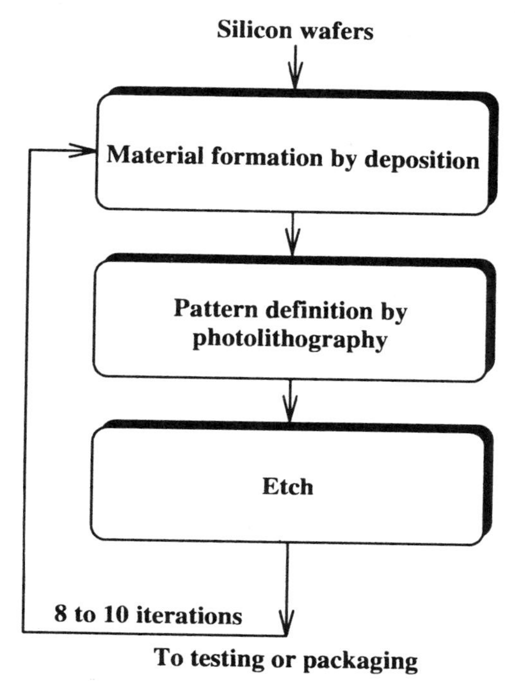

FIG. 2.11. Basic steps in the fabrication process.

2. **Define:** In this step, the entire surface is coated with a thin layer of light sensitive material called *photoresist*. Photoresists have a very useful property. The ultraviolet light causes molecular breakdown of the photoresist in the area where the photoresist is exposed. A chemical agent is used to remove the disintegrated photoresist. This process leaves some regions of the wafer covered with photoresist. Since exposure of the photoresist occurred while using the mask, the pattern of exposed parts on the wafer are exactly the same as in the mask. This process of transferring a pattern from a mask onto a wafer is called *photolithography* and is illustrated in Figure 2.12.
3. **Etch:** Wafers are immersed in acid or some other strong chemical agent to etch away either the exposed or the unexposed part of the pattern, depending on whether positive or negative photoresist have been used. The photoresist is then removed to complete the pattern transfer process.

This three-step process is repeated for all the masks. The number of masks and the actual details of each step depend on the manufacturer as well as on the technology.

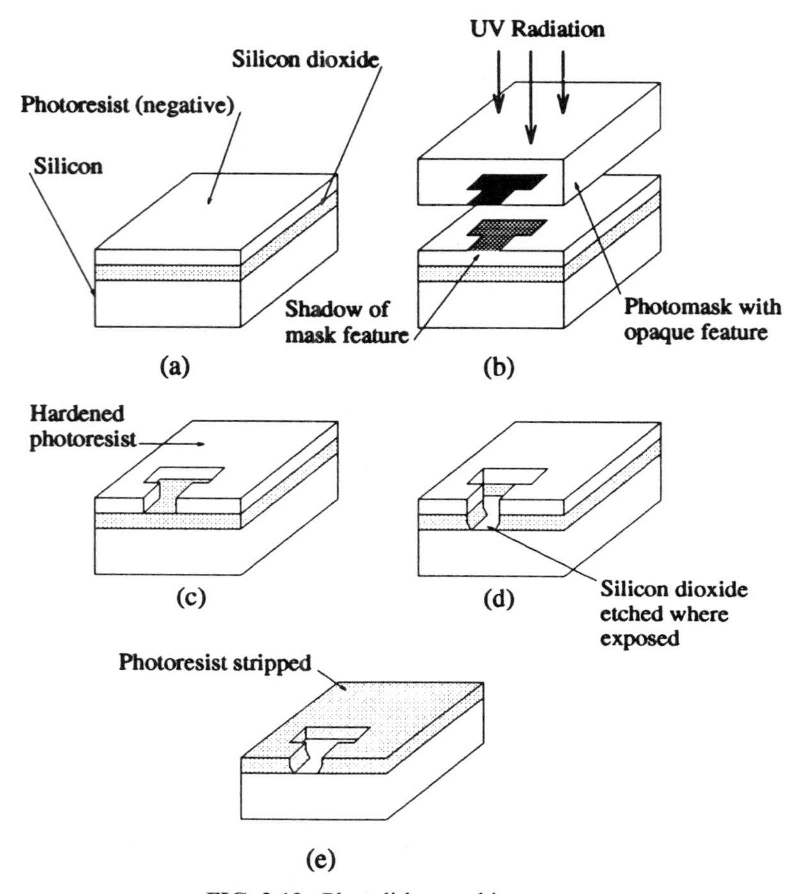

FIG. 2.12. Photolithographic process.

2.3.3.2 Materials for MCM-Ds Though similar fabrication processes are used there are differences in the various MCM-D technologies. The differences are characterized by the conductive and dielectric materials.

Cu, Al, and Au can be used as conductive materials. A comparison of these metal materials is given in Table 2.7. Polymers are used widely as a dielectric material, and plasma deposited silicon dioxide is the second possible candidate. The polymers are deposited from a solvent by spin or spray coating. The polymers used in MCM-D should be chosen such that they have low dielectric constant, good thermal conductivity, and good CTE match to the other materials used in the MCM.

It is not possible to find a material that achieves all the above goals optimally.Trade offs between the desired properties have to be made while choosing the dielectric material. A comparison of the properties of different polymers is given in Table 2.8.

TABLE 2.7. Comparison of Different Conductive Materials

Metal	Advantages	Disadvantages
Au	Chemically inert Good corrosion resistance High electrical conductivity	Poor adhesion to polymer dielectrics Sputtering targets are expensive
Cu	High electrical conductivity Good solderability Inexpensive Can be electrolytically or chemically plated	Poor adhesion to polymer dielectrics May require barrier or adhesion layers
Al	Good adhesion to polymer dielectrics Froms an inert native oxide Well understood	Low conductivity

TABLE 2.8. Comparison of Different Polymers

Vendor	Material	Photo-sensitive	Wet Etchable	Dielectric Constant	CTE
Amoco	UD-4212	No	Yes	2.9	50
Ciba-Geigy	PROB-400	Yes	No	3.3	40
Dow	BCB-3022	No	No	2.7	52
	BCB-19010	Yes	No	2.7	52
Du Pont	PI-2555	No	No	3.3	40
	PI-2611D	No	No	2.9–3.9	20
	PI-2722	Yes	No	3.3	40
	PI-2741	Yes	No	2.9	10
	PI-1111	No	Yes	2.8	19
Hitachi	PIQ-13	No	No	3.4	45–58
	PIQ-L100	No	No	N/A	3
	PL-2315	Yes	No	3.3	40
National Starch	EL-5010	No	No	3.2	38
Toray	UR-3800	Yes	No	3.3	45

Single-crystal silicon wafers are used as base substrates for deposition of wiring coatings of different types. Silicon of this kind is readily available from the semiconductor industry. It is exceptionally smooth and stable and has the same thermal properties as the semiconductor chips mounted on it. Accordingly, the wiring and dielectrics coated on crystal silicon can have the same high resolution (4,000 lines per cm) and submicrometer thickness as on the chip themselves. However, substrate sizes are limited to those available from the semiconductor industry, currently wafers 20 cm or less in diameter. It must be noted that crystalline silicon is brittle, and only a few metals have been developed for use on silicon substrates, with consequent limitation on abilities to handle large currents. Single-metal substrates have been tried but have not become popular, probably because they react to heat differently from silicon.

Many other materials such as ceramic and PCB laminate can also be used in the base substrate. The base material should have strong mechanical strength, excellent CTE match to the dielectric material and to the silicon dies, good thermal conductivity, and low dielectric constant. Aluminum nitride has been used as the base material due to the close match of CTE of the silicon [63]. Co-fired ceramic and glass-ceramic have also been used as the base substrate [202]. A comparison of different substrate materials is given in Table 2.9.

Combinations of gold conductors and low dielectric ceramic dielectris have been developed mainly for deposition on ceramic substrates. The main advantage here is a complete absence from the end product of the organic materials, eliminating moisture absorption problems and providing a large, strong, stable, high density assembly. A main concern has been the low thermal conductivity with which the ceramics conduct heat; but that appears to have been addressed by the use of thinner dielectric coatings around wiring, as well as by new base materials with acceptable thermal properties. When silicon dioxide coatings overlay silicon wafer substrates, wafer-wide conductive planes for distribution of power and ground are readily fabricated as aluminum layers with very thin insulating layers using conventional semiconductor processing. There is no need for decoupling capacitors of any kind in many applications, the power planes are themselves excellent capacitors with extremely low series inductance.

TABLE 2.9. Comparison of Different Base Materials

Material	Dielectric Constant	CTE	Fabrication Temperature	Thermal Conductivity
Almunia	9.4	6.6	1,600	20
Aluminum nitride	8.8	35	1,600	200
Mullite + glass	5.9	35	1,200	6
Borosilicate + alumina	5.6	40	900	6
Lead borosilicate + alumina	7.8	42	900	6
Glass-geramic	5.0	30	950	7
Composite SiO_2 + glass	3.9	19	900	4
Epoxy resin	4.7	1,400	200	0.1
Polyamide resin	4.7	750	300	0.1
MS resin	3.7	680	250	0.1
Polyamide				
PMDA-ODA	3.5	250–400	400	0.2
BPDA-ODA	3.0	20–60	400	0.2
Benzocyclobutene	2.7	650	350	0.2
Polyphenyl quinoxaline	2.7	350	450	0.2

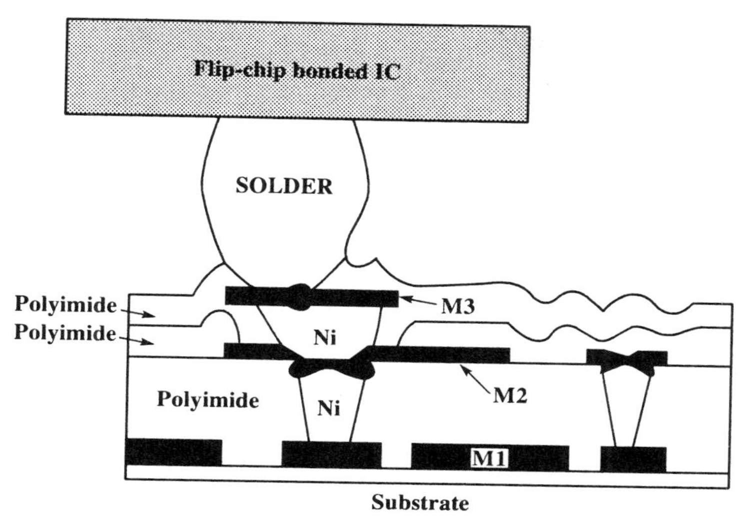

FIG. 2.13. Flip-chip attachment method.

2.3.3.3 Chip Bonding Technologies for MCM-Ds MCM-D demands chip attachment techniques that achieve the highest chip I/O density for the high performance system due to the necessity of shortening the length of the signal path. The most popular chip attaching methods that achieve this are flip-chip and TAB. Figure 2.13 shows the cross section of an MCM-D using the flip-chip attachment method. Three metal layers are used in this example. Wire bonding can also be used if the performance requirement is not critical. The TAB method allows preattachment testing of a die, whereas pre-attachment testing of a die is not possible in flip-chip solder-bump technology. Preattachment testing of a die can ensure that no defective dies are mounted on MCM-D. Testing of individual dies while still in the wafer can be done up to 40 MHz automatically, while testing above 40 MHz is costly. The TAB method provides a test fixture as a part of the die mount process. Therefore, dies can be tested using special hardware when they are mounted on the substrate. However, solder-bump technology can have better electrical performance since the soler-bump contacts may exhibit only a fraction of a pF of contact capacitance and low parasitic inductance as well.

2.3.3.4 Characteristics MCM-D, the VLSI based technology, has many advantages, such as high performance, high wiring density, low dielectric constant, and good electrical properties. The main advantage of MCM-D is the performance edge it has over either MCM-C or MCM-L due to its finer line pitch, smaller vias, and lower dielectric constants. For example, IBM has developed modules with a 10 μm line width and a 13 μm line pitch using MCM-D technology [28]. The vias in the MCM-D technology are also significantly smaller than the ones in any other technology. The reduced feature size results in smaller substrate area and less number of layers. The proportion of the number of interconnect layers required for a given application in MCM-D and in MCM-C can be as low as 1:4 [194]. MCM-D can achieve better electrical properties. The shorter signal paths allow lower capacitive load and reduced circuit noise. Signal-to-signal capacitance is reduced, resulting in the reduction in crosstalk noise. MCM-D technology provides low dC or aC interconnect resistance, rigidly controlled interconnect impedances, and high density decoupling and termination components. Combined with lower dielectric constants, a faster clock rate can be achieved using MCM-D. The dielectric constant of polymers is in the range of 2–4 at about 10 MHz, allowing for a 25%–40% improvement in signal propagation delay over the traditional co-fired MCM-C modules. In fact, when the system clock rate exceeds 100 MHz, severe loss of signal or severe signal distortion has been observed when other packaging and interconnect technologies are used [47, 160]. The use of MCM-D seems mandatory for high performance systems with clock rates of 100 MHz or higher [47].

The wiring density to support a generalized assembly of random logic ICs is approximately twice the device I/O count divided by the pitch of devices in the assembly. Thus a 240 I/O device arrayed at a 12 mm pitch would require a substrate wiring capacity of 400 cm of wiring per cm^2 on the substrate. This can be provided by two layers of routing on a 50 μm pitch. This gives a very useful indication of the interconnection geometry requirements for the MCM-D and many present day MCM-D technologies with two layers for signal routing with the conductor traces at a pitch between 40 and 100 μm. To achieve low trace resistances in such fine pitch wiring, low resistivity metals such as aluminum and copper are employed. The interconnection traces are also increased in thickness over those on chip level, with trace thickness of 2–5 μm. Linewidths are typically between 10 and 25 μm. The linewidth to line pitch ratio is usually selected to be greater than 2:1 to ensure low signal crosstalk. The power and ground connections in an MCM-D substrate may be provided within the signal layers in a two layer construction. Higher performance modules employ four or five metal layer substrates, with separate power and ground planes, and two layers of signal routing. The fifth layer may be added for pad connections. This ensures low impedance power and ground structures and avoids routing constraints. A polyamide or other low dielectric constant polymer is employed as the interlayer dielectric. The dielectric thickness separating the metal layer is typically 5–20 μm, thus providing low capacitance per unit length for the interconnection traces, with good interlayer isolation. This geometry also allows impedance controlled 50 Ω lines. While these lines are lossy, work has shown that MCM-D scale of transmission line has very useful performance well into the radio frequency microwave region. Signal propagation delays for the MCM-D interconnection structure are 55–60 ps/cm, giving a typical cross module delay of below 1 ns. This combination of power/ground and signal interconnection structure provides good power and clock distribution, with significantly improved margins compared with more conventional packaging and interconnection technologies.

The MCM-D approach offeres a high performance interconnection and packaging route to achieve the full potential of VLSI silicon devices and in particular provides a viable path to 100 MHz processor clock rates, while achieving significant improvements in size, weight, and relaibility. Since MCM-D technology is relatively new, it suffers from a lack of sufficient infrastructure to provide materials, equipment, and completed MCM-D substrates.

One of the most important issues is the availability of known good dies. *Known good die* (KGD) refers to the bare die that has been tested and burned-in at frequency specifications. KGD remains one of the largest hurdles to overcome for the entire MCM industry, especially for MCM-D technology, since repair and rework in MCM-D significantly increase the overall system cost.

Low total volume, low yield in the process, and high cost are significant disadvantages of MCM-D. Many MCM-D processes are directly derived from IC processes, generating overkill capacities for most current applications. IC processes require high resolution and expensive machinery and unnecessary clean room, resulting in the high cost of MCM-D technology. In summary, the major advantages and disadvantages of MCM-D technology are as follows:

1. Advantages of MCM-D technology
   a. High performance
   b. High wiring density
   c. Low dielectric constant
   d. Good electrical properties
2. Disadvantages of MCM-D technology
   a. High cost
   b. KGD problems

2.3.3.5 Applications MCM-D offers the best performance for system designers. It has been well established for performance in terms of clock frequencies, noise reduction, and propagation delay requirements. MCM-D has been mainly used in mainframes such as the IBM 390/ES9000 [202], the NEC SX-3[2], the Hitachi M880 [103], and the DEC VAX-9000 [53]. Since highly complicated IC fabrication technology and equipment are required for producing MCM-Ds, most current MCM-D fabricators are completely vertically integrated.

MCM-Ds are used in high capacity and high speed telecommunication switching and transmission systems due to the performance and density characteristics of MCM-Ds. In fact, AT&T has used MCM-DC in its Merrimack Valley facility since 1987 [139], and NTT has announced that it will implement switches on MCM-D technology on a worldwide basis by the mid-1990s [154]. Several avionic [81], HDTV [180], and workstation applications [28] are in prototype stages and will be in commercial products in the near future.

The use of MCM-Ds will become mandatory for high performance systems with clock rates of 100 MHz or higher. It has been projected that the performance of typical workstations will be in the range of 150–250 MHz in the near future

[31]. It is also predicted that MCM-Ds will take a major portion of the MCM market in near future.

2.3.4 Hybrid MCMs

Hybrid approaches such as MCM-DL (MCM-LD) and MCM-DC (MCM-CD) use combinations of different technologies. Both MCM-DL and MCM-LD refer to the technology of deposition of thin film on the laminate substrate, which not only works as a support but also contains interconnections for the power, ground, and some signals. Similarly, both MCM-DC and MCM-CD refer to the technology of deposition of thin film on th co-fired ceramic substrate, which contains interconnections for the power, ground, and some signals.

The base substrate contains much of the power, ground, and noncritical signals in the case of MCM-DL and MCM-DC. This is different from traditional MCMs, in which the base material supplies the support. The hybrid approaches may optimize the cost performance trade offs. For example, MCM-DL provides both improved via densities and PCB processing economies. Thus the hybrid approaches may be the most economic methods for upgrading the performance.

2.3.5 Comparison of Substrate Technologies

Table 2.10 compares the MCM families in terms of line widths, line density, line separation, turnaround time and years for which these technologies have been available.

MCM-L is a mature technology and has the highest availability and lowest cost of all MCM technologies. The new MCM-Ds and low dielectric ceramic MCM-Cs outperform the plastic package laden PCB, MCM-L, and co-fired MCM-Cs in many ways. The time delay between chips is less, electrical noise and crosstalk are less, and size is less. The chips can be larger, approaching 2.5 cm^2, and the I/O lead counts approach 500 per MCM. Discrete bypass capacitors are unnecessary for power distribution. Test and rework can be done in a more detailed fashion. With some designs, on-the-spot programming of interconnects by the used decreases lead times and should lower costs. Until the demand for these new MCMs becomes much stronger, product costs will remain fairly high. In the following, we describe MCM-L, MCM-C, and MCM-D in more detail.

TABLE 2.10. Multichip Module Classification

Characteristics	MCM-C	MCM-L	MCM-D
Line density (cm/cm^2)	20–40	30	200–400
Line width/separation (μm)	125/125–375	750/2,250	10/10–30
Turn-around time	1 month	9–13 weeks	10–25 days
Years of availability	> 10	>50	>5

In summary, MCM-D achieves the highest performance and wiring density, whereas MCM-L has the lowest cost. On the other hand, MCM-D requires the most sophisticated fabrication process and equipment, resulting the highest cost, whereas MCM-L has the lowest performance and wiring density among the three technologies. MCM-C lies in the range between MCM-L and MCM-D in terms of both performance and cost.

Generally speaking, MCM-Ls achieve the lowest cost and lowest wiring density of the MCM technologies, whereas MCM-D offers the highest wiring density and the highest performance at the expense of the highest cost. MCM-C lies in between MCM-L and MCM-D in terms of cost and wiring density. In Table 2.11, a comparison of other characterisitics of these technologies is presented.

TABLE 2.11. Comparison of Substrate and Interconnect Technologies

Parameters	MCM-L	MCM-C	MCM-D
Base substrate materials	N/A	N/A	Al_2O_3, *AlN*, *BeO*, *Si*, *CU*, ceramic
Conductor materials	*Cu*	*Au*, *Ag*, *PdAg*, *Cu*, *W*, *Mo*	*Au*, *Al*, *Cu*
Dielectric materials	Epoxy-glass Polyamide BT-epoxy Cyanate ester Sycar laminate sheets	Glass/ceramic Ceramic (Al_2O_3)	SiO_2 polyamide benzocyclobutene
Minimum line width (μm)	60–100	75–100	8–25
Minimum line pitch	625–2250	50–450	25–75
Dielectric constant	3.7–5	5–9	2.4–4
Via grid (μm)	1,250	125–450	25–75
Via diameter (μm)	150–500	50–100	8–25
CTE	4–16	3–8	3–7.5
Thermal conductivity (w/M °C)			
Base substrate	N/A	N/A	25–260
Dielectric	0.15–0.35	1.5–20	0.15–1.0
Maximum number of layers	46	63	8
MCM pin-out (mm)	Array 2.54	Array 1.00–2.54	Peripheral 0.63
Maximum substrate size (mm)	700	245	50–225
Wiring density (cm/cm^2)	30	20–40	200–400
Repair/rework	Easy	Easy	Difficult
Relative cost	1	1.5–2.0	2–10
Power dissipation	Poor	Moderate	High

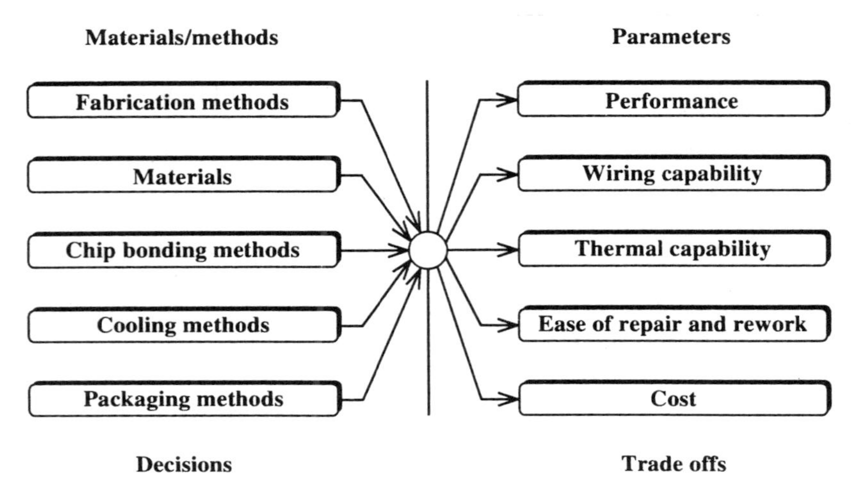

FIG. 2.14. MCM parameters are affected by different methods and materials.

2.4 LIMITATIONS OF THE TRADITIONAL CLASSIFICATION METHOD

The traditional classification method offers a simple way to classify the MCMs and is widely accepted. However, as different methods and materials are used in the same type of MCMs and hybrid approaches combining the different types of MCMs are used, it becomes more difficult to identify the characteristics of a type of MCM due to widely different parameters of the same types of MCMs.

The problem of the traditional classes that they are based on substrate fabrications methods only. However, the properties of an MCM depend on other methods and materials as well (see Figure 2.14). Therefore, the classification of MCMs should be based on all the important methods and materials. A good classification scheme should specify

1. Fabrication process
2. Dielectric material
3. Conductive material
4. Base material
5. Chip bonding method
6. Thermal management method

Since the classification parameters such as performance and cost are mainly determined by the factors discussed above, the new classification can provide a much more comprehensive picture for each type of MCM than the one provided by the traditional classification. In fact, once all the factors are specified, the range of each MCM parameter can then be estimated quite accurately.

2.5 SUMMARY

The classification of MCMs is important in order to study MCM properties, design methods, fabrication methods, and applications carefully. The parameters that characterize the MCM technologies and form the basis of classification include different fabrication methods, materials, and chip bonding methods. The traditional classification of MCMs is based on fabrication methods and classifies MCMs into MCM-Ls, MCM-Cs, and MCM-Ds. MCM-Ls have been available for a long time. They have a well established infrastructure. MCM-L is the lowest cost MCM technology and is suitable for many portable computers and equipments. However, the MCM-Ls are restricted by low wiring density. Thus they are not suitable for high performance systems. MCM-Ds offer the highest wiring density and performance of the MCM technologies. They are the best choice for future high performance electronic systems. The disadvantages of the MCM-D technology are high cost and lack of good infrastructure. MCM-Cs fill the gap between MCM-Ls and MCM-Ds. Their performance is similar to that of MCM-Ds of but at a much lower cost. MCM-Cs have been used in mainframe computers for many years.

2.6 PROBLEMS

1. There are two materials that can be used as dielectric materials in an MCM. The dielectric constant of the first material is 3.0, whereas the dielectric constant of the second material is 6.0. The minimum space between two chips that can be achieved by using the first mateiral is 6 mm, whereas the one that can be achieved by using the second material is 4 mm. Which mateiral should be used in the MCM to achieve the best perfrmance? Why?

2. There are several constants of a material used in an MCM that affect the parameters of the MCM, such as dielectric constant, thermal conductivity, and coefficient of thermal expansion. What parameters and functions of the MCM are affected by each of these constants?

3. Equation 2.1 is obtained by assuming that there are three interconnections for every four terminals, the average wiring length for an interconnection is 1.5 × chip pitch, and the efficiency in routing is 50%.

$$N \leq \frac{0.44W}{PI} \tag{2.1}$$

   Are these assumptions realistic? If the assumptions are to be changed so that there are four interconnections for every four terminals and the routing efficiency is improved to 70%, what will be the equation for estimating the number of chips that can be contained in an MCM?

4. What methods are available to remove the heat efficiently in an MCM? Discuss the advantages and disadvantages of each method.
5. We have discussed three chip bonding methods. Among these methods, which is the best one for a high performance MCM, considering the signal delay caused by the chip bonding method and the effect of the method on the wiring density of the MCM?
6. Among the three chip bonding methods, which one can remove heat most efficiently?
7. Pretest is an important ingredient of an MCM. The ease of pretest in an MCM depends on the chip bonding method used in the MCM. Identify the chip bonding method used in an MCM that allows ease of pretest of the MCM.
8. Reworking an MCM referes to removing a chip or chips from the MCM. Does the chip bonding method affect the reworking of an MCM? If so, which chip bonding method should be used in an MCM so that the reworking will be easiest?
9. If one is designing an MCM for automobiles, what are the factors one must consider? Which type of MCMs (MCM-L, MCM-C, or MCM-D) should one use? Why?
10. One of the limitations of the traditional classification method is that it cannot distinguish between many parameters among different groups. Suggest a classification method that can distinguish the parameters of performance, wiring capacity, and thermal capacity among different groups.

BIBLIOGRAPHIC NOTES

The three most popular bonding methods, wire bond, TAB, and flip-chip are compared by Collier [41] and Hodson [97]. Collier [41] describes the requirements of chip requirments of chip attachment techniques for silicon hybrid circuits; compares flip-chip, TAB, and wire bonding; and discusses the significance of assembly yield and test. Although area array flip-chip technology provides perhaps the ultimate electrical interconnection density, TAB can also be applied in a very high density format. Silicon devices can be bonded face down with short beam leads and equal inner and outer lead bonding (OLB) pitches. Assembly of high lead count TAB test chips onto fine line aluminumpolyamide and copper-polyamide connections on silicon substrates has been evaluated in the RISH (Research Initiative into Silicon Hybrids) program. The MCMs have been categorized as MCM-L, MCM-C, and MCM-D by the Institute for Interconnecting and Packaging Electronic Circuits (IPC). An overview of MCM-L technology is given in several papers [26, 93, 145], as is a review of present and emerging MCM-L technologies [93]. The properties of MCM-C technology have been reviewed [4, 91, 147]. Amey [4] provides an overview of ceramic technologies for MCM applications. Examples of modules applying advanced techniques for high density circuit patterning, impedance control, thermal management, and change and repair are described. There are several papers [68, 187, 195] on MCM-Ds. Tessier and Garrou [195] describe the current status of MCM-D substrate technology. The driving forces for MCM-D interconnects are reviewed, and the myriad of dielectric and metallization material options and substrate fabrication approaches described in the literature are compared. Hermetic and nonhermetic packaging alternatives for MCM-D substrates are also reviewed.

CHAPTER 3

DESIGN OF MULTICHIP MODULE BASED SYSTEMS

As MCM technology matures, an increasing emphasis has to be given to the design of MCMs. The design of MCMs or MCM based systems is a very complicated process, and requires an interdisciplinary approach. The design involves not only the usual manufacturing and reliability issues but also issues such as materials, thermal and electrical parameters, simulation, and testing. In the following, we first discuss the fundmental issues in MCM design such as design objectives, design process, and electronic design automation. Then, each individual phase in the design process is discussed. The characteristics of MCM design tools are also presented.

3.1 OVERVIEW OF MCM DESIGN

In this section, the fundamental issues of MCM design are discussed. First, we discuss the classification of MCM design objectives which is the starting point of any MCM design. Then the design flow of MCM based systems is presented and the automation of each design phase is discussed.

3.1.1 Design Objectives

MCMs are used in electronic systems to achieve one or more of the following objectives: high performance, small size, light weight, high reliability, fast, time-to-market, and low cost. As the primary reasons for using MCMs in an electronic system differ, the objectives or the priorities of the objectives of MCM design vary from one system to another. In the following, the objectives of MCM design and their potential impact on the final products are discussed.

The typical objectives of an electronic system include high performance, effective thermal management, reliability, low weight, low size, low cost, and short time-to-market. The objective of a particular system depends on the type of application for which the system is intended. For example, in high speed mainframe or telecommunication equipment, the primary concern may be the performance whereas the most important goal for portable equipment such as laptop computers may be the weight and size. Military applications require high reliability, while the cost may be the most important factor for consumer electronics. At the beginning of the MCM design process, all these objectives should be well understood and prioritized so that the best solution balancing all the objectives can be reached. Depending on priority, there are different design strategies. For example, if the highest priority for the design is size reduction, then the following factors need to be emphasized for MCM design:interconnect density, I/O counts, component placement, capping technology, thermal management, and a second level interconnection technology [60]. On the other hand, if performance is the top priority, the electrical issues such as interconnect delay minimization and noise control should then be of utmost importance. If low cost and short time-to-market are the primary goals, the use of standard parts is important since production in high volume can reduce the cost. Generic assembly and testing tools and standards should be used as much as possible to reduce the cost and time to-market. The choice of the MCM fabrication technology may also depend on the design objectives. MCM-D may be a better choice if the goal is to develop high performance systems. MCM-L is the most cost effective approach if it can saatisfy the design requirements.

An MCM must provide a proper operating environment for its dies by meeting all the relevant constraints imposed on it. The environmental constraints include physical constraints such as size or weight requirements, electrical constraints such as working frequency of the system or noise level, and thermal constraints such as the heat removal capability. These constraints arise due to the performance requirement of the design. In addition, the design of MCMs must also guarantee the manufacturability, testability, and repairability, since from the manufacturing view point these are most important concerns.

3.1.2 Design Process

The design of a large electronic system such as a mainframe is a complicated process. It includes the tasks of mechanical analysis, thermal management, and electrical design as well as the overall system level planning involving trade off analysis. The levels of abstraction also differ during the design process and impose different requirements on the design. At the highest level, only behavioral information is available, and the designer has many options. At the lowest level, the exact physical layout is to be decided and exact information is available. However, the range of options is quite limited. To pursue the best design strategy for each type of design activity and to concentrate on each area, the design process is divided into several phases. This allows the research and development

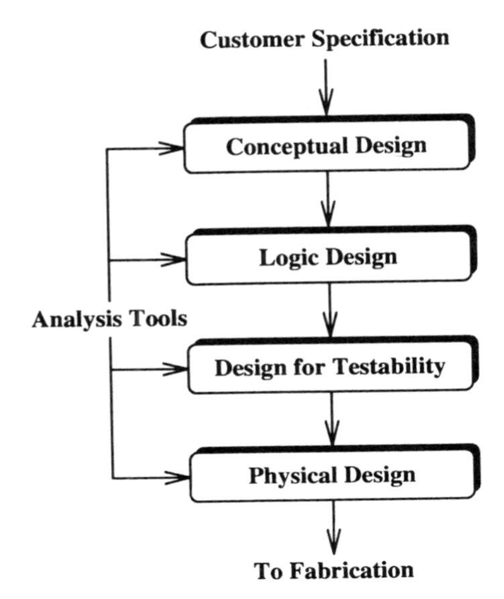

FIG. 3.1. Design process steps.

efforts to concentrate on specialized problems in each phase. This also helps in development of CAD/CAE tools for each phase.

The design process can be divided into system design, logic design, and physical design. The typical design process is shown in Figure 3.1 and discussed below.

1. **System Design:** System design refers to the design activities that produce the system specifications. It includes the selection of technology and materials to be used in MCMs, trade off analysis, and the specifications of clock frequency, size, power, functionality, and cost of the system. System design is of great importance in reducing the total time and cost of the system. The reduction in design time and cost is possible because a correct system design can reduce the number of design iterations.
2. **Logic Design:** Logic design is the process of converting behavioral specifications, which are the result of system design, into circuit representations. First, behavioral aspects of the system are considered. The outcome is usually a timing diagram or other relationship between subunits.Then, the logic structure that represents the functional design is derived and verified. Finally, this logic design of the system is simulated and verified to ensure its correctness. Logic design is discussed in detail in Section 3.3.
3. **Physical Design:** In this step, the circuit representation of each component is converted into a geometric representation. Physical design is a very complex process; therefore it is usually broken down into various sub steps. The physical layout is then verified to ensure that all the design require-

ments are met. The physical design phase is discussed in detail in Section 3.4. Algorithmic issues of physical design are discussed in Chapter 8.

These different design phases are not isolated. The design process involves iterations, both within a phase and between different phases. The entire design process may be viewed as transformations of representation in various phases. In each phase, a new representation of the system is created and analyzed. The representation is iteratively improved to meet system specifications. For example, a layout is iteratively improved so that it meets the timing specifications of the system. An undetected error that occurred in an early phase and discovered at a later phase causes a feedback to be submitted to the earlier phases. Several iterations may be necessary for all these phases to correct the error.

After a MCM is fabricated, it is tested to ensure that it meets all the design specifications and that it functions properly. As the testing of MCMs becomes increasingly complicated, the design of testing circuits is not a trivial task. As a result, design for testability should be an important factor in the early design phases. This allows efficient and cost effective testing of MCMs.

3.1.3 Electronic Design Automation

As semiconductor technology develops, the complexity of an electronic system increases dramatically. A complex electronic system can contain tens of millions of components. Therefore, is impractical to design such a system without automation. *Electronic design automation* (EDA) refers to the engineering field of providing tools and support for electronic design. EDA tools can help reduce design time and design cost and increase the reliability of design. We also use the term *CAD/CAE* tools for EDA tools in this book.

The EDA tools were first introduced for detailed phases such as physical design and logic design. In addition to the complexity imposed by the large number of components, physical design involves complex design rules, exacting details required by the fabrication process. As a result, almost all phases of physical design extensively use EDA tools, and many phases have already been partially or fully automated. This automation of the physical design process has increased the level of integration, reduced the turn-around time, and enhanced MCM performance. Logic design involves a large number of logic functions. The task of handling these functions is automated to ease the iterative improvement process.

3.2 SYSTEM DESIGN

System design, also called conceptual design, high level design, or early design, is the first step in the design of MCM based systems. The system design phase starts with an understanding and prioritization of the design objectives. The objectives should be clearly defined and investigated for different technologies. During the system design, a selection procedure is used to select the fabrication method, dieletric material, conductive material, chip bonding method, and thermal management method to be used. As specifications for the hardware and software elements of the architecture develop, different logic partitioning, chip functions, timing requirements, and clock distribution strategies will be evaluated, and trade-offs will be made to optimize system functionality and various factors such as speed, power, manufacturability, and materials available. Moreover, many design objectives need to be evaluated and trade offs need to be made. As a result, the selection of the technology and materials used in an MCM is not a trivial task. In addition, no detailed information is available at this stage, which makes a correct system design even harder to achieve. More attention has been drawn to this field recently [178]. System design is discussed in detail in Section 3.2.

During the selection process, trade off analysis is carried out to optimize the design as different objectives such as performance and cost cannot be optimized at the same time. The major decisions in the system design phase also include whether to develop an internal capability or to use the MCM products of another company. If MCMs are to be developed internally, whether an existing design is used or to be changed to suit the requirement of the new products or whether a completely new design is to be carried out also needs to be decided.

There are initial requirements for the MCMs or MCM based systems that are specified by customers or by market survey. These requirements may concern the performance, size, and cost of the products. These requirements need to be met which dictates many choices in selecting the MCM technologies. For example, the use of an existing design or manufacturing infrastructure can reduce the time-to- market and the cost. Another consideration is the long-term objective. For example, even if an MCM-L can satisfy the requirement for the current design and is a more cost effective approach than an MCM-D, an MCM-D may be a better choice if the company is pursuing a long-term goal of developing high performance systems. Yet another consideration is whether to use standard parts since this production in high volume can reduce the cost. Generic assembly and testing tools and standards should be used as much as possible to reduce the cost and time-to-market.

During system design, in addition to the selection of technologies, the clock frequency, size, power, and functionality of the system are also specified. As specifications for the hardware and software elements of the system develop, many partitionings of the system, ship functions, timing requirement, and clock distribution strategies will be evaluated. Trade-offs are made to optimize system functionality and various factors such as performance, power, manufacturability, and cost. The result of system design is the complete behavioral specifications of the system.

A good system design is important for the success of the entire design, because the choice of technology may determine the ultimate performance and cost of the system. In addition, in order to modify anything late in the design process, all the phases that have been completed must be carried out again. These iterations can be time-consuming. As a result, it is crucial to achieve a correct system design in one pass.

3.2.1 Trade Off Analysis

The selection of technologies has a major impact on the properties of MCMs and on the parameters of the final product. There are many parameters to be considered when choosing the technology and materials. The parameters include size and weight, reliability, wiring capacity, thermal capacity delay and electrical noise, performance, time-to-market, and cost. To evaluate the tradeoffs, the appropriate figure of merit (FOM) needs to be decided, which has higher weight among the parameters even though attention has to be given to every parameter. In the case of a high performance system, the FOM may be system cycle time, whereas the FOM may be size or weight if the final product is a portable system.

After deciding the appropriate FOM, the trade off analysis is used to compare the advantages and disadvantages of each choice of technology. An example of using trade off analysis to select the MCM technology is shown in Tables 3.1 and 3.2. In this example, five options (A, B, C, D, and E) in the design of a notebook computer are evaluated. The five options are rated according to each parameter such as size and cost. Then a total score is calculated for each option, based on the weight for each factor. The largest difference among the alternatives is due to the

TABLE 3.1. Summary of Sizes, Package-Related Energy Consumption, and Production Costs for the Packaging Options

Option	Type	Size (mm)	Power (W/MIPS)	Cost ($/part)
A	PCB with SCMs	120 × 120	0.027	536
B	PCB with TAB- mounted dies	120 × 110	0.026	736
C	MCM-L with wire bonded dies	82 × 64	0.015	785
D	MCM-C with TAB-mounted dies	62 × 60	0.015	950
E	MCM-D with wire bonded dies	44 × 45	0.014	1,060

TABLE 3.2. Ratings of Each Packaging Option Based on Several Factors

Goal	Weight	A	B	C	D	E
Size	10	3	3	8	8	10
Production cost	10	10	9	8	5	3
Weight	8	5	5	6	6	6
Power	7	5	5	6	6	6
Design cost	6	10	8	8	5	5
Thermal dissipation	5	3	3	8	10	8
Total weighted ranking		280	258	338	300	290

size factors. The thin film MCMs offers the largest savings in the size. The laminated MCMs has 75% savings over the PCB options. There are significant differences in the production costs as well. The PCBs are the least expensive to produce, followed closely by the laminated MCMs. On the other hand, the thin film MCMs are the most expensive option. The overall score shows that the laminated MCM is the best choice to satisfy the design requirement. The details of this example are presented in Franzon and Evans [67].

Methods of trade off analysis can be classified in two ways: according to the methodology or according to their contents. In the following, we discuss both schemes.

According to the methodology, methods of trade off analysis can be classified as advisory, estimation-based predictive, or simulation-based predictive methods. Advisory methods are usually implemented using knowledge-based systems or expert systems. On the other hand, predictive methods are based on estimation or simulation of the important parameters of the final products. The actual trade off analysis may be accomplished by combining the advisory approach with the predictive approach. In the following, the characteristics of advisiory methods and predictive methods are discussed.

1. **Design Advice:** The qualitative advantages and disadvantages of each design option are presented. In many cases, the qualitative description is enough to make the decision. However, in some cases, detailed evaluation, for example, how much better one design option is over another one, is required before the decision can be made. In this situation, predictive methods should be used.

   The characterization in advisory methods include type, basis, and ranking [54]. *Type* refers to the design issue that the advice is concerned with. Within a particular type or design issue, design advice is characterized by the *basis* by which the options will be compared. The basis may include performance, wiring capacity, thermal capacity, and cost. Finally, within a particular basis, the design advice is characterized by the *ranking.* Ranking indicates the relative importance of the advice. Design advice may be ranked according to quality, generality, or constraint-driven. If ranking is based on quality, the parameter that can be estimated most accurately will be given high weight. If the ranking is constraint-driven, the weight of each parameter is an input to the advises. The design advice can be a very complicated expert system since it deals with difficult trade offs.
2. **Prediction:** The advisory approach can only provide preliminary analysis. More detailed analysis can be accomplished by predictive methods. Since no detailed design has been made and no detailed information is available, the analysis can only be carried our in an " approximate" way.

Let us now concentrate on different types of trade off analyses used in MCMs. The trade off analysis includes wiring, sizing, reliability, thermal, and electrical analyses.

3.2.1.1 Wiring Analysis Wiring analysis estimates the wiring utilization of electronic systems. Since the available wiring capacity (AWC) must satisfy the required wiring capacity (RWC) in any design, it is very important to predict the RWC and AWC accurately. The fabrication technology has a major impact on the AWC. Therefore, wiring analysis is necessary for the selection of fabrication method.

As we have seen in Chapter 2, the RWC of an MCM can be estimated by

$$N \leq \frac{0.44W}{PI}$$

or

$$W \geq \frac{PNI}{0.44}$$

where N is the number of chips, W is the wiring capacity, P is the chip pitch, and I is the number of I/Os per chip. I is determined by the number of circuits per chip as predicted by Rent's rule, discussed in Chapter 1 [125].

On the other hand, the AWC can be estimated by

$$W = w \times S \times L$$

where W is the wiring capacity, w is the wiring density, S is the routing area available on each layer, and L is the number of layers.

To interconnect all the chips on an MCM, the AWC of the MCM must be greater than the RWC. If the AWC of the MCM is less than the RWC, then one of the following choices must be made.

1. Reduce the number of dies on each MCM by repartitioning the system into more MCMs.
2. Reduce the chip pitch by selecting a different bonding method or substrate material.
3. Select a fabrication method that produces less vias and/or smaller line pitch, resulting in larger wiring density.
4. Increase the number of layers.
5. Increase the size of the routing area.

Note that any of the above choices will change other system parameters. For example, choosing a fabrication method that can produce less vias and smaller line pitch can increase the cost. Increasing the number of layers in an MCM may reduce the performance.

In addition to the consideration of the wiring capacity of an MCM, there is the issue of the wiring capacity between different package levels such as between bare dies and an MCM or between MCMs and a PCB. The wiring capacity between different package levels depends on the type of connectors between them. Usually area connectors with pins distributed over the entire area of a chip or an MCM provide much higher wiring capacity than peripheral connectors. The wiring capacity between different package levels is usually much smaller than the wiring capacity within a package level. For example, the pitch between the pads on the substrate that provide the interconnection between bare chips and MCM substrate is larger than the wiring pitch on a die. As a result, in many cases the wiring capacity between different package levels is the bottleneck. Moreover, using the high pin count connectors between different levels of packages is costly. Therefore one should minimize the use of connectors in MCMs and PCBs since the use of connectors also reduces the reliability of the sysstem. However, there is a penalty for using large MCMs. The fabrication cost of the MCM is directly proportional to the size of the MCM substrate. As a result, a trade off has to be made regarding the size of the substrate.

By carrying out wiring analysis, designers are able to determine the impact of different fabrication techniques, connector methods, and sizes of MCMs and PCBs on the wiring capacity. Accordingly, it helps the designers to make appropriate trade offs among fabrication techniques, connector types, and sizes of MCMs an PCBs to implement the required functionality.

3.2.1.2 Sizing Analysis The sizing analysis of an MCM is critical to the prediction of performance and manufacturability of th MCM. The size may determine the length of critical nets, which in turn may determine the clock frequency.The distance between chips also depends on the size of the MCM. The size of an MCM also affects its reliability and yield, which contributes to its cost.

Size and weight are important parameters in portable equipment and aerospace systems. The size limitation of a system can be an area or volume limitation or both. Notice that during the system design, the size and weight of a subsystem can only be estimated. If an error occurs in the estimation of the size and weight of any subsystem, more advanced packaging should be used for the sub system to avoid a complete redesign of the whole system. In addition, when the size and weight are estimated, the calculation should also consider the heat removing system. This is due to the fact that when more components area packaged into a certain space to reduce the size, more heat will be generated. As a result, a more complicated heat removal system is required that takes more space. Therefore, there is a limitation on how much the space can be reduced. When the power consumption increases as more components are integrated into the system, the size and weight also increases, since the size and weight of the power supply increase. In CMOS circuits, the capacitive energy is the main form of power consumption. Generally speaking, the power consumption is directly proportional to the total capacitance of the interconnection and the clock frequency [58]. The capacitance of the interconnection is directly proportional to the interconnection length. Therefore, reducing the interconnection length leads to less power consumption and smaller and lighter systems.

As discussed above, the size of an MCM is related to its performance, cost, heat removal, an power consumption. By using sizing analysis, the proper size of

MCMs which satisfies performance and cost requirements of the system can be determined.

3.2.1.3 Reliability Analysis Reliability refers to the length of time during which an electronic system can operate properly. Reliability of an electronic system depends on circuits, chip packages, chip bonding technique, interconnects, cooling schemes, and power supplies [178]. The reliability of an electronic component is measured statistically since electronic components are rarely worn out.

The reliability of any electronic system increases when the number of package levels is reduced. Therefore, more components should be packaged at a package level to increase the reliability. In addition, the reliability of the system relates to the mismatch between CTEs of different materials used in the system. The significant mismatch between CTEs of different materials used in a system such as the mismatch between die material and substrate material reduces the reliability of the system. Reliability analysis helps to determine the impact of thermal characteristics of the MCMs on the reliability of the endproduct.

3.2.1.4 Thermal Analysis Electronic components must operate in a certain temperature range due to reliability concerns. The acceptable operational temperature of an electronic component is determined by the performance and reliability requirements as well as the expected environmental conditions of the system. The details of thermal analysis are discussed in Chapter 5.

Each die generates a certain amount of heat. On the other hand, the temperature of a die should be maintained below the specified temperature for high reliability. The lifetime of an IC die depends on its operating temperature. As a result, the heat generated by the dies in an MCM must be removed efficiently so that the temperature is maintained below the specified temperature. The amount of heat that can be removed in unit time in an MCM is referred to as the *thermal capacity* of the MCM. The maximum number of circuits contained in an MCM depends on not only the wiring capacity but also the thermal capacity of the MCM. Therefore, the thermal capacity of an MCM is an important criterion of the MCM.

The limitation imposed by the thermal capacity of the MCM is

$$N \leq \frac{T}{H}$$

where N is the number of dies in the MCM, T is the thermal capacity of the MCM, and H is the amount of heat generated by a die per unit time.

3.2.1.5 Electrical Analysis The basic elements in any electronic system are electronic components that need to communicate with each other. The electrical signals that travel between components create noise. In addition, the signals that travel between components cause signal delay. The signal delay directly affects the performance. The noise, if it becomes too large, can cause serious signal distortion,which in turn can cause a malfunction of the system or degradation of the system performance. As a result, the noise and delay must be accurately predicted and carefully controlled.

As discussed in Chapter 2, the typical delay in an MCM can be estimated by the following equation:

$$T_d = 0.033L\sqrt{\varepsilon_r}$$

where T_d is the signal delay in an MCM (in nanoseconds), L is the chip pitch (in cm), and ε_r is the dielectric constant of the substrate material. The *chip pitch* is the distance between two chips.

As can be seen from the above equation, both the chip pitch and the dielectric constant of the substrate material contribute to signal delay and should be minimized to achieve the highest performance.

The noises includes reflection noise, crosstalk noise, and simultanenous noise. The reflection noise is that noise caused by the change of impedance. The crosstalk noise is the noise caused by signal lines that run parallel to each other, whereas the simultaneous switching noise occurs when a large number of off-chip or on-chip drivers switch simultaneously. The electric noise has to be controlled within a certain level to ensure the proper operations of the circuits.

By analyzing delay and noise in system design, designers can ensure that the performance requirement of the design can be satisfied.

3.2.2 System Partitioning

In addition to trade off analysis, system partitioning is also an important part of system design. A complex electronic system is usually constructed hierarchically, from trannsistors to chips to MCMs to boards to backplanes. This hierarchy is typically defined bases on functionality during system design process. The areas of the system that are the performance bottlenecks and the features that could be integrated to reduce the number of components must be addressed in the system specification. One way of system partitioning is to apply MCM technology to a particular area of the specification and system architecture. Particularly circuits requiring high speed digital processing or high bandwidth data transfer such as CPU and signal processing functions are ideal applications of MCMs [141]. The number of I/Os may also be an important objective for the system partitioning. Systems with a high pin count and high interconnect density could be partitioned into modules to reduce multi level interconnects among single chip packages and PCBs. A system may also be partitioned by function. For example, analog functions may be targeted at a particular MCM. Noise sensitive circuits such as low voltage or low current devices may be effectively isolated by implementing MCM technology [141].

The system partitioning poses conflicting demands on the algorithms used to generate system hierarchies automatically. Optimal partitioning of large systems is very expensive in terms of requiring large amounts of memory and CPU time

[151]. On the one hand, the problem size makes the speed and memory effciency of the algorithms critical. On the other hand, the design must be modeled with enough detail so as to satisfy the constraints imposed by the packaging or system designer and to optimize the trade offs. One way of solving such conflict is to use a sequence of algorithms that are optimized to perform specialized tasks rather than using a single algorithm. The initial algorithms in the sequence are used to partition and cluster the components of the design into a smaller number of tightly connected components. As a result, the size of the problem no package or usesr constraints are violated [151].

To partition a system effectively, it is importnt to understand the objective or FOM of the partitioning. One FOM can be the complexity absorbed by the MCM, which is defined as the ratio of the number of chip I/Os to the number of modulel I/Os. A small number of module I/Os can be accommodated with larger module I/O pitch, achieving a more manufacturable assembly. In addition, a small number of module I/Os can also reduce the required interconnect density on the board. Another FOM of partitioning can be the cost/performance ratio. The area of design partitioning [151] has been extensively studied. The objective function most offen used is to optimize the number of I/Os and /or the number of gates per partition. One approach is to use an upper bound on the number of I/Os, and the size of each partition is restricted by certain bonds. While atempts have been made to consider all the complex constraints imposed in MCM design such as I/O limits, timing, and area, this remains one of the most important, open problems for MCM design.

3.2.3 ESDA Tools

As the variety of the fabrication methods, materials, chip bonding methods, and thermal management methods used in an MCM grows, it becomes increasingly necessary to automate the selection process and the trade off analysis in the system design. The automation of system designing which is called *electronic system design automation* (ESDA), has evolved rapidly to meet this requirement. The ESDA tools allow designers to explore options and trade offs before a design is committed to hardware, software, or a combination of both in the system design phase. In ESDA tools, the most important, features are the selection of MCM technologies and the system partitioning.

The selection of technology and materials to be used in an MCM is based on the trade off analysis. The goal of MCM design is to optimize various parameters such as performance and cost of the final product. Trade off analysis compares the gains in some parameters against the losses in other parameters. Figure 3.2 shows the relationship between the MCM technologies and the parameters of the final product. Trade off analysis and selection of MCM technologies are difficult tasks:

1. **Search Space is Large:** Since there are manydifferent fabrication methods, dielectric and conductive materials, chip bonding methods, and

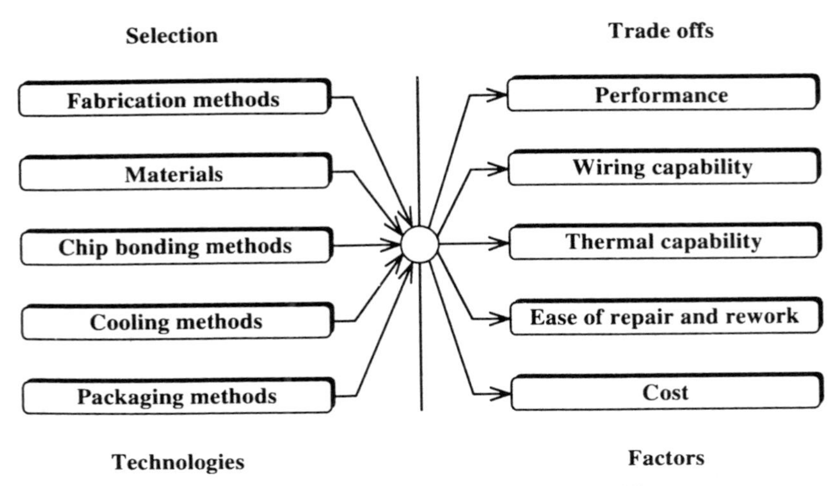

FIG. 3.2. Selection of technology affects several factors.

thermal management methods to be considered, the search space can be very large.

2. **Parameters are Interrelated:** As many parameters are interrelated, a change in one parameter may cause changes in other parameter. This makes it difficult to compare different selections.
3. **Detailed Information is Not Available:** Since the detailed design has yet to be carried out, details are not available at this stage. It is, therefore, difficult to analyze the design accurately.
4. **There is a Lack of Global View Because of Subsystems:** It is difficult to evaluate the subsystem since any advantages gained in one subsystem may cause a disadvantage in the complete system because the subsystems are interrelated.

In addition to the trade off analysis and selection of technologies, the design-entry method is also one of the key elements in an ESDA tool. This is due to the fact that at the high level of design, designers need to study the interplay of concepts. The design-entry method should be intuitive so that the designers can work as they think. Graphical-entry and textual-entry are the most commonly used design-entry methods. Graphical-entry methods may include truth tables, state charts, and block diagrams, whereas VHDL and Verilog hardware description languages are used in textual-entry methods [135].

ESDA tools are used in the early phase of design during which the system specifications, in terms of either hardware or software, are yet to be determined. Therefore, it is necessary for the ESDA tools to have hardware–software codesign capabilities so that the hardware and software can be designed and evaluated at the same time. This requires that both hardware and software behaviors be represented and validated in one system-level model. The model should be consistent for both hardware and software design so that behavioral descriptions can be implemented into either the hardware or the software design environment at any time during the system design process.

Furthermore, the ESDA tools should also be able to partition the hardware and software and optimize the hardware–software partitions. The ESDA tools should also be able to analyze, simulate, and verify the results. The trade off analysis may be carried out to select the technologies and materials to be used in MCM that are most suitable for the design.

3.3 LOGIC DESIGN

After system design, behavioral specifications of the system are determined. Logic design is then carried out to convert behavior description of an electronic system into circuit description. There are two methods of logic design, namely, schematic capture and the use of behavioral languages [61]. In the schematic capture method, the design is specified and captured hierarchically in a graphical form by a schematics editor [152]. On the other hand, a system can also be specified by using a behavioral language. The widely used behavioral languages include ELLA, HDL, and VHDL. After the circuits have been generated, they should be verified by logic simulaltion and timing simulation. Logic design is to convert the behavioral specifications given in terms of performance specifications into the circuit description in terms of logic gates. Traditionally, the logic design for small systems was straightforward and could be completed manually with few automatic design tools. However, as the complexity of electronic systems increases, it becomes increasingly difficult to manually perform the logic design without extensive use of design automation tools. As a result, many logic design tools have ben developed. Logic design automation is gaining acceptance in the electronic industry as it provides opportunities to explore variations in logic design to achieve the optimal trade offs between cost and performance.

The behavioral specifications of an electronic system can be given as various levels of detail [62]. *System level* is the highest level of the system and desscribes the operational characteristics of the system. The following level, called the *algorithmic level,* describes the algorithms performed by each process and their associated data structures and procedures. The *register transfer level* defines the set of data manipulation operations and data transfers between registers, which are called the *data path*, as well as the orderings of the operations and transfers, which are called the *control path.* The *logic level* describes the combinatorial logic functions and finite state machines. *Circuit level*, which is the lowest level of a system, defines the dc and ac electrical characteristics of each logic unit.

An electronic system can be described at different levels of abstraction from the exact physical layout level to a behavioral description of the system that carries no specific information about how the functions of the system are to be

	Behavior	Structural	Physical
System	Performance specs.	Processors, Memories	Physical partitions
Algorithmic	Algorithms	Hardware subsystems	Clusters
Register transfer	Data path, Control path	ALUs, Registers	Floorplans
Logic	Logic functions	Gates, Flip-flops	Cell estimates
Circuit	Transfer functions	Transistors	Cell layouts

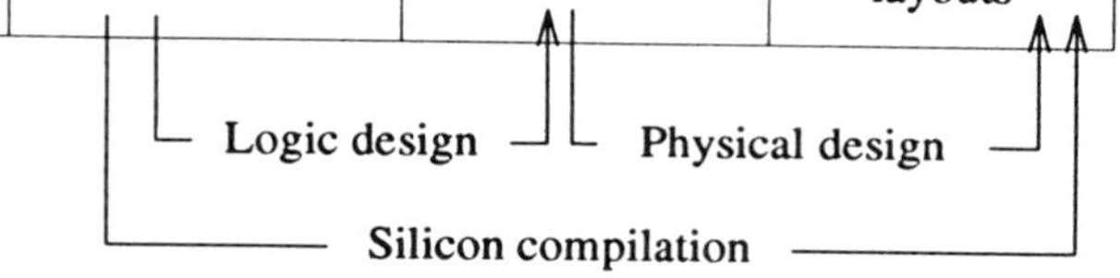

Logic design — Physical design

Silicon compilation

FIG. 3.3. Design levels.

implemented. The behavioral descriptions of the system can be converted into structural descriptions, which in turn can be converted into physical descriptions. The sturctural description defines a network of abstract components that realizes the specified behavior in an implementation independent manner, whereas the physical description defines the physical layout of each abstract component. The various levels for each description are illustrated in Figure 3.3.

Logic design includes all the design activities that generate the structural description of a design based on its behavioral description or creates a more detailed behavioral or structural description from a less detailed one. More precisely, logic design consists of the following design activities:

1. Conversion from one level of behavioral description to the same level of structural description
2. Conversion from one level of behavioral description to a lower level of structural description
3. Conversion from one level of either behavioral or structural description to a lower level of the same description

Physical design, also called *physical synthesis*, is concerned with the generation of one level of physical layout from the same level of structural description or from a higher level of structural dsscription or from a higher level of physical description.

Another possible design process is to derive the physical layout directly from the behavioral description either of a higher level or the same level. This design technique is called *silicon compilation*.

Furthermore, logic design can be further partitioned as high-level synthesis, logic synthesis, logic optimization, and logic verification. High-level synthesis is the technique of generating the structural description at the register transfer level from the behavioral dsescription at the algorithmic level. Detailed discussios of high-level synthesis and its tools have been published [143, 207]. *Logic synthesis* deals with converting the behavioral specifications given in terms of switching circuits and finite state machines into abstract structures. The details of logic synthesis are discussed by Edwards [61]. *Logic optimization* plays an important role in both the high-level synthesis and the logic synthesis. *Optimization* refers to the technique used to improve the quality of the design, beginning and ending at the same level of the same kind of description. Logic optimization is to improve the quality of logic design. After the circuit representation of an electronic system has been generated, it is important to verify that the sturctural dsescription of the system is logically consistent with the behavioral dsescription, since the errors in logic design can result in an expensive delay in design process if it is not detected and corrected promptly. The technique of verifying logic design is called *logical verification.*

The details of logic design can be found in any standard textbook on logic design or circuit design for digital systems [61, 204]. Therefore, logic design is not covered in detail in this text.

3.4 PHYSICAL DESIGN

Once the logic or circuit dsscription of a system is determined, it must be translated into the geometric patterns so that it can be fabricated. The process of converting the logic specifications of an electrical circuit into a geometric layout is called *physical design.* The input to physical design is a circuit diagram, and the output is the layout of the circuit that can be fabricated.

Physical design is accomplished in several phases such as partitioning, placement, and routing. Each of these phases is discussed in detail in Chapter 8. Due to the complexity of physical design, almost all its phases include extensive use of EDA tools, and many phases have already been partially or fully automated. This automation of the physical design process has increased the level of integration, reduced the turn-around time, and enhanced chip performance. To present a global perspective, these are briefly described.

1. **Partitioning:** A system may contain one hundred million transistors. The layout of the entire circuit cannot be handled due to the limitation of memory space as well as computation power available. Therefore, it is normally partitioned by grouping the components into blocks (subcircuits/submodules). The blocks can be boards, MCMs, or chips, depending on the level of partitioning. The actual partitioning process considers many factors, such as the size of the blocks, number of blocks, and number of interconnections between the blocks. The output of partitioning is a set of blocks along with the interconnections required between blocks. The list of interconnections is referred to as a *netlist.* The netlist provides the information about the logical connections among devices. The netlist also includes the performance constraints. Performance constraints may be minimum, maximum, or matched lengths, delay, and/or noise type limitations. The netlist also needs to be consistent with chip and functional test pattern nomenclature. Otherwise, there will be significant additional test development and debug times [60].

   In large circuits, the partitioning proces is hierarchical, and at the topmost level a chip may have 5–25 blocks. Each module is then partitioned recursively into smaller blocks.

2. **Placement:** During placement, the blocks are exactly positioned. The goal of placement is to find a minimum area arrangement for the blocks that allows completion of interconnections between the blocks. Placement is typically done in two phases. In the first phase an initial placement is created. In the second phase, the initial placement is evaluated and iterative improvements are made until the layout conforms to design specifications.

   The quality of the placement will not be evident until the routing phase has been completed. Placement may lead to an unroutable design, that is, routing may not be possible in the space provided. In that case, another iteration of placement is necessary. To limit the number of iterations of the placement algorithm, an estimate of the required routing resources is ussed during the placement phase. Good routing and good circuit performance heavily depend on a good placement algorithm. This is due to the fact that once the position of each block is fixed, very little can be done to improve the routing and the overall circuit performance.

3. **Routing:** The objective of the routing phase is to complete the interconnections between blocks according to the specified netlist. The goal of a router is to complete all circuit connections using the shortest possible wire length and using the smallest number of layers.

Since placement may produce an "unroutable" layout, the chips or MCMs might need to be replaced or repartitioned before another routing is attempted. In general, the physical design is iterative in nature, and many steps such as routing and placement may be repeated several times to obtain a better layout. In addition, the quality of results obtained in a step depends on the quality of solution obtained in earlier steps. For example, a poor quality placement cannot be "cured" by high quality routing. As a result, earlier steps have more influence on the overall quality of the solution. In this sense, partitioning and placement problems play a more important role in determining the numbers of layers and performance, than routing.

After the physical layout has been generated, it should be verified against any fabrication rule violation or design error. The design verfication consists of two parts:

1. Nonsimulation based:
   a. Design rule checking
   b. Sizing
   c. Power
   d. Package fit
   e. Optimization
   f. Functionality
2. Simulation based:
   a. Functional simulation
   b. At-speed simulation
   c. AC-text simulation
   d. Parametric testing

The nonsimulation based verification is simple and requires little time to run. However, simulation based verfication may be quite complicated and may require large computing resources.

The functional simulation is used to detect a single stuck-at-1(SA1) or stuck-at-0 (SA0) fault in the circuit if one exists. This requires sufficient vectors to "cover" all possible SA1 and SA0 fault locations. However, in reality, not all possible SA1 and SA0 fault locations may be covered. In this case, the percentage of coverage is the fault grade of the vector set. For a high fault grade score (99% or up), it is considered as all the locations are "covered." At-speed simulation is to perform a simulation that is executed at the specified maximum operational frequency of the system with timing checks enabled. An AC test is used to check either the propagation path delay in a nonmemory path or the external set-up and hold time for memory elements. Both rising and falling edges should be checked. The parametric testing is used to check the correctness for V_{IH} and V_{IL}. V_{IH} refers to the high-level input voltage, which is the maximum voltage that is required to be applied to the input of a device for a logical 1 voltage level. V_{IL} refers to the low-level input voltage which is the maximum voltage that is required to be applied to the input of a device for a logical 0 voltage level. Further details on simulation based verification can be found in Walker and Camposano [204]. After the physical layout is verified, it is sent to the fabrication to produce the prototype.

3.5 DESIGN ANALYSIS

MCM design is a long process, and any reworking at a later phase due to an error at an earlier stage is expensive. Therefore, verification needs to be carried out at each step to verify the correctness of the design at each step. Design analysis is used to verify the design throughout the entire design process, including system design, logic design, and physical design. Design analysis tools include electrical and thermal analysis tools. To develop these analysis tools, it is important to understand the electrical and thermal behavior of MCMs and develop the correct electrical and thermal models for MCMs. As a result, the studies of electrical and thermal properties of MCMs are important subjects. The details of electrical and thermal considerations in MCMs are discussed in Chapter 4 and Chapter 5, respectively. In this section, we discuss the characteristics of the analysis tools used in different phases of design process.

There are several general models that can be used in any analysis tools, as discussed below.

1. **Heuristic Methods:** General guidelines or empirical studies are used to model a system in heuristic methods. The advantage of general guidelines is their simplicity, whereas its shortcoming is that it becomes outdated as technology advances. Empirical results can also be used as the basis of prediction. Quantitative empirical results usually take the form of mathematical expressions or tables obtained from fitting experimental or detailed simulation data. A typical example is Rent's rule. Heuristic methods are usually simple and fast. Therefore, they are suitable in the early phase of design in which many different options have to be considered. However, their major disadvantages are lack of accuracy and the fact that the scope in which the model is valid is not always clear since the model is fitted using only limited data.
2. **Analytical Methods:** In analytical methods, the models are derived from principles or statistical results. Usually several assumptions are made for certain conditions, and then mathematical formulas are developed under the assumptions. Since the assumptions are clear, the scope in which a particular model is valid can be easily determined. Another way to develop analytical models is to use probabilistic techniques.The electrical models such as transmission line models discussed in Chapter 4 are typical examples of analytical models. The disadvantage of using analytical models is that they may involve solving complicated differential equations that requires much computing power.
3. **Simulation Methods:** In simulation methods, the physical behavior in an MCM is simulated. For example, the electrical behavior in an MCM can be simulated. As a result, one can predict whether delay in the MCM can satisfy design requirement and if there is any significant electrical noise that violates the signal integrity requirement. The thermal behavior of an MCM can also be simulataed to study the heat transfer effect or temperature

pattern in the MCM. Usually simulation involves solving a large number of equations simultaneoulsy. The simulation methods can provide the details regarding the design. Simulation can be performed at different levels, for example, for a single gate, an ALU, or an entire system. Simulation is used extensively in design verification. To reduce the expensive verification time, test vectors required for simulations should be generated automatically, and the simulation results should be compared with the expected responses automatically.

Among the three aproaches, the heuristic methods are the fastest to produce the prediction, whereas the simulation methods can produce the predictions of the best quality. As a result, heuristic methods are generally more suitable in the early phase of design where there are many options to consider. The simulation methods are best suited in the later phase of design where detailed analysis is required (see Fig. 3.4)

1. **Preroute Analysis:** Thermal analysis is used to predict junction and case temperature. A thermal analysis tool should be able to extract physical design information directly from an MCM drawing for analysis, and the result from thermal analysis can be used to run reliability analyses for the design or the system being designed. Upon the result of such analyses, placement can be modified and heat sinks or thermal vias can be designed. at this stage. The components and package types should be changed now if necessary to determine the impact on the mean time between failure (MTBF) of the system design or just a component [56].

   Signal analysis can be carried out without the information of the location of all the traces. The unrouted interconnects can be modeled by specifying Manhattan distance, impedance values, and propagation speed [56]. Preroute signal analysis can be beneficial for time-to-market, since many signal integrity problems can be detected and corrected by modifying

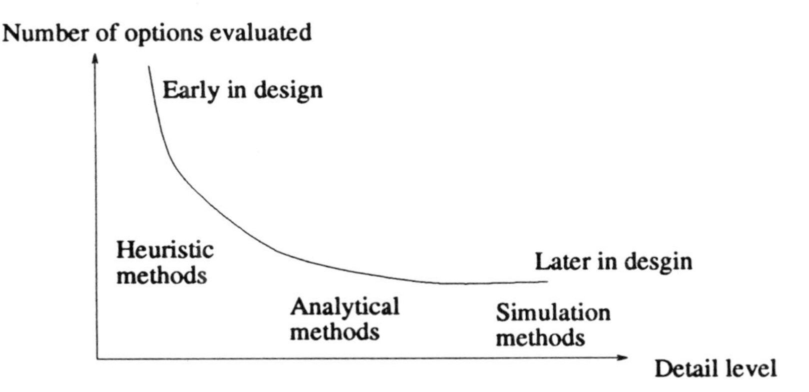

FIG. 3.4. The analysis becomes more detailed in the latter phases of design [178].

the placement before time and effort are spent in the expensive routing. The reflection-related problems such as overshoot and undershoot can be analyzed quite accurately at this stage. Modifying the placement or net schedule can be much easier at this stage than at a later stage when the design is fully routed.Virtually all signal-related issues except crosstalk can be analyzed at the placement stage [56]. Modifying the placement, for example, moving high speed drivers and receivers closer together, can alleviate some of the ringing and delay problems. Terminators can be added to minimize overshoot and undershoot to correct reflection problems. Once reflection problems have been corrected, delay violation for the critical nets can be avoided by analyzing the delay and editing the layout accordingly. After reflection and delay-relataed issues have been solved, a more detailed analysis is used to evaluate simultaneous switching noise (SSN). The impact of SSN on the noise margin can be determined by examining the critical componets. If SSN levels are unaceptable, different packages with less package parasitic may be used, or the switching times of the drivers are adjusted to reduce the number of drivers that switch at the same time [56].

After the design is verified for correctness, a back annotation file, which contains minimum and maximum delays for each driver/receiver pair is generated. The annotation file can be used to consider interconnect delays for functional simulation and timing analysis. Timing analysis is carried out for critical paths, considering both gate and interconnect delays. Any problem in the timing analysis has to be solved before the design can be fully routed to reduce the design time.

2. **Post Route Analysis:** After routing, thermal analysis is carried out again so that the routed conductors are taken into account to increase the accuracy of the predicted temperatures. The new predicted temperatures in turn also increase the accuracy of signal analysis, since the thermal shift and interpolating VI curves can be more accurate. Reliability analysis can also be more accurate due to the more accurate temperature prediction.

   Signal analysis is usually repeated by replacing all the original interconnect estimates based on the Manhattan distance with actual routed traces and vias.

   The annotation file is generated for the routed design and is sent to a timing analysis tool for a final timing analysis. The detailed modeling of routed interconnects and vias is included in this file. If timing requirements have been met, the design is sent to fabrication.

As performance becomes more critical, CAD tools will be required for detailed, three-dimensional interconnect electromagnetic analysis and detailed microwave design for critical path. Electromagnetic interference effects caused by radiation from or to chips, MCMs, or boards must also be considered during analysis and design. Three-dimensional thermomechanical analysis will be required to analyze the impact of board or MCM warpage caused by board processing or board ambient temperature excursions. The thermal analysis is used to estimate component temperatures given physical layouts and component packaging information and heat dissipation so as to control the reliability impacts of temperature. In addition to the reliability concern, the operating temperature of an MCM affects its performance as well. All the electrical parameters of the components depend on temperature. Moreover, temperature influences electrical characteristics of MCM substrate material. The parameters being affected by temperature include timing characteristics, impedance characteristics, dielectric constant, and electrical conductivity of MCM substrate. As a result, both performance and reliability impacts of thermal issues must be fully understood in the design process.

3.6 TESTING

As the design complexity increases, the cost of testing goes up quickly. As a result, the value of design for testability (DFT) increases as the design becomes more complex. Therefore, a complex MCM design cannot succeed without careful DFT. In fact, the testability issues must be considered from the very beginning of the MCM design.

The electrical test access may be severely limited in some MCMs, restricting the use of traditional manufacturing test approaches. DFT is important for all electronic systems. However, it is only for MCMs that DFT becomes mandatory to ensure an economically viable product and satisfactory yields.

The testing of an MCM includes the testing of bare ICs, substrate, and the package of the MCM. Bare ICs are much more difficult to test than packaged ICs. Full functional tests can be performed on packaged ICs that can be run at full operating frequency and over the entire temperature range. The testing capacity for packaged ICs has been developed to a point that the probability that they are "known good" can range to 0.99999 or higher [85]. The "known good" ICs refer to the ICs that are known with high probability to be free of any as-received defects and that will remain fully functional and defect free under the following situations [85]:

1. After assembly onto a board or substrate
2. Over the entire design temperature range and working frequency
3. After next level burn-in or environmental stress screening
4. After having been fielded for a time at least as long as the minimum guaranteed field life

Bare ICs cannot achieve the same high probability as "known good" as packaged ICs due to the lack of benefit of individualized testing in sockets. Some testing for bare ICs can be done at wafer probe. However, only a limited number of parameters are typically checked in these tests, and, traditionally, full functional testing at working frequency and full design temperature range is not carried out at the wafer level. This is due to the difficulties such as maintaining the entire wafer uniformly at high temperatures, frost problems at cold temparatures, dissipated heat removal from the IC under test, providing adequate power and ground, and bringing signals to and from the ICs without excessive degradation from probe losses, contact resistance, or interference from the surrounding wafer. As a result, only limited probe testing is carried out for bare dies, resulting in probabilities of being "good" of only 0.50–0.99 for bare dies [85]. Pretesting of bare dies prior to the assembly onto the MCM depends on having sufficient electrical test access. Many approaches such as pressure contact methods and metallugical techniques can be used to ensure electrical connection to the dies under test [85]. On the other hand, on-chip circuitry can be added to improve internal test access or even allow the dies to be largely self-testing. Boundary-scan and built-in-self-test (BIST) are such approaches.

Another difficulty of MCM testing arises due to the small size of the test nodes. Access to the test nodes ranges from very limited to nonexistent. In the case of PCB testing, the components (packaged chips) on the PCBs can be thoroughly tested before mounting. Therefore, most failures are limited to the interconnect problems. However, the key components on MCMs are bare dies, which are difficult to test. In addition, the repairability of MCMs creates another problem. If a packaged chip is defective, the chip is simply discarded. However, when an MCM is found to be defective, it usually cannot be thrown away due to its high cost. Instead, it may be economically possible to repair the device. As a result, the designers must include diagnostic capability as a part of the testability measures. These characteristics make MCM testing far more challenging than traditional VLSI or PCB testing.

In addition to the problems discussed above, MCMs are expected to suffer from virtually all of the problems associated with assembled PCBs:

1. Shorts and opens in the interconnections
2. Wrong components
3. Missing components
4. Physically damaged components
5. Components with functional defects
6. Components suffering from performance defects
7. Components suffering from parametric defects

These defects can be functional or parametric. To detect these defects, the functional and parametric tests for MCM testing are required just as with PCBs, and many testing approaches have been developed for functional testing and parametric testing as well as DFT in PCBs [69, 71, 211].

Although the whole process of MCM testing is very similar to that of PCB, the MCM presents some challenging difficulties to detecting and diagnosing these problems:

1. MCM substrates often have much higher density pads than PCBs. For example, an MCM with node counts ranging from about 64 to 600 can be built on a substrate as small as 35 mm^2, with pad size ranging from 4 to 20 mil. Therefore, the bed-of-nails tester that is used commonly in board testing is not feasible in MCM substrate testing. On the other hand, it is also difficult to access the internal nodes through the limited I/O pins in MCMs.
2. Many of the dies currently available for MCM assembly do not come with the same performance or reliability guarantees as packaged ICs. Thus, an additional burden is placed on module-level testing.
3. There is poor availability of equipment for MCM testing.

To ensure high MCM yields, MCM testing should be used throughout the fabrication process to verify the quality of each processing step and component that goes into the module. Thus MCM testing can be divided into three basic areas: substrate testing, die testing, and module testing. Substrates are tested during fabrication and prior to component attachment. Dies are tested prior to mounting on the substrate. The assembled module is tested prior to final sealing to permit repair of faulty components.

The testing of MCMs is a difficult and expensive process. As a result, a test strategy has to be carefully developed in the DFT phase to enhance the testability of the MCMs [153, 208]. One way of enhancing the testability of MCMs is to use know good dies (KGD). However, KGD is an extremely difficult issue by itself, and currently much effort is being spent on it. Another way of enhancing the testability of MCMs is to design built-in test (BIT) capability. For example we may design bare dies with boundary-scan circuitry built in. Yet another way to make the MCMs easier to test is to use an overlay process in which interconnects are built over the top of the bare die in contrast of the normal process of placing dies on an interconnect structure. The overlay process generates a top surface that is available for both I/O pads and test pads, which can be used to access the internal nodes. The added probing capability helps designers to evaluate the MCMs, functionality and other interconnect structure performance [153].

DFT tools are used to automate scan test-point selection. They aim at helping designers to create high quality tests while reducing test-program development time and test-fixture cost. Different strategies such as in-circuit, cluster, BIST, and scan-test should be supported in these tools.

To guarantee the testability of a design, designers and test engineers have to agree to work together and create a testing environment. More automation for DFT is required in order to achieve the maximum efficiency. The circuitry required for testing should be automatically added while maintaining the optimal placement and timing.

3.7 MCM DESIGN TOOLS

Now we focus on the characteristics of MCM design tools and compare them with both VLSI and PCB design tools.

3.7.1 Characteristics of MCM Design Tools

As high speed becomes a focus of MCM design, the design process is no longer based on schematic-driven layout. Instead, designers tend to make more and more design decisions during iterations between the physical layout and the simulation environment. This shift of focus is caused by the impact of the physical layout of the design on the electrical behavior of the circuits. When signal frequency is higher than 50 MHz, the impact of the physical layout (such as parasitic elements, interconnect dimensions, and the physical characteristics of the MCM) becomes significant to the functionality and performance of the design [14]. Signal integrity issues such as transmission line effects, crosstalk noise, and timing issues introduced by interconnect delay must be dealt with early in the design process. As a result, EDA tools are required to integrate logical and physical representations of a design to facilitate the analysis of physical effects within the simulation environment. In addition, place and route tools should be performance-driven so as to control the delay and noise effects in the physical layout. The combination of logical simulation/timing and physical analysis tools and performance-driven physical tools can help to solve timing and signal-quality issues in high performance design. Today's complex system, high clock speed, fast rise times, and high device pin counts not only affect the physical layout of an MCM but also have a dramatic impact on its electrical performance. Transmission line effects are no longer negligible for high speed MCMs. Transmission line effects bring ringing and reflection effects as well as undershoots, overshoots, pedestal, and crosstalk effects. Their impact on circuits include unwanted logic transitions, compromises in setup and hold times, and device damage. As a result, the tight integration between high speed analysis tools used to extract physical layout parasitic and the simulation environment used by designers is required for CAD tools. Effects such as interconnect delay, inductance, capacitance, and characterisitc impedance must be back annotated into the logic simulator and timing analysis tool to ensure the required performance.

Another important feature of MCM CAD tools is constraint-driven. The performance and manufacturing constraints such as thermal distribution, delay, crosstalk, ringing, simultaneous switching, and electromagnetic interference noise must be controlled at each step of the design process. These constraints should be used to guide the design rather than be a part of design verification or a postprocessing cycle as in the traditional design process, which causes many time-consuming iterations. For example, in the physical design phase, the place-and-route tools and the analysis tools need to be coupled together. On one hand, the place-and-route tools are driven by constraints generated by technology files, timing simulation tools, and thermal and electrical analysis tools. These tools convert electrical and thermal constraints into physical constraints such as the maximum length of a net or minimum separation between two parallel signal traces. On the other hand, these physical design tools are armed with the analysis tools that provide quick feedback on the correctness of the physical design.

Physical layout tools (CAD tools) should be integrated with design analysis tools (CAE tools) and computer-integrated manufacturing (CIM) tools, which are used to automate fabrication, assembly, and test equipment. In summary, the desired features for the seamless CAD/CAE/CIM tools include

1. Input
   a. Schematic capture
   b. Netlist generation
   c. List of components generation
   d. Capture of component operation vectors
   e. Library of component drawings/parameters
2. Substrate design
   a. Automatic assignment of pad patterns for components
   b. Autoplacement
   c. Thermal-driven placement
   d. Autorouting
   e. Performace-driven routing
   f. Arbitary angle and gridless routing
   g. Automatic generation of power and ground grids
   h. Handling of multiple power and ground planes
   i. Design rule check
   j. Z-axis checking
   k. Handling of different chip bonding methods such as wire bond, TAB, and flip-chip
   l. Handling of dies in cavity
3. Electrical simulation
   a. Calculation of line resistance/capacitance/inductance
   b. Circuit simulation
   c. Clock distribution/timing analysis
   d. Power/ground distribution analysis
   e. Calculation of decoupling capacitance required
   f. Calculation of termination resistance required
4. Producibility
   a. Generation of mask making data in correct format
   b. Generation of test data for bare substrates
   c. Generation of CIM data for automated assembly

   d. Generation of test vectors for assembled modules
   e. Design for testability check
   f. Design for manufacturing check
   g. Design for repairability check
5. Reliability
   a. Temperature prediction
   b. Reliability analysis
   c. Vibration/shock analysis
   d. Chemical/corrosion check
   e. Automatic prediction of MTBF

3.7.2 Design Kit

As MCM technology advances, many different choices of technology become available. As a result, it becomes increasingly difficult for EDA tools to be general enough to handle each different case. A design kit is developed to solve this problem. A *design kit* refers to a design-tool-specific approach to designing MCMs. It is an effective approach to the problem of too many different technologies in MCMs. Usually a design kit is based on a single fabrication supplier. Initially substrate materials, processing technique, die assembly approach, and an MCM supplier are selected based on the performance goals, cost objectives, volume production requirements, operating environment, reliability demand, and other issues. The entire set of issues must be carefully analyzed to select a technology and an MCM supplier. Once the MCM supplier has been selected, the design kit can be used to assist the design process. Library information and custom software are provided to assist in the design process. Since there are so many different technologies available for MCMs, it is difficult to have a tool that is generic and at the same time optimal for each technology. As a result, it is desirable to have a tool tailored to a particular set of technologies, correct design system parameters, specific design rules from the chosen MCM supplier, and libraries for these particular technologies; this leads to an optimal solution for the customers. Design of MCMs is driven by the fabrication and assembly manufacturing rules. As a result, the design of MCMs can only start with the selection of an MCM supplier and settlement on the design rules. Design rules can include [141]

1. Trace widths and clearances
2. Trace angle allowance
3. Via size, shapes, and stacking
4. Metal plane construction
5. Manufacturing process build order
6. Component attachment technology
7. Special rules applied to an area, a layer, or a set of nets

In 1992, MicroModule System was one of the first vendors to introduce a commercial MCM design kit [115]. After that, many other MCM vendors also developed MCM design kits. These design kits configure a generic or commercial MCM design tool to become an MCM supplier-specific design system by loading a design kit's software into a generic or commercial MCM design tool. In doing this, designers can ensure that their design will meet the supplier's requirements with a minimal risk for error. It also helps them to avoid spending time manually translating the supplier-specific data into the design system, potentially an error-prone procedure. Documentation, design process flow, data to define the supplier's capabilities, and tools to verify a design's integrity are usually included in an MCM design kit. Technology data and specialized tools to support the supplier's technology are also included in the design kit. Technology data include design rules and constraints that are used to drive the layout and routing tools to create high performance designs. An MCM design kit can also contain the physical design process guidelines, with software describing predetermined substrates, spacing, widths, and vias. In addition, template generators help the designer to build foundry-compliant components, such as thermal vias and bonding pads. The design kit can also contain software to help analyze crosstalk and transmission line effects for design checks; a tester interface to extract data on where to place test points; and a release utility to collect and format the information required by the MCM supplier.

3.7.3 Platforms for MCM Design Tools

The design process of MCMs or MCM based systems involves many steps. As a result, many design tools are developed to handle the different phases or different problems of the design. The support for the design of electronic systems includes not only design tools but all software and infrastructure support such as libraries and their related software support, as well as the software for storing and managing design programs, design data, user interface technology, and design decision support systems (e.g., ASIC gate, megacell, and die library support). There is a demand for a platform that can carry out all the design activities. That platform is the CAD framework, which is a platform on which different design tools can be integrated.

A *CAD framework* is a software platform that provides a common operating environment for CAD tools, possibly from different vendors. Therefore, standardization such as data format or process compatibility among different vendors is required for any framework. The CAD Framework Initiative (CFI) Inc., a nonprofit organization in promoting in the standardization of framework, published CFI's Release 1.0 Pilot Program. The document provides standards that address design representation, inter-tool communication, tool encapsulation, an execution-log format, base system-services guidelines, and error handling. Several vendors, including IBM, Hewlett-Packard, and Sun Microsystems, are participating in this program [134]. In addition to the requirement of openness and standardization, a framework should enable users to launch and manage tools as well as to create, organize, and manage data. A framework should have the option of supporting the process management or data management or both, selectable by the users. It should be able to view the entire design process graphically. The intuitive graphical interface should give users quick feedback on design changes to speed up the design process. It should be able to perform design management tasks like configuration and version management that change the data format of one design to the data format of another. Among the key elements of a CAD framework are platform-independent graphics and user interface, inter-tool communications, design data and process management, and database services.

3.7.4 Comparing MCM Design With VLSI and PCB Designs

In many ways, such as size and density, MCMs are halfway between VLSI chips and PCBs. MCMs are larger than VLSI chips and smaller than PCBs. MCMs have greater routing density than PCBs but less than VLSI chips. There are also distinct differences between MCMs and PCBs/VLSI chips. The thermal management in MCMs is significantly more difficult than in either PCBs or VLSI chips. Another issue is the testing of MCMs. The thorough testing of bare ICs is much more difficult than the testing of chips that have been sealed into packages, creating a KGD problem. Besides the KGD problem, the DFT becomes crucial for the success of MCMs due to both the difficulty of MCM testing and the high cost in producing MCM protocols.

Unlike VLSI or PCB design, there is an important concern in MCM placement: the heat dissipation of chips. The heat dissipation should be as uniform as possible. However, this goal often conflicts with electrical objectives such as minimizing the delay or critical nets. It is therefore common to optimize the placement based on electrical performance and solve the thermal problem by other means.

Another problem in MCM design that does not exist in PCB or VLSI design is the generation of mesh reference planes. Due to processing constraints, the ground plane of some thin film substrate MCMs need to be a mesh. As a result, a mesh ground plane has to be generated by CAD tools. The mesh plane is usually generated in two steps. During the first step, the mesh plane pattern is generated with spacing and line width dictated by the substrate processing constraints. Then holes are cut for the feedthroughs from the routing layers to the top metal layer for substrate bonding and for thermal bumps if flip-chip technology is used to attach the chips to the substrate. As a result, the feedthrough cuts depend on the location of the individual chips and on the I/O pins of the chips and are independent of the mesh in the plane. Therefore, the resulting small slivers of metal can cause design rule violations and mask tooling problems [33]. These design rule violations must be removed by some post-processing. Hence, it is desirable to integrate a polygon editor with the ground plane generator to achieve the post-processing of the mesh plane [33].

Besides ground plane generation, output data format issue is also important in MCM design. Gerber format is preferred in the PCB industry, while GDSII stream format is the standard for the IC industry. The conversion between these two formats is straightforward. However, the Gerber format design is usually photographic based whereas the database produced in GDSII format is intended for a pattern generator, which is used to drive the electron beam machine. As the feature size of the MCM substrate continues to decrease, the mask tends to be written by either electron beam or laser rather then being generated by photographic based methods. As a result, many MCM designs require the output format in GDSII format.

MCM design may share some characteristics with PCB design and some characteristics with VLSI design. However, there are new characteristics in the MCM design that make it a very different task than either PCB design or VLSI design. For example, a part of the MCM design is the routing of signal nets. As in PCBs and VLSI chips, the routing of signal nets in an MCM is the task of placing electrical traces in a multilayer substrate to interconnect the pins of each signal net. The routing of an MCM bears some characteristics of PCB routing and some characteristics of VLSI routing. The finer geometries in an MCM are similar to the VLSI environment, whereas the transmission line effects due to the long traces in an MCM are similar to the phenomenon in a PCB. A VLSI router or PCB router may well be the first tool to look at when the routing tool for MCMs is to be designed. However, an MCM may contain thousands of nets and has finer pitch than a typical PCB. Thus, the complexity of an MCM routing problem may be much higher than that of a PCB routing problem. Furthermore, the relaxation of the normal routing geometry rules may be necessary to achieve performance and cost goals. These new requirements may include the relaxation of one-layer, one-direction rules, arbitrary-angle routing, vairable-width traces and spacing on individual net basis, and evenly distributed wiring. On the other hand, the traces in an MCM are typically much longer than in VLSI chips, where a simple RC delay model is sufficient. As a result, the lossy transmission line model has to be used for the electrical traces in an MCM. In VLSI design, the transmission line effects are not considered due to the small size of a chip, whereas the transmission line model is used in PCB design since the interconnects in PCBs are long.The frequencies at which the MCMs have to operate are increasing rapidly. As a result, frequency dependent losses in MCMs are more severe than in PCBs and must be modeled accurately.

To model the transmission line effects accurately, the transmitter and receiver must be accurately modeled. The components are modeled as black boxes in PCB tools, and the detailed electrical characterisitics of the drivers driving the pins are not stored. As a result, the PCB tools cannot accurately model the transmitter and receiver. On the other hand, VLSI tools store the entire design hierarchy. One can expand to the next level recursively until the desired driver circuit is included.

Due to the complex nature of MCM design, the whole design has to be partitioned recursively until each unit can be efficiently designed. In PCB design, the components (chips) are treated as black boxes and only the properties related to the pins are stored. As a result, it is impossible to expand to the next level. On the other hand, the VLSI tools store the entire design hierarchy, and hence they are better suited for the MCM design. In the schematic design, however, no boundaries exist. Therefore, the schematic- and netlist-driven tools can be used directly in MCM design.

Design rules depend on the substrate technology. Most design rules in PCBs are single-layer design rules such as simple spacing and clearance rules. On the other hand, the design of MCM-D requires multilayer design rules that can be verified using VLSI tools.

Special purpose masks may be required when the chips are assembled onto the MCM substrate. For example, the metal masks for the evaporation of metal bump pads on the MCM substrate and on every chip to be mounted on the substrate may be needed in center flip-chip processes. The PCB or VLSI tools cannot handle these masks.

In high speed digital systems, large output drivers are used. This increases the switching current or the transient current that flows in and out of the chip through the power supply and ground pins, since the drivers not only drive the load but they also drive the MCM substrate traces, the substrate to package wire bond, and the package parasitic. These components all have inductance. The inductance causes the ground (or power supply) to bounce when the drivers switch. This is also referred to as *simultaneous switching noise* because it is most pronounced when many off-chip drivers switch simultaneously. The simultaneous switching noise problem becomes more severe in MCMs because of the large number of drivers and the addition of MCM substrate traces. No VLSI or PCB tools can model these effects.

TABLE 3.3. VLSI and PCB Tools Cannot Support All Requirements of MCM Design

Issues	VLSI Tools	PCB Tools
Physical design hierarchy handling	Yes	No
Schematic design hierarchy handling	Yes	Yes
Placement and route	Has major	Has major
Physical design verification	Yes	Yes, but cannot handle multilayer rules
Graphical database ready for mask tooling	Support pattern generators	Problematic
Special purpose mask support	No	No
Transmission line support	No	Yes, but needs improvement
Accurate transmitter/receiver modeling	Yes	No
Simultaneous switching noise modeling support	No	No
Substrate testing support	No	No
After-assembly testing	No	No

Another feature required for MCM design is the high-level design tools. The high-level tools of design verification, synthesis, and partitioning must be developed to allow designers to work at the system level, developing designs in terms of functions to be implemented rather than the actual physical implementations themselves. In addition, a homogeneous environment in which multilevel design tools that are equally applicable at the IC, MCM substrate, and board levels must be created.

MCM design is a concurrent engineering process. MCM design involves intense interaction from all engineering disciplines throughout the design cycle. Tools that are specialized in different areas need to be integrated on a common framework that supports common databases and interfaces for a wide range of tools from different vendors.

Due to the above characteristics of MCM design as compared with PCB design and VLSI design, it is important to develop MCM tools based on the experience of both PCB tools and VLSI tools while stressing the unique requirements of MCM design. In Table 3.3, the features of VLSI tools and PCB tools are summarized.

3.8 SUMMARY

The design of MCMs or MCM based systems is a very complex process. As a result, it is usually partitioned into several phases such as system design, logic design, and physical design. System design is concerned with the selection of technology and the complete behavioral specifications of the system. System design must be carried out very carefully to prevent expensive iterations of later phases. Almost all the aspects of logic design and physical design have been automated due to the fact that these two phases usually deal with millions of components and it is virtually impossible to carry out the design manually. To verify the correctness design at each design step, extensive use of design analysis tools is important. Design analysis includes electrical analysis and thermal analysis. The testing and design of testability is one of the most difficult problems in MCM technology. Until the KGD problem can be solved satisfactorily, the testing of MCMs remains the major obstacle to further development of MCMs.

3.9 PROBLEMS

1. If the endproduct is used for military applications and its reliability is of the utmost importance, what would be the objectives of MCM design? What factors of MCMs need to be considered?
2. What is sizing analysis? What effects does the size of an MCM have on other parameters of MCMs such as performance?

3. What are the types of electrical noise present in MCMs? How does the noise affect MCM design?

4. A CAD framework is used as a platform to integrate different CAD tools. List different functionalities of a CAD framework.

5. If a constraint-driven physical layout tool is to be developed, what constraints should be considered?

6. What are the major differences between an MCM CAD tool and a VLSI CAD tool? What are the differences between an MCM CAD tool and a PCB CAD tool?

7. What is a design kit? Why is a design kit a useful approach?

BIBLIOGRAPHIC NOTES

MCM is a comparatively new technology, and very little literature is available on MCM designs. Sandborn and Moreno [178] cover the system design (conceptual design) of MCMs. They also discuss details of the methodology of conceptual design, trade off analysis, and several CAD tools for the conceptual design of MCMs. An early analysis method characterized as physical simulation based is presented by LaPotin et al. [127]. A detailed case study is also included in that paper. The logic design of MCMs is similar to the logic design of any VLSI or PCB. As a result, any standard textbook on this subject can be used to study the logic design of MCMs. White [207] discusses the structured, orderly, logical approach to circuit design. This approach concerns discrete board design, SSI and MSI logic design, bit-slice design, structured software and hardware programming, and systems concepts. Edwards [61] covers the automatic logic synthesis and logic optimization techniques for digital systems. Though there is no book written for the physical design of MCMs, Sherwani [184] covers almost all of the aspects of the physical design of VLSI circuits. Both the basic principles as well as advanced research topics on the physical design automation of VLSI circuits are discussed.

CHAPTER 4

ELECTRICAL DESIGN OF MCMs

Many MCMs are specifically designed for high performance systems. In such cases, performance may be the most important design objective. In a given architecture, the performance is determined by the maximum operating frequency. To maximize operating frequency the delay in the *critical path,* i.e., the maximum delay path, must be minimized. In addition to high performance, the system must perform correctly and reliably. This is related to the issue of noise in the signals. Electrical noise may cause inadvertent logic transitions, thereby introducing errors in the system. Hence, to design a reliable high performance system the electrical design should optimize both delay and signal noise. It should be clear from the above discussion that a reliable high performance system cannot be designed without a proper electrical design. Thus electrical design is one of the most important elements in the overall MCM design. In this chapter we discuss the basic concepts used in electrical design.

4.1 DELAY

Total delay in system is a sum of various delays in different parts in the system. In this section, we consider different types of delays. Delays are caused by device delays and by interconnect delays. Device delays are reduced by improvements in the farbication technology. Since fabrication technology has made significant strides in recent years, interconnect delays are becoming the dominant factor in the overall delay of the system. MCM interconnects are very long compared with the IC interconnects; As a result, transmission line analysis is required. On the other hand, IC interconnects can be conveniently analyzed using a significantly simpler model.

Noise is caused by the shape of the interconnect and the interaction between two or more signal lines. If an interconnect is improperly designed, it may cause reflections and degrades the signal shapes. If two fast switching lines are placed too close to each other, crosstalk may cause inadvertent logic transitions.

To analyze the various factors involved in signal delay, let us consider the example of two flip-flops mounted on an MCM, which are driven by the same clock source and are on a critical path. The total dealy between the flip-flops must be less than the clock period to ensure that the MCM operates correctly. The total dealy t between the flip-flops can be determined by

$$t = t_l + t_b + t_i + t_u + t_s \tag{4.1}$$

where t_i is the delay introduced by the interconnection between the two flip-flops; t_l and t_b are delays of the latch and the buffer at the source, respectively, t_u is the setup time required by the latch on the receiving end; t_s is the clock skew (see Fig. 4.1) [58].

We focus on the delay introduced by MCM interconnections. The delay inside VLSI devices is a complex topic in its own right and is outside the scope of this book; Information on this topic can be found elsewhere [6].

The interconnect delay t_i is given by

$$t_i = t_p + t_{se} \tag{4.2}$$

where t_{se} is the *setup time* and t_p is the *propagation delay*. *Setup time* is the time required by the signal to settle down before the signal can be sampled safely by the receiver and determined to be consistent at a logic-0 or a logic-1. *Propagation delay* is defined as the period from the 50% point of the output waveform at the source end of an interconnection to the 50% point of the input waveform at the receiving and (see Fig. 4.2).

Delay modeling involves two other important concepts, namely *rise time* and *time of flight.* The *rise time* of a signal is the time from the 10% point of a waveform to its 90% point (Fig. 4.2) and the *time of flight* is the time for the light to travel along the interconnection. The time-of-flight is also called *time-of-flight delay*.

4.1.1 Interconnections

Interconnect refers to the medium used to connect any two or more circuit elements. Interconnections include the package pins, lead frames, bonding wires, TAB frames, solder bumps, wires inside chips, wires on MCM substrates, or wires on the PCBs. Interconnections differ widely in their electrical performance. Their performance is related to the length and cross section of the interconnect and the driving frequency of the signal, among other factors. As a result, different models have to be used to estimate accurately the delay in different types of intercon-

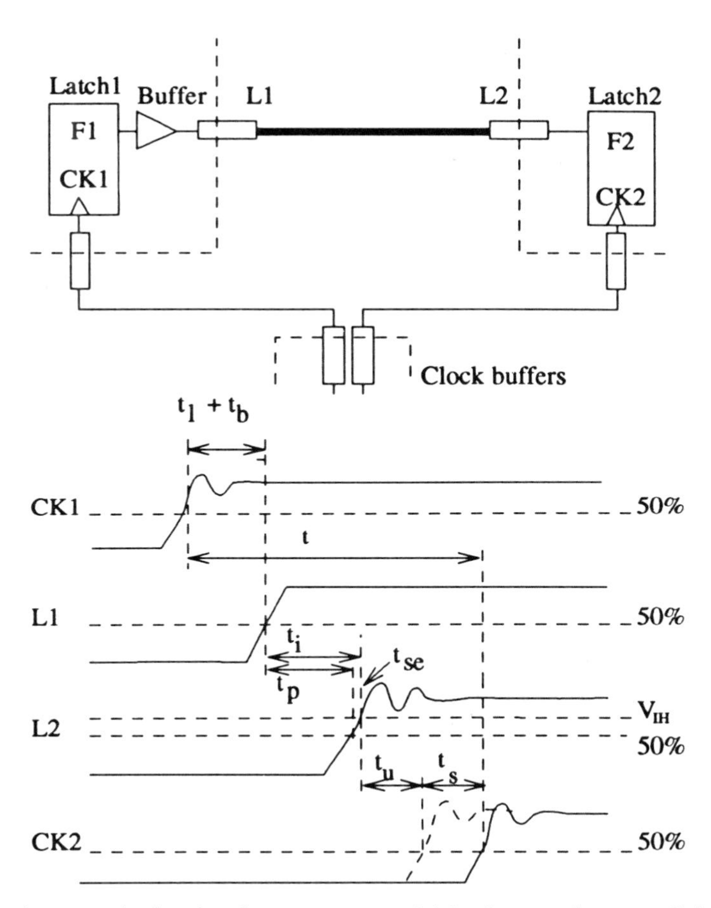

FIG. 4.1. An example showing the components of delay in a synchronous digital system.

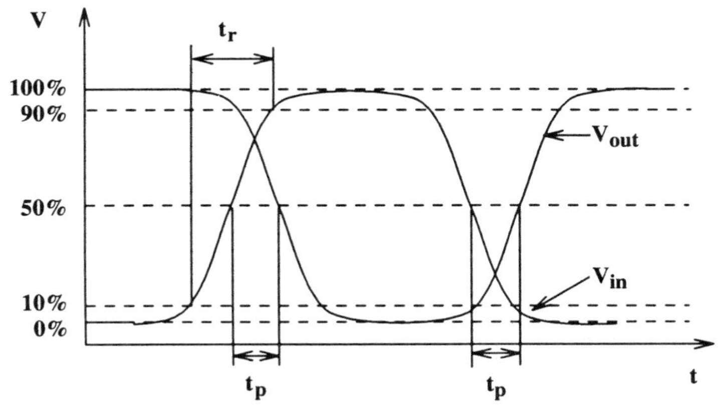

FIG. 4.2. Definition of rise time and propagation delay.

nects. Two of the most frequently used models, the RC model and the transmission line model, are presented in this section.

At low intergration levels (SSI and MSI), transistor delay is the dominant factor in the overall delay. As the feature size of transistors is reduced by improvements in the fabrication technology, the transistor delay is significantly reduced and the interconnect delay becomes the dominant factor. This is especially true for MCMs and PCBs in which the interconnect lengths are much longer than in VLSI chips. As a result, interconnnect delay plays an important role in determining the maximum clock frequency of a system. With increasing interconnect lengths in MCMs and PCBs, interconnection capacitance may dominate the gate capacitance and hence determine the overall performance of the whole system. The large interconnection capacitance also increases power consumption and causes noise. In addition to large capacitance and inductance loads resulting from long interconnections, the resistance of the lines may become a major concern. In a system where packaged chips are mounted on a PCB, the clock frequency inside the chip can be much higher than the clock frequency outside the chip. To adjust the difference, a large number of buffers are used in VLSI input and output stages. In a PCB system, the distance between chips is very large, and long wires are required to interconnect chips. The buffers in VLSI chips and the long wires in the PCB increase system delay. By using naked dies directly and integrating many dies on a single substrate, fewer and smaller buffers are required in each die, and thus high clock frequency inside an MCM can be obtained. Moreover, dies are placed close to each other. Therefore, interconnect distances on MCMs are much smaller than the interconnect distances on PCBs. These two factors allow MCM based systems to operate at higher frequencies than PCB based systems. However, very tight chip spacings and high clock frequency effects the performance of the MCMs. As more dies are integrated onto one substrate, the length of the interconnection between dies becomes longer. As a result, the interconnection delay in an MCM dominates the total delay since the internal delay on dies reduces dramatically as VLSI technology develops. Another side effect is related to very high clock frequencies, where the simultaneous switching noise becomes more severe. In addition, capacitive and inductive couplings between neighboring circuits become more significant due to the closer spacing between neighboring lines and longer interconnect wires in general. As a result, the crosstalk noise can be more significant. As the MCM technology develops, the delay and noise of interconnections become more significant. Therefore, the delay and noise in MCM interconnections must be carefully studied, and much effort is needed to minimize such delay and noise in the system.

MCM interconnections can be represented in several different models. Two of the most frequently used models are RC and transmission line models. In the RC model, only resistance and capacitance effects are considered, whereas in the transmission line model the effect of inductance is also considered. While it is easier and faster to calculate the delay using the RC model, the transmission line model is more accurate. Nevertheless, the RC model is widely used since in many cases it gives fairly accurate results in a much shorter time.

The RC model can generate fairly accurate results when the clock frequency is low or the interconnection is short. However, if the interconnection is sufficiently long or the circuits are sufficiently fast such that the rise time of the signal is comparable to the time-of-flight delay, the RC model is no longer sufficient and the interconnections should be modeled as transmission lines. More precisely, when $t_r < 2.5t_f$, the transmission line model should be used. On the other hand, when the rise time is much larger than the time-of-flight, for example $t_r > 5t_f$, a lumped RC model suffices. When t_r is the range of $2.5t_f$ and $5t_f$, either the lumped RC or the tranmission line model can be used, depending on the accuracy required [6].

4.1.2 Transmission Lines

The transmission line model is mandatory when analyzing delay and noise in many high performance MCM applications due to the short rise time and long interconnnections. Figure 4.3 illustrates the cross sections of transmission line structures that can be used to model the interconnections in electrical systems. The wire above ground model (Fig. 4.3a) can be used to represent a bonding interconnections; the coaxial model (Fig. 4.3b) is used for interconnects on PCBs and in local area networks; the on-chip interconnections can be modeled as a microstrip structure (Fig. 4.3d), and a wire sandwiched between two reference planes in PCB or MCM can be described by the triplate strip line model (Fig. 4.3d).

There are two types of transmission lines: lossless and lossy. In a lossless transmission line, resistance is considered negligible. As a result, the signal can propagate from one end to the other end of the line without any signal degradation. In a lossy transmission line, signal degradation (attenuation) due to the

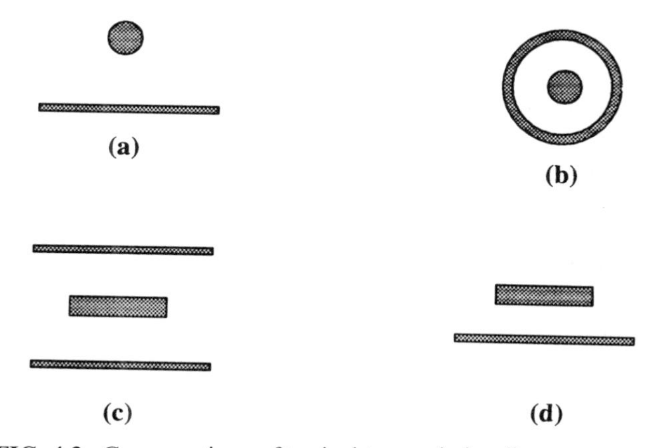

FIG. 4.3. Cross sections of typical transmission line structures.

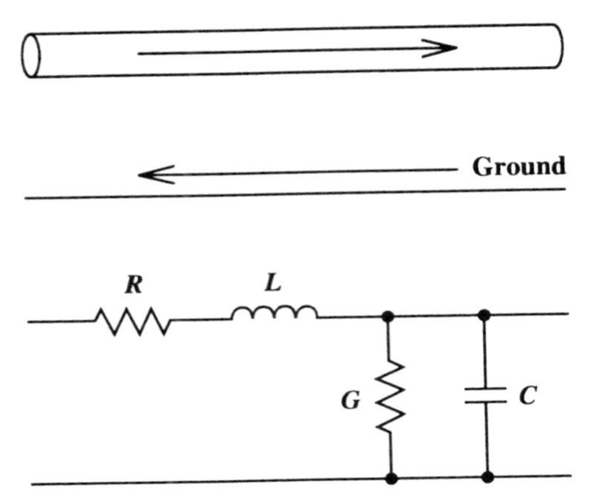

FIG. 4.4. Electrical model of a uniform transmission line.

resistance of the line must be taken into consideration as signal propagates along the line. Many interconnections such as the wires on PCBs, package pins, lead frames, bonding wires, TAB frames, and solder bumps all have low resistance due to large cross sections. As a result, these interconnections can be treated as lossless transmision lines. On the other hand, on-chip interconnections and wires on thin film MCMs have significant resistance due to small cross sections. Therefore, they should be treated as lossy transmission lines [58].

To discuss the properties of lossless and lossy transmission lines, we introduce some basic electrical parameters describing the properties of transmission lines. The primary electrical parameters of a transmission line are resistance along the line, *inductance* along the line, conductance shunting the line, and capacitance shunting the line (denoted as R, L, G, and C, respectively in Fig. 4.4). Each of these parameters are specified as per unit length (Fig. 4.4). Series components are the cumulative effect of the signal and ground planes. For example, in calculating the resistance of a microstrip, not only the resistance of the signal line but also the resistance of the ground plane must be taken into account.

Other basic parameters of a transmission line include the characteristic impedance Z_0, propagation speed v, time-of-flight t_f along the transmission line, total capacitance C, total inductance L, and total resistance R. In this section, we discuss the transmission line theory and the relationships among the basic parameters in both lossless and lossy transmission line models.

4.1.2.1 Lossless Transmission Lines

Many interconnections in MCMs and PCBs can be modeled as lossless transmission lines. To analyze the properties of a lossless transmission line, we assume that there are a voltage signal and a current signal propagating along a lossless transmission line as shown in Figure 4.5.

Note that, at every point to the right of the step, both the voltage between the signal and ground lines and the current into the lines are zero. At every point to

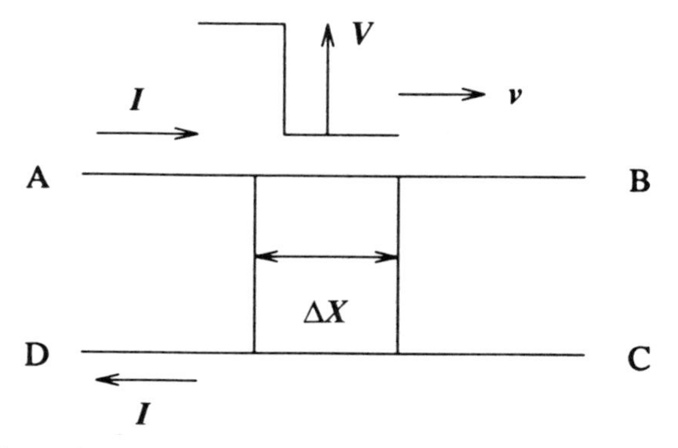

FIG. 4.5. A voltage signal and a current signal propagating along a transmission line.

the left of the step, the voltage between the two lines is V, and the current in the direction AB is I and the current in the direction of DC is $-I$.

The *magnetic field flux* Φ is defined as

$$\Phi = \int \mathrm{B} \cdot d\mathrm{S} \tag{4.3}$$

where B is the magnetic field vector, dS is the incremental area vector, $\cdot$ is the vector dot product, and the integral is taken over the entire surface for which Φ is defined [6].

Inductance of a device is defined as the proportionality constant between the magnetic flux it holds and the electric current it carries. It is given by

$$L = \frac{\Phi}{I} \tag{4.4}$$

The inductance of a transmission line is then defined as

$$\mathscr{L} = \frac{\phi}{I} \tag{4.5}$$

where $\mathscr{L}$ is inductance and ϕ is the magnetic field flux, both per unit length of the line.

The propagation speed of the voltage/current signal is defined as

$$v = \Delta x / \Delta t \tag{4.6}$$

where Δx is the distance the signal propagates in time Δt. Thus, the change of magnetic flux is

$$\Delta\Phi = \Delta(IL) = \Delta(I\mathscr{L}x) = I\mathscr{L}\Delta x \tag{4.7}$$

According to Faraday's law, the change in magnetic flux will generate a back electromotive force in the line, and this electromotive force equals to the input voltage. As a result, we have

$$V = \frac{\Delta\Phi}{\Delta t} = I\mathscr{L}\frac{\Delta x}{\Delta t} = I\mathscr{L}v \tag{4.8}$$

Transmission line phenomenon is a result of the interaction between the inductive (magnetic) field and the capacitive (electric) field effects in the transmission line. Capacitance of a line is defined as a fraction of the electric charge it holds and the potential difference across it [6]:

$$C = \frac{Q}{V} \tag{4.9}$$

Therefore the magnitude of the current wave traveling down the line can be calculated as

$$I = \frac{\Delta Q}{\Delta t} = \frac{\Delta(VC)}{\Delta t} = \frac{\Delta(V\mathscr{C}x)}{\Delta t} = V\mathscr{C}v \tag{4.10}$$

By combining Equations 4.8 and 4.10, we obtain the propagation speed as

$$v = \frac{1}{\sqrt{\mathscr{L}\mathscr{C}}} \tag{4.11}$$

Then the characteristic impedance Z_0 is obtained as

$$Z_0 = \frac{V}{I} = \sqrt{\frac{\mathscr{L}}{\mathscr{C}}} \tag{4.12}$$

It can be seen from the above equations that a uniform lossless transmission line can be completely defined by any two parameters of $\mathscr{L}$, $\mathscr{C}$, Z_0, I, V, and v [6].

The propagation speed can also be calculated from the material properties of the medium by

$$v = \frac{1}{\sqrt{\varepsilon\mu}} = \frac{c_0}{\sqrt{\varepsilon_r\mu_r}} \tag{4.13}$$

where ε and μ are the dielectric cosntant and magnetic permeability of the propagation medium; ε_r and μ_r are its relative permittivity and permeability; and c_0 is the speed of flight in vacuum. For nonmagnetic materials, μ_r is approximately 1.

Time-of-flight delay from one end of the line to the other end can be calculated as

$$t_f = \frac{l}{v} \tag{4.14}$$

4.1.2.2 Lossy Transmission Lines Many interconnections such as on-chip interconnections and traces of some thin film interconnections have significant resistance, and therefore they must be treated as lossy transmission lines. The properties of lossy transmission lines are discussed in this section.

To analyze the properties of a lossy transmission line, we assume that there is a unit step signal in the input end of the line. The response of this line at a distance x from the beginning of the line is given by

$$V(x,t) = V_A + V_R \tag{4.15}$$

where $V_A = e^{-\alpha x}$ is an attenuated step function exponentially dependent on the attenuation constant α and the distance x (see Fig. 4.6). V_A is also called the *fast-rising portion.*

$$V_R = \frac{R_x}{2Z_0} \int_{t=x\sqrt{\mathcal{L}\mathcal{C}}}^{t} \left[\frac{e^{-\mathcal{R}t/2L}}{\sqrt{t^2 - x\sqrt{\mathcal{L}\mathcal{C}}}} \frac{I_1 \mathcal{R}}{\in \mathcal{L}} \sqrt{t^2 - x\sqrt{LC}} \right] dt \times u\left(t - x\sqrt{\mathcal{L}\mathcal{C}}\right)$$

represents a RC-like behavior with a slow rise time. The term $u(t-x\sqrt{\mathcal{L}\mathcal{C}})$ is a unit step function that is zero for $t < x\sqrt{\mathcal{L}\mathcal{C}}$ and 1 for $t > \sqrt{\mathcal{L}\mathcal{C}}$. V_R is also called the *slow-rising portion.*

The attenuation constant α consists of two components: conductor loss due to line resistance (α_R) and skin effect (α_S), and dielectric loss due to shunt conduc-

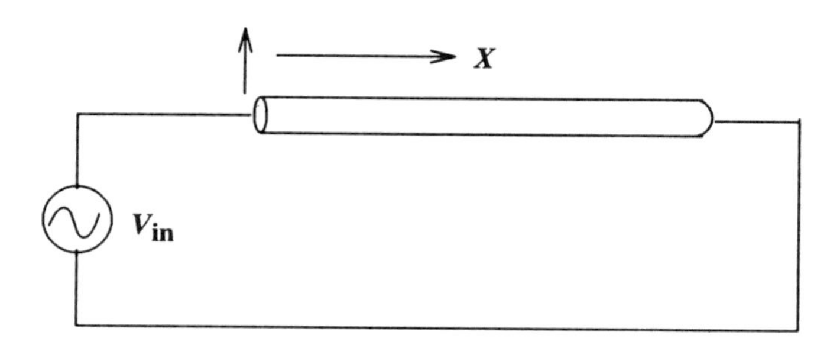

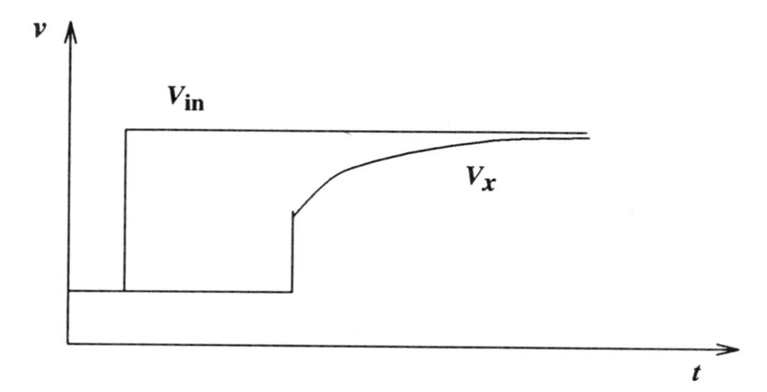

FIG. 4.6. Signal propagates along a lossy transmission line.

tance (α_D), all per unit length. The attenuation constant is expressed as

$$\alpha = \max(\alpha_R, \alpha_S) + \alpha_D \tag{4.16}$$

Conductor Losses

DC RESISTANCE At low frequencies, the signal loss is mainly caused by dc resistance of the transmission line. The low frequency loss factor per unit length α_R is given by the dc resistance of the line [6]:

$$\alpha_R = \frac{\mathcal{R}}{2Z_0}$$

$$= \frac{\rho}{2W_{int}H_{int}Z_0}$$

where W_{int} and H_{int} are the conductor width and thickness.

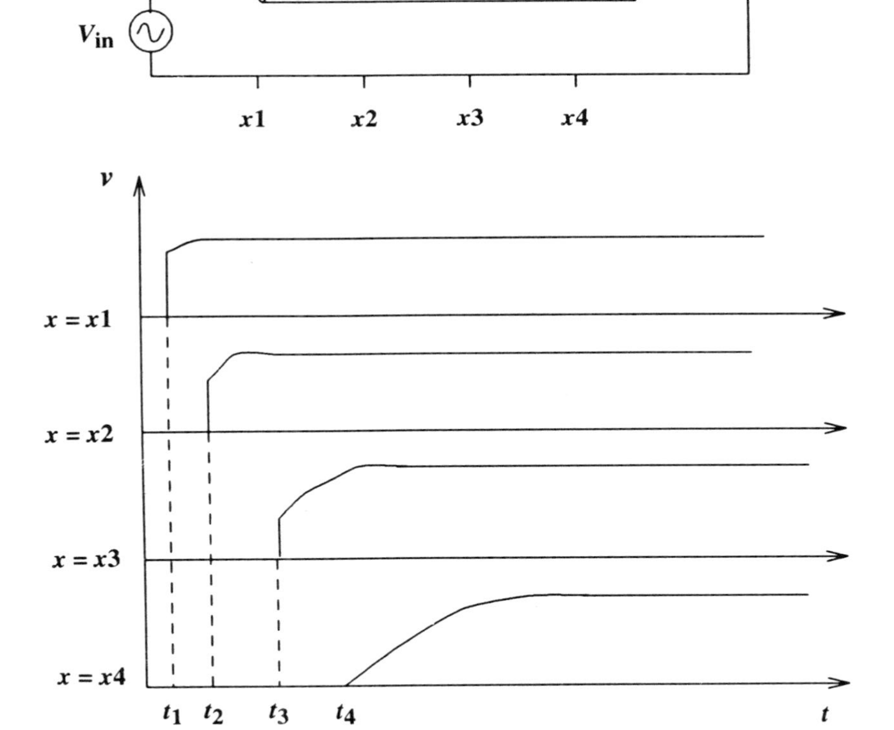

FIG. 4.7. Waveform propagating along a lossy transmission line.

Because the ground plane is much wider than the signal line and the current can spread sideways at low frequencies, the contribution of the ground plane can be ignored, resulting in the factor of two in the denominator in the above calculation of α_R.

To achieve high performance, the initial step V_A should be sufficiently high. In other words,

$$\frac{V(x=l, t=0)}{V(x=0, t=0)} = e^{-\mathcal{R}L/2Z_0}$$

must be close to 1 so that a good portion of the initial step reached the end of the line.

SKIN EFFECT At low frequency, the current is distributed uniformly throughout the cross section of the conductor through which it flows; however, at higher frequency, the current is concentrated on the surface area of the conductor. This is known as the *skin effect* [6].

An incident electromagnetic field generates currents on the surface of a conductor in the direction of the electric component of the field. If the conductor is perfect, the current is confined to an infinitesimally thin layer at the surface, and the electric field set up by this current cancels the incident field, resulting in a zero field in the conductor. In reality, all conductors have a finite resistance, which causes the field and the current to penetrate in the conductor, and this gives rise to resistive losses. As the conductivity and frequency increases, the thickness of penetration decreases. Because of this effect, at high frequencies a hollow tube is as good a conductor as a solid rod of the same diameter. Therefore, the effective resistance of an interconnection for high frequency signals does not decrease by increasing the thickness more than a critical value. The skin effect phenomenon is shown in Figure 4.8.

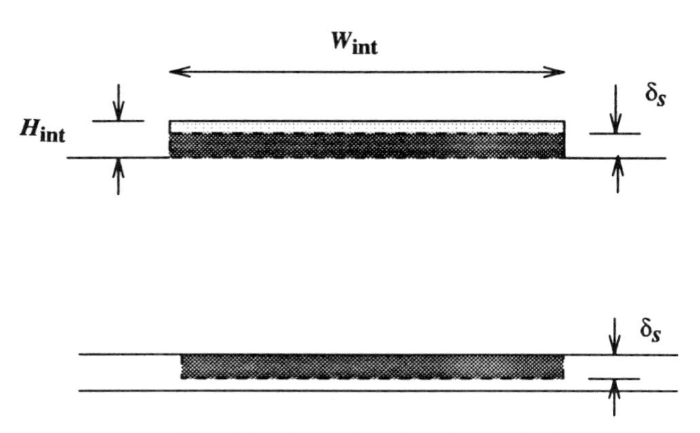

FIG. 4.8. Skin effect in microstrip line.

As a consequence of this electromagnetic induction phenomenon, the magnitude of the current density drops exponentially with distance away from the surface; thus the effective area that the current passes through is reduced. As a result, the resistance of the transmission line increases as the signal frequency increases.

The skin effect can be measured by a constant called *skin depth*. The distance at which current density becomes a fraction $1/e$ of its value at the surface is called skin depth, denoted by δ_s. The skin depth is expressed as [6]

$$\delta_S = \sqrt{\frac{\rho}{\pi \mu f}} \tag{4.17}$$

where f is frequency and μ and ρ are the permeability and resistivity coefficients of the material. For nonmagnetic materials, μ is approximately equal to $\mu_0 = 4\pi \times 10^{-9}$ H/cm. When conductor thickness is approximately $2\delta_s$ or more, increasing conductor thickness does not reduce the effective resistance of the interconnection.

Because of the skin effect, the high frequency components of the current are confined to the surface areas, and the high frequency attenuation factor per unit lenght α_s is given by

$$\begin{aligned}\alpha_S &= \frac{2\mathscr{R}_{SKIN}}{2Z_0} \\ &= \frac{\rho / W_{int}\delta_S}{Z_0} \\ &= \frac{\sqrt{\pi \mu_0 f \rho}}{W_{int} Z_0}\end{aligned}$$

Note that the resistance in the high frequency attenuation factor is multiplied by two because the signal and ground plane are assumed to contribute equally to the series resistance.

The conductor loss α_C is determined by the larger one of α_R and α_S.

$$\alpha_C = \max\left(\alpha_D, \alpha_S\right) \tag{4.18}$$

Dielectric Losses The dielectric attenuation constant is defined as follows [6]:

$$\begin{aligned}\alpha_D &= \frac{\mathscr{G} Z_0}{2} \\ &= \frac{2\pi f \mathscr{C} \tan \sigma_D \sqrt{\mathscr{L}\mathscr{C}}}{2} \\ &= \pi f \tan \sigma_D \sqrt{\mathscr{L}\mathscr{C}} \\ &= \frac{\pi \sqrt{\varepsilon_r} f \tan \delta_D}{c_0}\end{aligned}$$

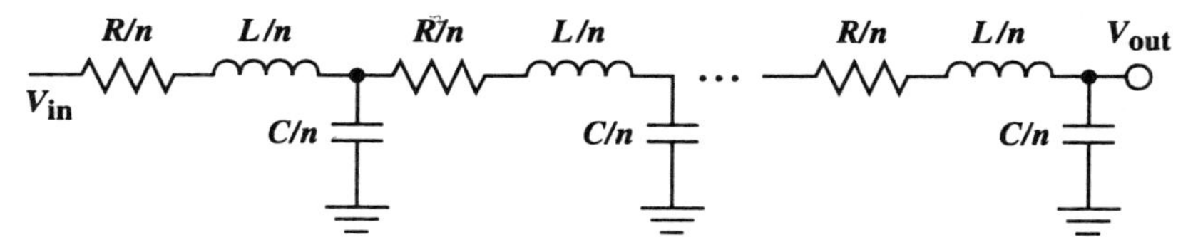

FIG. 4.9. Approximate circuit for a lossy transmission line to be used in a circuit simulator.

The loss tangent tan δ_D is defined as

$$\tan \delta_D = \frac{\mathcal{G}}{\omega \mathcal{C}} = \frac{\sigma_D}{\omega \varepsilon_r} \tag{4.19}$$

where δ_D is the conductivity of the dielectric, and ω is the angular frequency ($\omega = 2\pi f$). If the dielectric materials are chosen correctly, dielectric losses are negligible at most frequencies of interest [6].

A lossy transmission line can be simulated with a circuit simulator such as SPICE. Figure 4.9 illustrates the approximate transmission line circuit within sections. Except signal losses, lossy transmission lines have the same properties as lossless transmission lines.

The characteristic impedance and time-of-fight delay can be calculated as follows [6]:

$$Z_0 = \sqrt{\frac{L}{C}}$$

$$t_f = \sqrt{LC}$$

where L, C, and R are the total inductance, capacitance, and resistance of the line. Time-of-flight delay can be calculated from the line length and dielectric constant of the insulator:

$$t_f = \frac{l\sqrt{\varepsilon_r}}{c_0} \tag{4.20}$$

4.1.3 Interconnect Delay

Electrical devices from basic gates to large chips all have a certain amount of delay. Delay is a result of the cumulative effects of inductance and capacitance. The signal attenuation caused by resistance also increases the delay.

Interconnect delay plays a major role in MCM system delay, and it is determined by many factors, such as line length, conductor geometry, and material. In this section, we briefly present the interconnect delay in the lumped and transmission line models.

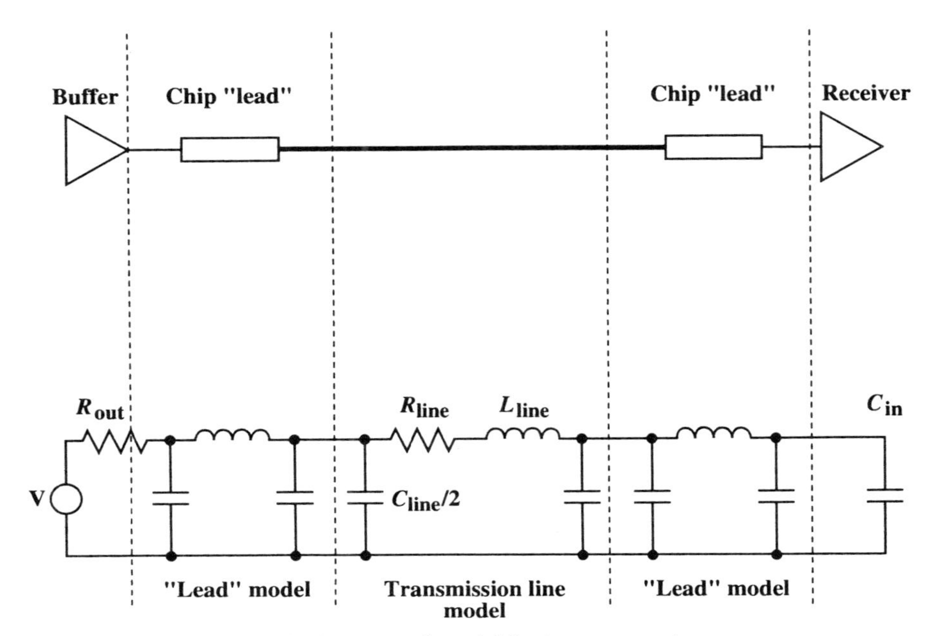

FIG. 4.10. Lumped model for interconnection.

4.1.3.1 *Lumped Model* We first consider interconnect delay using a lumped model for interconnection as shown in Figure 4.10. A lower bound estimate on the propagation delay of this model is the RC delay given by [58]

$$t_p \approx 0.7\left[R_{out}\left(C_1 + C_{line} + C_2 + C_{in}\right) + \frac{1}{2}R_{line}C_{line} + R_{line}\left(C_2 + C_{in}\right)\right] \tag{4.21}$$

where R_{out} is the equivalent resistance of the driver, R_{line} is the total line resistance, C_{line} is the total line capacitance, C_1 and C_2 are the lead capacitances, and C_{in} is the input capacitance of the die. The effect of any line inductance, L_{line}, is to increase this delay by 10%–30% or more.

4.1.3.2 *Transmission Line Model* The delay of a transmission line depends on the type of the line. The delays in a surface microstrip line and in a buried microstrip line are discussed in this section.

The empirical (and approximate) equations for the characteristic impedance Z_0 and for the time-of-flight in a surface microstrip transmission line are given as follows [58]:

$$Z_0 = \frac{87}{\sqrt{\varepsilon_r + 1.41}} \ln \frac{5.98h}{0.8w + t} \tag{4.22}$$

$$t_f = 3.337\sqrt{0.475\varepsilon_r + 0.67} \tag{4.23}$$

where ε_r is the dielectric constant of the dielectric and the dimensions h, w, and t are height, width, and thickness, respectively, of the microstrip line.

Similar equations for a buried microstrip are as follows [58]:

$$Z_0 = \frac{60}{\sqrt{\varepsilon_r + 1.41}} \ln \frac{5.98h}{0.8w + t} \tag{4.24}$$

$$t_f = 3.337\sqrt{\varepsilon_r} \tag{4.25}$$

4.2 METHODS OF MINIMIZING INTERCONNECT DELAY

MCM interconnections are generally long, and interconnect delay dominates gate delay. As a result, it is vital to minimize interconnect delay to improve the performance. Due to the generally long interconnections and high performance requirements, transmission line models have to be used for MCMs. We discuss here the techniques of reducing delay on a transmission line.

To get the minimum propagation delay on a transmision line, the first signal to arrive at the end of the line must have sufficient voltage to switch the receiver, e.g., it should exceed V_{IH} on a $0 \rightarrow 1$ transition or V_{IL} on a $1 \leftarrow 0$ transition. (V_{IH} refers to the high level input voltage, which is the minimum voltage that is required to be applied to the input of a device for a logic-1 voltage level. V_{IL} refers to the low level input voltage, which is the maximum voltage that is required to be applied to the input of a device for a logic-0 voltage level.) In other words, the fast-rising portion of Equation 4.15 V_A should have enough amplitude. When this is achieved, the situation is called *first incidence switching* [58]. First incidence switching requires small signal attenuation in the transmission line. If the arriving signal does not have sufficient voltage to switch the receiver, it is called *undershoot* or *overshoot*, depending on the type of transition.

The first incidence voltage at the end of a matched terminated line is calculated by

$$V_{first} = V_{in} \frac{Z_0}{R_{out} + Z_0} e^{-\mathscr{R}l/2Z_0} \tag{4.26}$$

where V_{in} is the open circuit output voltage swing of the driver, l is the length of the line, and only dc resistance is considered here. Thus, to achieve first incidence switching, line losses must be small and the characteristic impedance of the line must be significantly greater than the Thevenin equivalent output resistance of the driver. For this reason, in a high speed system, the line characteristic impedance tends to fall in the range between 40 Ω and 75 Ω, higher values being preferred, particularly for CMOS drivers with their high values of R_{out} [58].

Because the cross section of wires can be very small in MCM-Ds, controlling line losses to achieve first incidence switching requires careful design. For example, consider the following CMOS circuit. Assume R_{out}=15 Ω in the logic-1 state, $Z_0 = 65\ \Omega$, $V_{in} = 5$ V, and $V_{first} = V_{IH} = 3.85$ V, then, according to Equation 4.26, first incidence switching requires $\mathcal{R} \times l < 7\ \Omega$. For a $5 \times 20\ \mu$m thin film copper conductor with $\mathcal{R} = 165\ \Omega$/m, the longest line that can still achieve first incidence switching would be 4.2 cm [58].

The line losses increase with frequency because of the skin effect. This changes the shape of the signal, increases the rise time, and makes first incidence switching more difficult for long lossy lines for design bandwidths above 200 MHz [58]. The smaller output impedances of TTL and ECL circuits make first incidence switching easier to achieve than the CMOS circuits.

If the first incidence switching is achieved, the delay will primarily depend on the values of the dielectric constant and the capacitive load. Reducing delay is the main reason for using dielectric material with smaller dielectric constants.

First incidence switching also requires that the reflection noise be controlled. We discuss the reflection noise in the next section.

4.3 NOISE

In the last two sections, we analyzed the interconnection delay in MCMs and the methods to minimize the delay. Interconnections in MCMs not only cause delay but also generate noise. The density of MCM substrates is high. Thus, the capacitive, inductive, and resistive couplings between neighboring circuits become more significant for MCMs because of smaller spacings, resulting in higher crosstalk noise. Furthermore, with larger amounts of currents switching simultaneously, the inductive noise associated with power lines also increases. This type of noise is called *simultaneous switching noise*. In addition, the discontinuities in the transmission lines also cause reflection noise.

In this section, we discuss reflection noise, crosstalk noise, and simultaneous switching noise caused by MCM interconnections. We also discuss the factors that contribute to the generation of noise and the effect of noise on system performance. The methods of controlling the noise are covered in the next section.

Interconnect noise increases as a result of shorter rise times, larger total chip currents, longer interconnections on MCM, and smaller spacings between interconnections. Certain types of noise, such as crosstalk, that previously were significant mostly at the board level become prevalent also at the MCM level because of the higher density of MCM substrates.

As the size of MCMs increases and the wavelengths of the signals become comparable to interconnection lengths, the transmission line properties of interconnections also gain importance. Whenever a change in characterisitc impedance occurs, part of the incident electromagnetic wave is reflected, just like part of a light beam is reflected upon striking a sheet of glass.

As chip dimensions and clock frequency increase, the wavelengths of the signals become comparable to interconnection lengths, and this makes interconnections better "antennas." In addition, capacitive, inductive, and resistive couplings between neighboring circuits become more significant with higher packaging density in MCMs.

Chip-to-chip interconnections, package pins, and bond wires all have parasitic inductances. When a current that flows through changes, a voltage fluctuation is generated across the inductor proportional to the inducatance L and the rate of change of the current dI/dt. As a result, when the circuits switch on and off, the voltage levels at the power distribution lines fluctuate, resulting in simultaneous switching noise.

All the types of noise mentioned above can cause increased delay or inadvertent logic transitions. Thus, they should be minimized through careful design to avoid performance degradation or system errors.

4.3.1 Reflection Noise

Reflection noise is the noise caused by the discontinuities in a transmission line. In the previous sections, the transmission lines were assumed to be infinitely long. The effects of a transmission line of finite length have not been considered. However, the real transmission lines are of finite length. As a result, the end of a transmission line introduces a discontinuity that may generate reflection noise. There are also other cases of discontinuities on a transmission line. Some examples are shown in Figure 4.11.

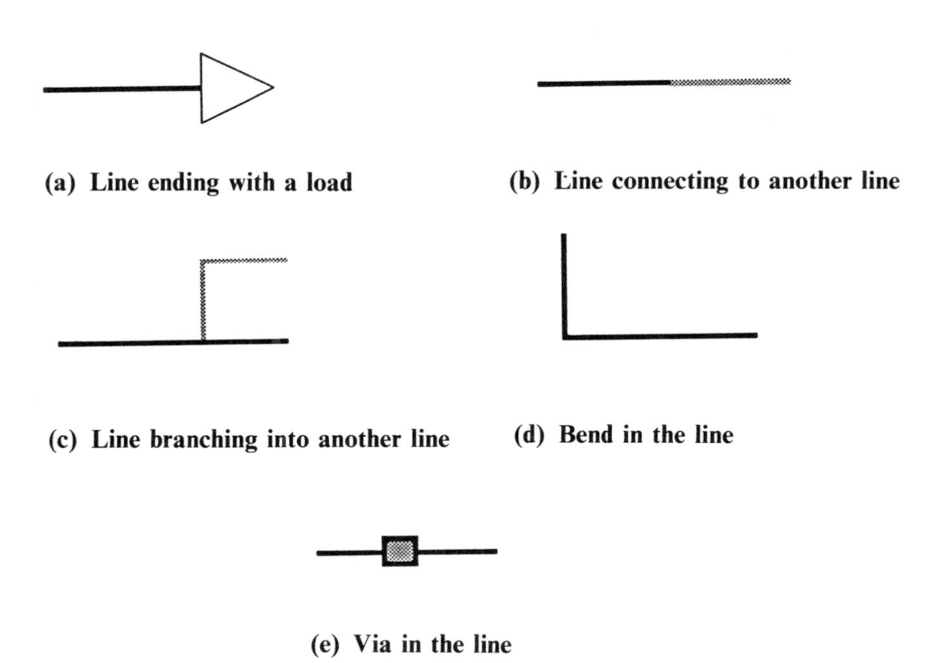

FIG. 4.11. Discontinuities in the transmission lines.

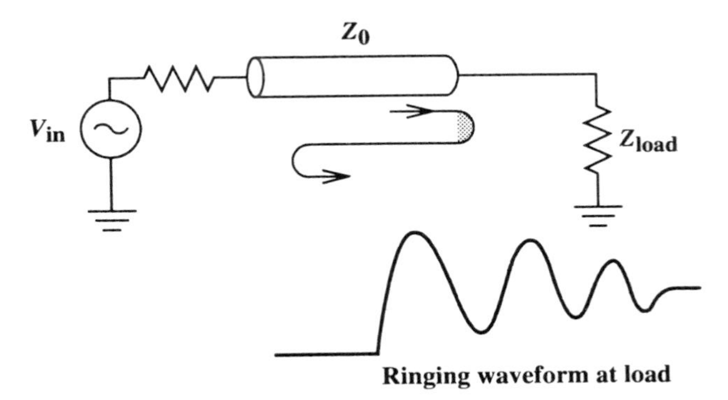

FIG. 4.12. Multiple reflections cause ringing.

The portion of the incident electromagnetic wave voltage that is reflected at a change of characteristic impedance is given as the reflection coefficient, Γ, where

$$\Gamma = \frac{Z_{load} - Z_0}{Z_{load} + Z_0} \tag{4.27}$$

Z_0 is the characteristic impedance of the transmission line on which the incident wave travels, and Z_{load} is the impedance of the load on the line. If the load is an open circuit, then $Z_{load} = \infty$ and $\Gamma = 1$, and the reflected wave has the same voltage as the incident wave . If the load is a short circuit, then $Z_{load} = 0$ and $\Gamma = -1$, and the reflected wave is inverted with respect to the incident wave. On the other hand, if the load is matched to the line impedance, $Z_{load} = Z_0$, then no reflection occurs.

The successive partial reflections of the wave from each end creates a damped ringing signal as shown in Figure 4.12 [58]. Ringing is a potential problem if the time it takes for the wave to travel down the line is longer than one fourth of the rise time [58],

$$t_p > t_{rise}/4 \tag{4.28}$$

Waiting for the ringing noise to settle down can take upto an additional $4 \times t_p$ [58]. Therefore, it is very important to control the ringing noise. In the case of a short line, ringing may not be significant because the ringing settles down before the end of the rise time. However, for a long line, the ringing noise can become a serious problem.

Reflections also occur whenever the line branches. The effect of the reflection is small, however, if the propagation delay along the length of the stub length is kept small when compared with the signal rise time.

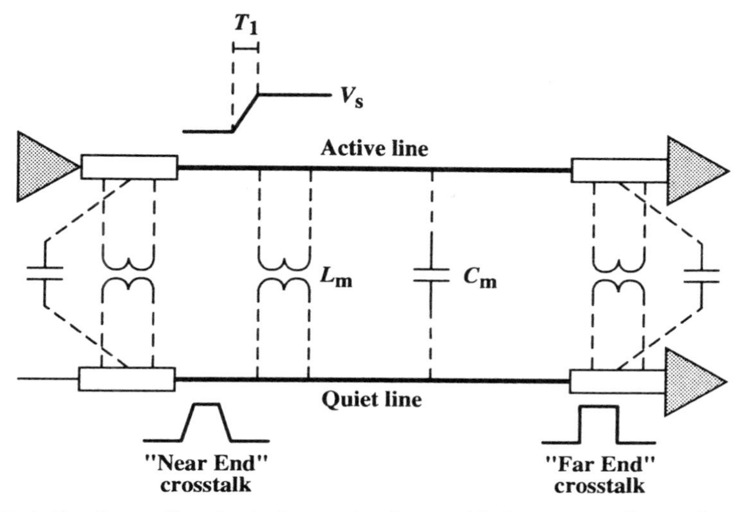

FIG. 4.13. Crosstalk noise is the result of mutual inductance and capacitance.

4.3.2 Crosstalk Noise

In addition to the reflection noise, mutual inductance and capacitance between different electrical signal paths cause unwanted electrical coupling known as *crosstalk noise*. Whenever a signal edge travels down a signal wire, chip attach lead or connector lead, both forward and backward noise pulses are induced in the neighboring wires, as shown in Figure 4.13.

Capacitive coupling $K_C = C_m/C$ and inductive coupling $K_L = L_m/\mathscr{L}$ between adjacent lines add at the near end of the quiet line and subtract at the far end. The maximum noise voltage at the near end can be approximated by

$$V_n \approx K_B \frac{2}{v}\frac{V_S}{T_1} l \qquad \text{if } l < \frac{vT_1}{2}$$

$$\approx K_B V_S \qquad \text{if } l < \frac{vT_1}{2}$$

where $K_B = (K_C + K_L)/4$ is the coupling coefficient, C_m is the mutual capacitance between the lines per unit length, C is the total capacitance of the line per unit length, L_m is the mutual inductance per unit length, L is the inductance of the line per unit length, V_s is the voltage swing on the active line, $v = c/\sqrt{\varepsilon_r}$ is the propagation velocity of electromagnetic waves in the dielectric, T_1 is the rise time of the signal, and l is the coupled line length [58].

The maximum noise voltage at the far end is given approximately as

$$V_f = K_F \frac{2}{v}\frac{V_S}{T_1} l \tag{4.29}$$

where $K_F = (K_C - K_L)/4$ is the coupling coefficient. If the medium in which the line is buried contains no other materials besides the dielectric, then the far end crosstalk is zero because $C_m/\mathscr{C} = L_m/\mathscr{L}$. This is only perfectly achieved when homogeneous dielectric stripline conductors are present. However, even with other conductors present, and for buried microstrips, K_F is usually small.

Crosstalk is a result of mutual capacitances and inductances between neighboring lines. It can increase circuit delay or even cause latches to switch to a wrong value. Therefore, the crosstalk noise must be minimized.

4.3.3 Simultaneous Switch Noise

In high speed digital systems, large output drivers are used. This growth in drive strength increases the switching current or the transient current that flows in and out of the chip through the power supply and gound pins since the drivers drive not only the load but also the MCM substrate traces, the substrate to package wire bond, and the package parasitics. These components all have inductance. The inductance causes the ground (or power supply) to bounce when the drivers switch. This noise is also referred to as *simultaneous switching noise* because it is most pronounced when many off-chips driver switch simultaneously. Simultaneous switching noise is also referred to as *power supply level fluctuations* or as *dI/dt* noise. For example, consider a 5 V, 32 bit driver chip with a rise time of 2 ns driving a load of 320 pF (10 pF/bit). This corresponds to a $dI/dt = C\Delta V/\Delta t =$ 0.8 A/s [58]. When this transient current passes through the inductive power distribution network, a noise voltage is produced.

The ground noise is more critical than the power noise because of the TTL compatibility. Input receivers have their switching point around 1.2–1.6 V. Therefore, more noise can be tolerated on the power bus than on the ground bus.

Controlling simultaneous switching noise is an important design issue that has received much attention for high speed and high pin count VLSI. The problem is worse in MCM due to the even larger number of drivers and the addition of MCM substrate traces. Switching noise can result in a number of problems if not handled properly [58]:

1. Simultaneous switching noise can cause inadvertent logic transitions to the output of what were intended to be quiet off-chip drivers.
2. The changes in internal chip supply voltage can change the static operating states of circuits and make the circuits operate more slowly and, thus, increase the delay in the switching drivers. The delay increase might be up to 2–3 ns or more for CMOS and TTL circuits, depending on the circuit details and the number of switching drivers. Overshoots and undershoots might also appear in these drivers.
3. For on-chip circuits acting as input gates, simultaneous switching noise acts to reduce the effective noise margin at the inputs.
4. For on-chip memory devices, such as latches, large amounts of ground-rail and power-rail noise might cause false changes in state.

4.4 METHODS OF CONTROLLING NOISE

Noise can cause increased delays or inadvertent logic transitions. Delays of synchronous signals may increase due to reflections (nodes may take a longer time to settle), crosstalk (coupling between the lines may increase the effective line capacitance and therefore the propagation delay), or simultaneous switching (a reduction in the supply level diminishes the current drive of a circuit and therefore increases its delay). As a result, noise should be minimized through careful design. In this section, some methods for controlling noise are presented.

4.4.1 Controlling Reflection Noise

Reflection noise is generated by the discontinuities of the transmission lines. Therefore the key to reducing the reflection noise is to remove the discontinuities, for example, let $Z_{load} = Z_0$. The act of equalizing the characteristic impedance of load and interconnect is called *impedance matching*.

The four most common techniques for impedance matching are shown in Figure 4.14 [58]. They are

1. Parallel termination
2. Thevenin equivalent parallel termination
3. AC termination
4. Series termination

4.4.2 Controlling Crosstalk Noise

Crosstalk noise is determined by the capacitive coupling $K_C = C_m/\mathscr{C}$ and inductive coupling $K_L = L_m/\mathscr{C}$. The coupling between the lines can be minimized by ensuring sure that no two lines are laid out parallel or next to each other for longer than a maximum length. Crosstalk requirements can determine the required spacing between the lines and thus the signal line pitch (signal line pitch = line width + line spacing). Often the required spacing is at least twice the minimum spacing that the fabrication technology would allow [58]. As the spacing increases, C_m and L_m decrease, and thereby crosstalk noise decreases. Sometimes crosstalk noise may also be reduced by re-routing parallel lines so that the length of the parallelism is minimized.

Crosstalk can also be reduced by keeping lines apart and by placing a ground plane between the lines. By inserting a ground plane, $C_m/\mathscr{C}$ is reduced because the total capacitance $\mathscr{C}$ increases whereas the mutual capacitance C_m remains approximately the same. The ground planes also reduces $L_m/\mathscr{L}$ because the current return path goes through the ground plane rather than neighboring lines. This forces most of the magnetic field flux to be between signal lines and the ground plane rather than between the neighboring signal lines.

The placement of a ground line between adjacent microstrip or stripline signal lines can be used to reduce crosstalk by almost 50% [58]. This is only beneficial if

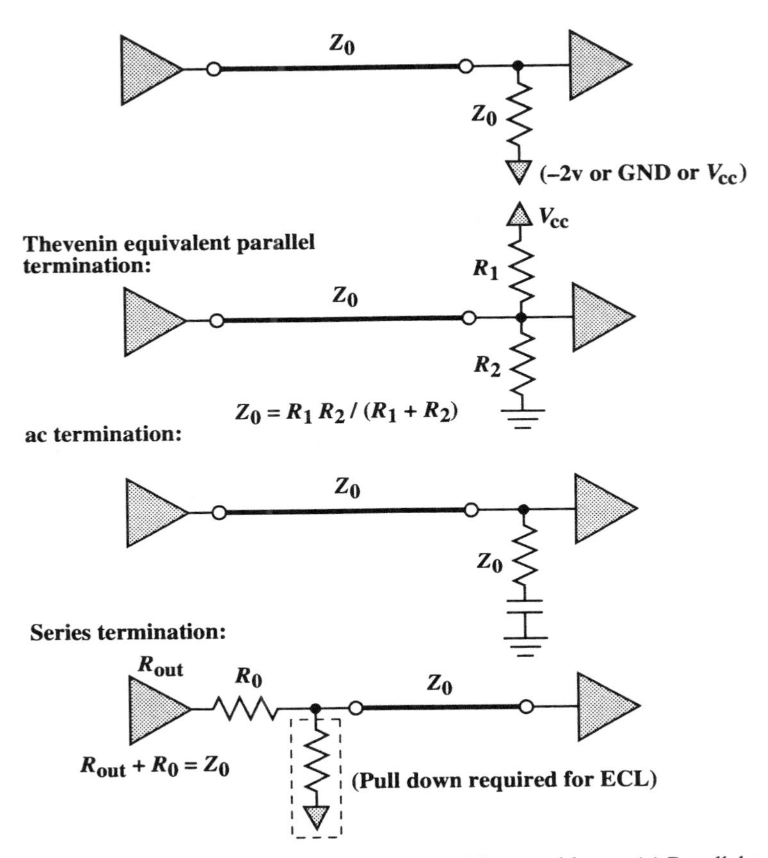

FIG. 4.14. The four most common impedance matching tecnhiques. **(a)** Parallel termination, **(b)** Thevenin equivalent parallel termination, **(c)** ac termination, **(d)** Series termination.

the spacing is already large enough to add a ground line.The ground line cannot be placed too close to the signal lines or it will change the characteristic impedance of the signal lines.

Choosing a stripline over a mircrostrip configuration reduces the required spacing for the same impedance Z_0, as the presence of two reference planes in the stripline reduces the ratio C_m/C (therefore, K_B and K_F). In either case, the greater the distance from the ground plane, the larger both the line width and spacing must be in order to keep Z_0 and crosstalk the same.

Matching terminations in both ends of wires can be used to help reduce the effects of crosstalk noise. If the neighboring wires do not have matching terminating resistors, the noise pulses are reflected from each end. As a result, part of the near end noise shown in Figure 4.13 will still arrive at the receiver.

Choosing a material with a lower dielectric constant also has positive effects on reducing the required spacing between two signal lines while keeping the coupling coefficients K_B and K_F unchanged. In addition, choosing a material with a lower dielectric constant increases the propagation speed v.

4.4.3 Minimizing Simultaneous Switch Noise

Simultaneous switch noise can be reduced through the following design techniques [6]:

1. Placing decoupling capacitors close to the chip reduces the simultaneous switching noise by charging up during the steady state and then assuming the role of power supply during the current switching period, as shown in Figure 4.15.
2. Minimizing the inductance of MCM interconnections. Inductance of an MCM increases as the length of any round or rectangular lead increases. By using solder bump technology, which has the least inductance compared with TAB and wire bound technologies [58], the simultaneous switching noise resulting from inductance can be minimized.
3. Using multiple power and ground connections may reduce the δI noise further. If there are multiple power/ground pins, each connection can supply a fraction of the total current transient, and together they can support a larger current spike.
4. It is common practice to separate the power/ground bus of the output drivers from that of internal circuitry to minimize the influence of the output driver switching noise on the internal circuits.

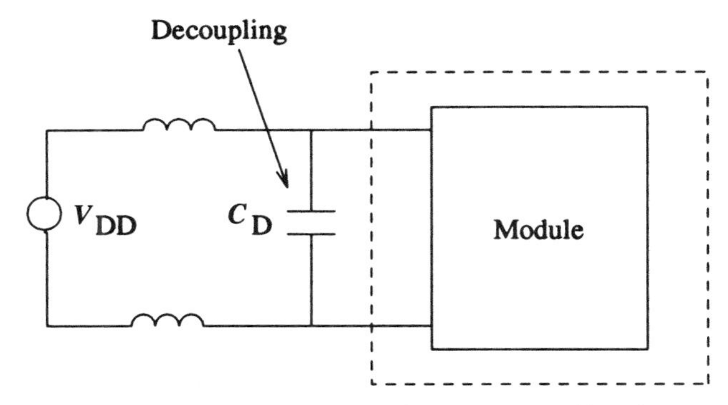

FIG. 4.15. A decoupling capacitor reduces *dI/dt* noise.

5. The δI noise can also be reduced by controlling the turn-on characteristics of output drivers. The drivers can be turned on slowly rather than sharply to reduce the size and slope of the current spike. The other approach is to reduce the number of drivers switching simultaneously by introducing additional delay on the outputs. Both approaches may add some delay to the system.
6. Using a high dielectric constant medium between the plane-pair, if there is no signal line between them.

4.4.4 Effects of Characteristic Impedance

The characteristic impedance Z_0 of interconnections determines the propagation delay, noise level, and power dissipation [6]. In this section, we discuss the effects of Z_0 on delay and noise.

From Equations 4.12 and 4.20, we know that the smaller the line impedance, the larger its capacitances ($Z_0 = 1/vC$). Accordingly, a stronger driver is required to reduce propagation delay when the line impedance is small because of the large capacitance. If the driver resistance is not much smaller than Z_0, the driver and line will act as a voltage divider, and the initial potential step on the line will be proportional to $Z_0/(Z_0 + R_{out})$, as shown in Equation 4.26, where R_{out} is the resistance of the driver. With $R_{out} > Z_0$, many round-trips and reflections from the end of the line will be required for the output to reach its final value. In CMOS, stronger drivers have more buffer delay because more stages of cascaded inverters are required to achieve a very small output impedance. Consequently, large Z_0 is desirable because it results in an easy-to-drive line with a small capacitance.

When the line impedance is too high, circuit speed may start to degrade. Delay due to capacitive discontinuities along or at the end of the line is proportional to Z_0 because large impedance lines (which do not require large currents) cannot supply as high a current as small impedance lines do. They added delays due to discrete capacitive loads or short stubs are proportional to $Z_0 C_L$. As a result, small Z_0 helps to reduce delays due to capacitive loads attached to the transmission line.

Switching noise decreases when Z_0 increases because a high impedance line has a small capacitance and requires smaller current surges ($\delta I \propto \delta V/Z_0$). Crosstalk, however, increases with higher Z_0 because a high impedance line has a small capacitance to the reference plane. When the line and the reference plane are not strongly coupled, the coupling between the lines gains importance and crosstalk increases. In addition, self and mutual inductances of the lines increase with Z_0 because larger Z_0 implies lines high above the ground plane with a large current loop area and high inductance ($Z_0 \propto \mathcal{L}$). Reflection noise also increases with higher Z_0 because the smaller the line capacitance is, the stronger the influence of capacitive discontinuities becomes.

In summary, large impedance lines have the following advantages [6]:

1. Lines have small capacitances.
2. Lines are easier to drive. As a result, they require smaller buffers with shorter buffer delay.
3. Lines generate less switching noise.

On the other hand, small impedance lines have the following advantages [6]:

1. Lines have large capacitance.
2. Lines can supply a large current to capacitive loads. As a result, they have less added delay due to capacitive loading of fan-out nodes.
3. Lines are strongly coupled to the reference plane. As a result, they generate less crosstalk with neighbors.
4. Lines receive smaller and narrower reflections from capacitive discontinuities because a line with a larger capacitance is more stable.

From the above discussion, we can see that there are both advantages and disadvantages of either a large value or a small value of characteristic impedance. As a result, a careful study of the tradeoffs is necessary for choosing the right value of impedance.

4.5 ELECTRICAL DESIGN OF MCMs

In previous sections, we have discussed two major electrical design considerations in MCMs: delay and noise. These considerations should be dealt with at the beginning of the MCM design process. The goal of the electrical design of MCMs should be the maximization of the performance of the system, as limited interconnect delay,and minimization of the possibility of the false operation in the field due to the electrical noise. In this section, we discuss some of the delay and noise issues in designing MCMs.

4.5.1 Technology Selection and System Planning

Different MCM technologies and conductor materials have different electrical and physical characteristics, thus using different technologies also could result in different interconnection delay and noise. To show this, we discuss some electrical and physical parameters of different MCMs [104].

Table 4.1 lists the physical and material parametres of some MCMs. We can see that AlO has a high dielectric constrant while 2-L, 4-L, and 5-L have a lower dielectric constant.

Table 4.2 lists the electrical parameters for MCMs. It shows that AlO has a high Z_0 and a large attenuation constant α, while GC MCM has the lowest attenuation constant.

TABLE 4.1. Physical and Material Parameters

Parameter	AlO	GC	2-L	4-L	5-L
Signal pitch (μm)	450	450	50	25, 50	25, 50
Signal width (μm)	100	75	12.5	10, 15	10, 15
Signal thick (μm)	25	20	5	4.5, 6	4.5, 6
Dielectric thick (μm)	175	65	4	5.5, 9	5.5, 9
Dielectric const.	9.5	5.3	3.5	3.5	3.5

TABLE 4.2. Electrical Parameters

Parameter	AlO	GC	2-L	4-L	5-L
Z_0 (Ω)	51	50	44	36, 41	36, 41
T_0 (ps/cm)	110	80	79	65	65
R_{dc} (Ω/cm)	0.3	0.2	3.2	4.4, 2.2	4.4, 2.2
α 1GHz (1/cm)	0.013	0.009	0.04	0.06, 0.03	0.06, 0.003
KbH (%)	3.4	0.98	3.0	3, 1.7	3, 1.7
KbV (%)	0.98	1.4	–	–	–
KbD (%)	0.49	0.52	–	–	–
KfH (%)	–1.3	–6.7	3.0	3, 1.7	3, 1.7
KfV (%)	–0.36	–0.42	–	–	–
KfV (%)	–0.40	–0.17	–	–	–

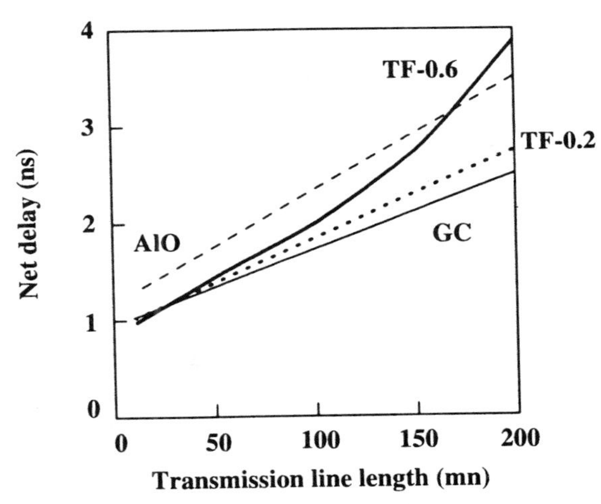

FIG. 4.16. Net delay comparison of the interconnection media.

The comparative net delays as a function of the line length ranging from 10 to 200 mm for the three types of interconnection media are plotted in Figure 4.16 [104]. The lines corresponding to the alumina (AlO) and glass-ceramic (GC) substrates show a linear response due to the predominantly capacitive charging effects. The two curves for the thin film (TF) medium show the effects of the line resistance on the delay. The low resistance thin film line shows a nearly linear response, while the high resistance line exhibits a quadratic response owing to the RC charging effects. It is instructive to note that the thin film is not the optimal medium to drive long transmission lines due to the significantly higher attenuation and distortion.

Comparisons of crosstalk among different types of MCMs are shown in Figure 4.17. It can be seen that the AlO MCM has a higher crosstalk noise than the other MCMs.

From the electrical design point of view, a technology must be selected so that, with a suitable choice for partitioning and floorplanning, the delay and noise requirements are met.

There are a number of requirements for the control of noise. As circuits get faster and rise time becomes faster, more stringent requirements on the technology arise. In particular, chip connection and connector styles with lower inductances and less mutual coupling are preferred. Either short lead connection methods must be used or reference planes are needed beneath the leads. As rise time increases, striplines are preferred over microstrips to reduce the mutual coupling between crossing lines in adjacent signal layers. In addition, the design of decoupling capacitors becomes more critical for controlling the simultaneous switching noise.

The desire of producing fast first incidence switching imposes requirements on line impedance, dielectric constant and line losses. The latter become particularly difficult to handle with thin film interconnections if aluminum lines are used or if the design bandwidth is so high that the skin effect becomes important.

The desire for a high line impedance must be balanced against the desire to minimize the line pitch, as determined by crosstalk requirements. If the line pitch is large, then extra layers might have to be added to create more interconnect capacity.

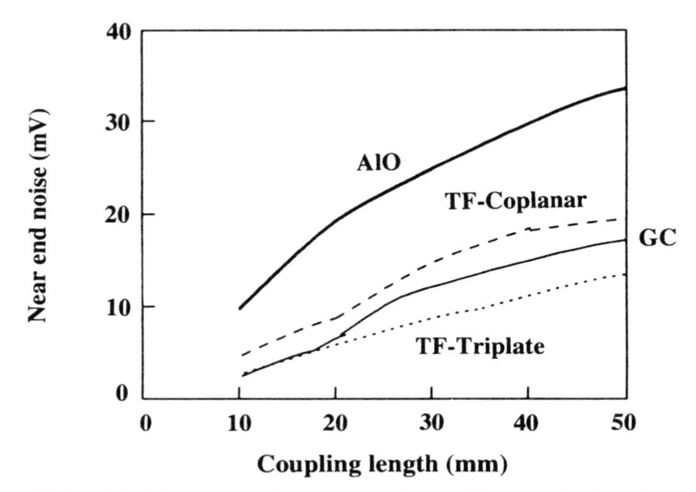

FIG. 4.17. Near end noise as a function of the coupled length.

4.5.2 Noise Budgeting

Three major types of electrical noise have been discussed. All noises arrive at the circuit receiver at random times. To ensure that the system works correctly, the noise should be controlled so that the sum of the noises will not exceed the noise margin of receivers. The goal of noise budgeting is to give the relative weight to the different types of noise according to the total noise margin and the level of difficulty in controlling each type of noise under the given package technology.

For clock signals in the system, noise control is more important; hence, noise budgeting for clock signals tends to be far more conservative.

4.5.3 Modeling, Simulation, and MCM Layout

The process of generating an MCM layout is described in Figure 4.18 [58]. Based on the requirements of timing design, delay requirements for each net are generated by subtracting the worst-case delays of the active components between each pair of latches from the clock period (see Fig. 4.1). An estimate of the wiring rules is produced from these delays. The wiring rules specify which nets need to

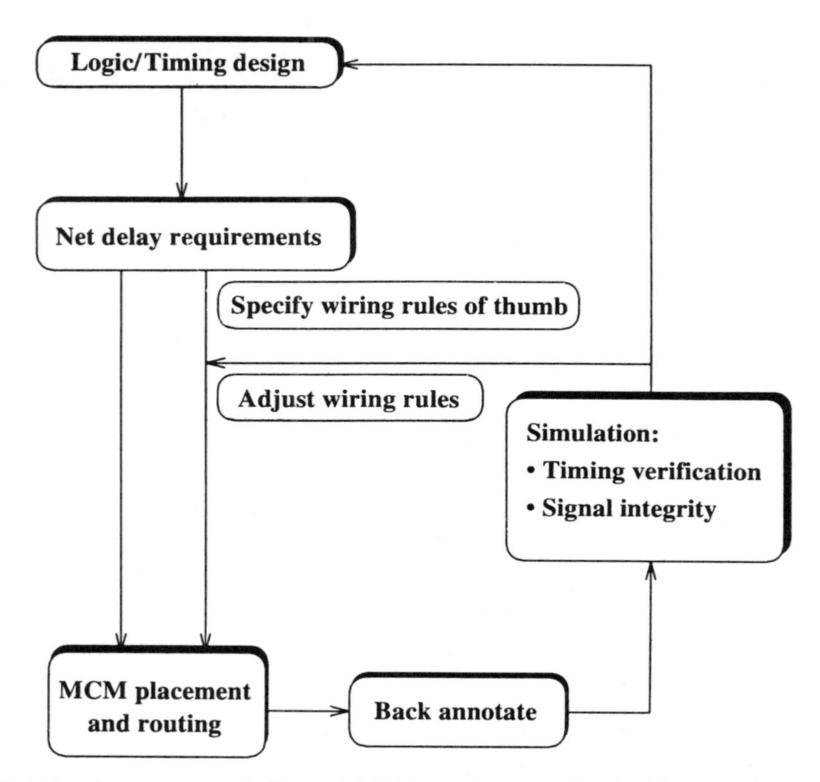

FIG. 4.18. The steps in producing an MCM layout from an electrical design point of view.

use controlled topologies and matching terminations, the limits on the lengths to each receiver, and the stub length limit for those nets, as well as the spacing requirements between nets.

The wiring rules are passed to the placement and routing tools, which generate the layout. It is necessary to verify that delay requirements and noise budget requirements (signal integrity requirements) are met. This requires that electrical models be obtained and the simulations be carried out. The results of these simulations are compared with the requirements. If requirements are not met, the wiring rules are adjusted and the processes are repeated.

Models must be developed for drivers, receivers, transmission lines, and line discontinuities including vias, chip connection leads, and connectors. Thin film MCM interconnects are resistive by virtue of their small cross-section areas. To model rigorously all the transmission line effects, the coupled lossy transmission line model must be used. Discontinuities are generally modeled by the lumped RLC circuit model.

4.6 SUMMARY

In this chapter, electrical design considerations for MCMs are discussed. The goal of electrical designing in MCMs is to improve the system's electrical performance, that is, to optimize the delay and noise performance of MCMs. Both delay and noise in MCMs are strongly influenced by the interconnections due to the high density and long length of interconnections and high working frequency of MCMs. To represent interconnections in MCMs accurately, transmission line models must be used. Both lossless transmission line model and lossy transmission line model are used in MCMs. The time of flight along a transmission line is basically determined by the line length and dielectric constant of the dielectrics. To minimize the interconnect delay, the first incidence switch must be achieved. The noise generated by interconnections in MCMs include reflection noise, crosstalk noise, and simultaneous switching noise. The noise can cause increased delays or inadvertent logic transitions. The noise should be minimized through careful system design, logic design, and physical design. Since noise and delay directly influence the MCM performance, they must be considered at the beginning of the design process. However, there are few EDA tools suitable for electrical design of MCMs currently available currently. Much work needs to be carried out to understand the electrical characteristics of the MCM design.

4.7 PROBLEMS

1. In an MCM system, suppose the polyimide dielectric constant is 3.5, and the rise time of signal is 1 ns with $N = 4$. Which model should be used to model a 2.5 cm interconnect line?

2. The dielectric of alumina material is about 9.5. Calculate the propagation speed of signal along an alumina transmission line and the time of flight if the length of the line is 10 cm.

3. Consider a $10 \times 15\,\mu$m aluminum conductor with $\rho = 2.8$ and the characteristic impedance of $50\,\Omega$. Calculate the attenuation constant of the conductor if the signal frequency is 200 MHz.

4. In a flip-flop MCM, the equivalent resistance and capacitance of the bonding wire are $0.8\,\Omega$ and 0.1 pF, respectively. The capacitance of the die and the lead both are 2 pF. Calculate the time delay of the bonding wire.

5. For an interconnect line in an MCM, suppose the $\varepsilon_r = 4.5$, and the minimum possible rise time of the signals along this line can be 4 *ns*. Calculate the maximum length of the line such that the ringing problem cannot happen along this line. If $\varepsilon_r = 8$, what will be the maximum length of the line?

6. For a $10 \times 15\,\mu$m aluminum conductor with $\rho = 2.8$ and $\varepsilon_r = 9.5$, the characteristic impedance of this line is $Z_0 = 60\,\Omega$, and a 5 V step signal is propagating along it. If the V_{IH} of this cicuit is 3.8 V and the equivalent resistance of the receiver is $20\,\Omega$, calculate the longest line length such that the first incident switching can still be obtained. In addition, calculate the minimum delay of this line.

7. Suppose there two parallel lines in the substrate of an MCM, the coupling coefficient is 0.5, and the dc noise margin is 0.8 V. If one line has a $0 \rightarrow 1$ transition within 5 ns, to avoid the inadvertent transition in the other line, calculate the maximum coupled line length between these two lines.

BIBILOGRAPHIC NOTES

Transmission lines have been discussed comprehensively in many books. Bakoglu [6] describes both the lossless and the lossy transmission line models in detail. Detailed simulations and analyses about interconnections [157, 163] delay analyses [6, 58, 157, 163], discussions about reflection noise [6, 58], discussions of crosstalk noise [6, 34, 58], and methods of controlling simultaneous switching noise [6, 33, 34] have also been published. Besides the three types of noise discussed in this chapter, there are some other types of electrical noise that may need to be considered in MCM design. Details about these types of noise [58] and the electrical design guidelines for MCM [58, 72, 104] can be found elsewhere.

CHAPTER 5

THERMAL DESIGN OF MCMs

In an electrical system, power consumed by the electrical devices is directly converted into heat, which causes a rise in the operating temperature of the device if it is not properly dissipated. The failure rate of electrical components is directly related to the operating temperature. In fact, the failure rate can increase sharply even if there is a small increase in temperature. For example, a transistor operating at a junction temperature of 180 °C has only one-twentieth the life of a transistor operating at 25 °C. The failure rate is even worse in chips and systems, since they contain a large number of transistors. Moreover, switching speeds, especially in CMOS systems, are degraded at elevated temperatures. Therefore, careful thermal design of an electrical system is required to keep the operating temperature below the upper temperature limit.

As the clock frequency and the number of transistors inside the device increase, the heat produced by the device increases. In the past several decades, successive revolutions in device technology from TTL and ECL to NMOS and CMOS has reduced the transistor switching energy from more than 10^{-9} J in 1960 to 10^{-13} J in the 1980s, and thus the heat generated by a single transistor has been reduced dramatically [10]. However, the effect of this reduction is overshadowed by a large increase due to an increase in the integrated components in the chip and the system working frequency. Therefore, today's fast and high density devices generate a large amount of heat and have high heat removal requirements. Heat removal becomes more critical in MCMs since hundreds of dies are mounted in close proximity on a single module. As a result, the heat generated in an MCM per unit area is very high. In many MCM designs, thermal management emerges as the most important aspect of the system.

Heat is removed mainly by conducting it away from the chips and allowing it to convect into a circulating coolant. Air-cooled and liquid-cooled thermal technologies using heat sinks, cooling pistons, cooling bellows, and micro-channels have been used for VLSI and PCB systems [6, 10, 16].

Although the principles of heat generation, thermal management, and design for MCMs are similar to those for SCMs, the thermal management in MCMs is more difficult and complicated than in SCMs. The difficulty and complexity in thermal control of MCMs stem from multiple chips on the substrate and typically high frequency of the system. Allowing many chips to be mounted on a single module causes a large amount of heat to be generated per unit area and makes heat removal very difficult from MCMs due to the small package size. Another reason for better thermal management of MCMs is due to the fact that MCMs are often used in applications that require higher reliability and thus more strict operating temperature requirements.

In the following, MCM thermal management and thermal design issues are discussed. First the basic concepts of thermal management in MCMs and SCMs are reviewed. Then MCM thermal resistance and thermal management technologies are presented. We classify these thermal management methods according to the equipment used, such as thermal vias, heat sinks, cooling pistons, cooling bellows, and cooling channels. Comparisons of the advantages and disadvantages of these technologies are also presented. Finally, we discuss some thermal design issues for MCMs. Additional material on the effects of thermal considerations on MCMs physical design, along with the CAD tools for thermal design and evaluation are presented in Chapter 7.

5.1 BASIC CONCEPTS

To understand the thermal management and thermal design of MCMs, one needs to understand some basic concepts related to heat generation, heat transfer, and properties of materials. These concepts, namely, power dissipation, heat flux, heat transfer, heat path, thermal resistance, heat removal in single chip, and the figure of merit for performance are introduced in this section.

5.1.1 Power Dissipation

The total power dissipated in a chip depends on the level of integration, technology (ECL vs. CMOS), operating frequency, and so forth. Today's high speed and highly integrated chips dissipate large amount of heat.

Assuming a circuit node is switching in every clock cycle, the power dissipation of this node can be expressed as

$$P = \frac{1}{2} f_c V_{DD}^2 \mathrm{C} \tag{5.1}$$

where f_c is the clock frequency of the circuit, V_{DD} is the power supply voltage, and C is the capacitive load of this node. The factor $(\frac{1}{2})$ is due to the fact that the capacitance needs to be charged and discharged to dissipate a total of CV^2 energy.

For the output driver, assuming it will switch every other cycle, an additional ($\frac{1}{2}$) factor is added. Thus the dissipation equation of a node is

$$P = \frac{1}{2}\frac{1}{2} N_D f_c V_{DD}^2 C \tag{5.2}$$

where N_D is the number of drivers in this node.

Consider a system with 100 output drivers. If the clock frequency is 50 MHz, V_{DD} and load capacitance are 3.4 V and 70 pF, respectively, then total power dissipation will be 1 W. Many VLSI chips have significantly high power dissipation. For example, Alpha chip from DEC dissipates about 30 W of power.

A typical MCM module may contain up to 100 chips; therefore, the total power dissipation in an MCM can be very large. For example, the MCM in the IBM 3090 contains 100 chips; although the maximum power for each chip is just 7.0 W, the total power dissipation for the whole module is as large as 500 W [10].

5.1.2 Heat Flux

The total power dissipation is not enough to describe the heat generation and heat removal of a device completely, since it does not consider the effect of area. For those devices that have a large surface area (such as PCBs), the heat generated per unit area is not high and therefore can be removed easily. On the other hand, for some small sized packaged devices (such as VLSI chips and MCMs), although the heat generated is small, the surface area is much smaller, and thus heat removal may be difficult. As a result, besides power dissipation, we need another parameter called *heat flux*. It is defined as the power dissipation per unit area. For example, the Mitsubishi SiC chip has 4 W total power dissipation with 8 × 8 mm chip size; thus its heat flux is 6.25 cm^2.

5.1.3 Heat Transfer Methods

There are three different kinds of heat transfer methods, namely, conduction, convection, and radiation. Heat transfer is described by heat flow equations.

1. **Conduction:** This is the transfer of thermal energy from the high temperature region of a solid to a lower temperature region. The heat flow is through the solid. The conduction for a block of solid material is described by the Fourier cooling law, which states that

$$Q = \frac{kA}{L}(T_h - T_c) \tag{5.3}$$

where Q is the heat flow in watts, k is the thermal conductivity of the block, T_h and T_c are the temperatures on the two opposite sides of the block, L is the length of the heat path, and A is the cross-sectional area. Conduction is a main mechanism of heat transfer inside electrical devices. As the heat conductivity of VLSI and MCM substrates is very low, the heat transfer through the substrate is not efficient.

2. **Convection:** This is the transfer of thermal energy from the high temperature surface of a solid body to a low temperature liquid or gas. Its heat flow is governed by Newton's cooling law

$$Q = hA(T_s - T_f) \tag{5.4}$$

where h is the heat transfer coefficient, A is the surface area exposed to the fluid, T_s is the temperature of the surface, and T_f is the temperature of the fluid. The transfer coefficient can be increased by increasing the coolant flow rate. Concection methods that use some additional means to increase the coolant flow rate are called *forced convection* methods. For example, a fan (or pump) may be used to drive the air coolant (or liquid coolant).

3. **Radiation:** This is the transfer of thermal energy from the surface of a solid body to the ambient surroundings by electromagnetic waves. The total heat transferred Q of radiation heat transfer is

$$Q = \sigma \varepsilon F_{hc} A(T_h^4 - T_c^4) \tag{5.5}$$

where σ is the Stefan-Boltzman constant, and ε is the emissivity of the radiating surface of area A. The temperature of the (hot) radiating surface and of the (cold) neighboring bodies is denoted by T_h and T_c respectively, in degrees Kelvin. F_{hc} is the view of the shape factor [58]. In space applications, radiation is the only mechanism of heat transfer.

Generally, efficient thermal management will attempt to make use of all three methods of heat transfer.

5.1.4 Heat Path and Thermal Resistance

In an electrical device (VLSI chip or MCM), heat is always generated inside the device by semiconductor junctions and then transferred to the ambient through the device surface. The whole path from junctions to the ambient can be viewed as consisting of two paths. One is an internal path, which refers to the path from junctions to device surface; the other is an external path, which is the path from the device surface to the ambient. The heat path is shown in Figure 5.1.

Heat passing through the heat path is like electrical current passing through a circuit path, where heat flow corresponds to electrical current and temperature corresponds to the voltage. Similarly, we can define the concept of thermal resistance:

$$R_t = \delta T / P \tag{5.6}$$

where δT is the device temperature rise, and P is the total power dissipation. The thermal resistance model is shown in Figure 5.2. In thermal management,

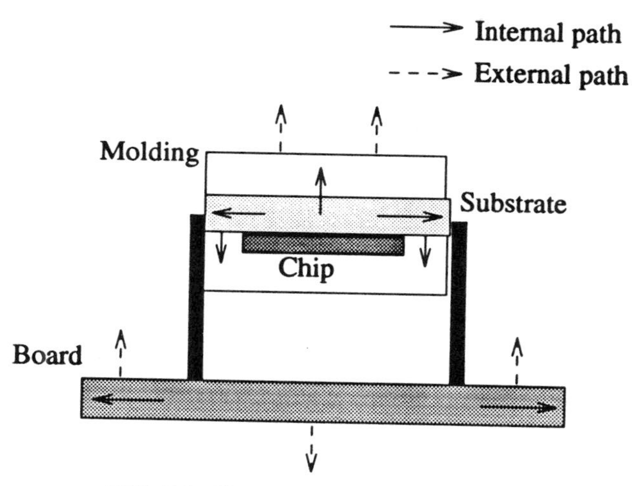

FIG. 5.1. Heat transfer path in a device.

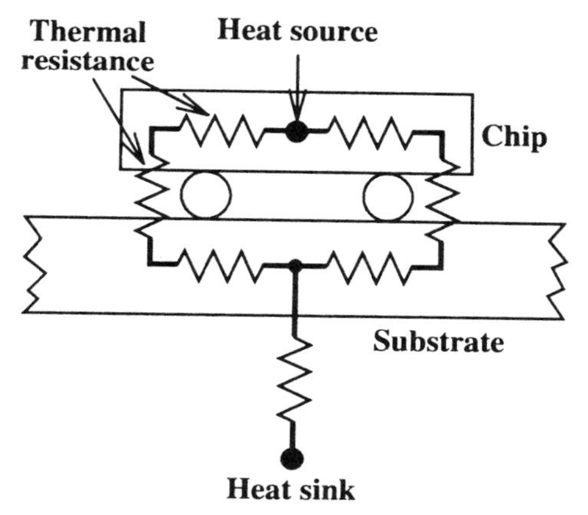

FIG. 5.2. Thermal resistance model.

the thermal resistance should be reduced so that the temperature rise is minimum.

5.1.5 Junction Temperature

The temperature at the interface of *p*-type and *n*-type semiconductors within an electrical device is called the *junction temperature*. Since the actual junction temperature is very difficult to measure, junction temperature also refers to the average temperature of the device. Although there is no industry set standard, junction temperature design limits vary from 80° to 180 °C. For most components, it is about 120 °C.

It is well known that the failure rate of an integrated circuit is accelerated by an increase in junction temperature; hence maintaining the junction temperature below the specified temperature limit is the main objective of thermal management [6].

5.1.6 Heat Removal From Single Chip Modules

Many heat removal technologies have been used for ICs, since heat generation in a VLSI chip depends on several factors, such as system structure, chip size, package technology, and material used. Selection of the proper thermal management technology for an IC is determined by these factors and by the thermal management cost.

In low-cost DIPs, dies are bonded to a lead frame and encapsulated in plastic molding. To improve the heat transfer, a heat spreader is used under the die. Thus, the thermal conductivities of the heat spreader, lead frame, and plastic are the main factors that determine the heat transfer. With this structure, thermal resistance of a 40 pin DIP can be as low as 38 °C/W during natural convection and 25 °C/W with forced convection of air. Ceramic chip carriers and pin arrays do not use a plastic mold. A substrate with an air cavity and a heat sink can be used. Thermal resistance can be reduced to 10 °C/W for a 68 terminal ceramic chip carrier.

For some high density and high power VLSI chips, air-cooled methods may not be effective and water-cooled methods must be used to remove the heat. One example of such a system is Fujitsu M-780 [10].

5.1.7 Figure of Merit for Thermal Performance

Several different parameters may be used as the figure of merit for thermal performance of a device. Two such parameters are thermal resistance and heat flux. Thermal resistance is a very useful figure of merit for thermal management, but it does not reflect the effects of device area on heat transfer. For two devices with the same power, the one with the larger area has less heat resistance than the one with a smaller area. As a result, heat flux is used as the figure of merit for thermal management.

5.2 MCM THERMAL RESISTANCE

The goals of thermal management in MCMs are to reduce the thermal resistance of MCMs and to keep the junction temperature under the limit. Reduction in either internal thermal resistance or external thermal resistance leads to the reduction in the total thermal resistance. In this section, we discuss the internal and external path thermal resistances in MCMs. Since in MCMs heat is transferred by conduction in the internal path and by convection in the external path, the heat flow equations of conduction and convection we discussed in previous sections can be used.

5.2.1 MCM Heat Path

Like all the other electrical devices, the heat path of an MCM also consists of an internal path and an external path. Different MCMs can have different heat paths, depending on the bonding technologies and thermal management technologies used.

1. **Internal Path:** MCMs may have two possible internal paths, depending on how the dies are bonded to the substrate. One internal path is *through the substrate*; the other is *through the chip side*. In the *through-the-substrate* heat transfer medhod, the heat is transferred from chips to substrate; then the substrate in turn transfers the heat to the heat sink. As the MCM substrate materials usually have low heat conductivities, the thermal resistance of the substrate can be very high. To reduce the thermal resistance, thermal vias can be used in the substrate or dies can be sunk into the substrate at the expense of wiring capacity. Wire bonded or TAB chips in MCMs are die bonded to the substrate, so their heat transfers through the internal path are through the substrate. Solder-bump or flip-TAB-chip MCMs can also have this kind of internal path.

   In *through-the-chip side* heat transfer method, the dies are placed with their active side facing the substrate, and the heat sinks are directly mounted on die surfaces. The heat is transferred from dies to heat sinks through an interface between dies and heat sinks. Since the internal path does not contain the substrate, the thermal resistance can be very low. Only flip techniques can use this internal path because of the potential damage to the surface of the chips.
2. **External Path:** The coolant and the surface area that the coolant contacts form the external path of MCMs. In many applications, MCMs use heat sinks to contact coolant. Thus the contacting surfaces are the fin structures of heat sinks. External thermal resistance is affected by several external factors, such as the heat sink structure, coolant material (air or liquid), coolant flow, and coolant temperature.

5.2.2 Internal Thermal Resistance

For the composite, rectilinear structure that is encountered in many IC chips and MCMs, the heat flow Equation 5.3 can be written as

$$Q = (T_h - T_c)\sum_p (kA / L_p) \tag{5.7}$$

where p is the layers of material in the internal path, and L_p is the thickness of each layer.

According to the definition of heat resistance in Equation 5.6, the internal chip module resistance can be described as

$$R_{jc} = (T_h - T_c)/q_c = \sum_p (L_p / kA) \tag{5.8}$$

where L_p is the thickness of each layer, k is the thermal conductivity, and A is the cross-sectional area. It can be seen that as the thickness of each layer decreases and/or the thermal conductivity and cross-sectional area increases, the resistance of the internal path decreases. It also shows that internal resistance can be altered not only by the choice of material and bonding technology, but also by a change in geometry.

5.2.3 External Thermal Resistance

The heat transfer in the external path of MCMs is mainly by convection. We obtain external thermal resistance

$$R_{ex} = \frac{1}{hA} \tag{5.9}$$

by substituting Equation 5.4 into Equation 5.6. It can be seen clearly from Equation 5.9 that the external thermal resistance varies inversely with the heat transfer coefficient h and the wetted area A. The heat transfer coefficient is usually found from the available empirical correlations and/or theoretical relations for a particular geometry and flow regime. For flow along plates and in the inlet zones of parallel-plate channels, usually encountered in many electronic cooling applications, the average convective heat transfer coefficient is calculated differently depending on the flow velocity [10].

In the low velocity coolant flow, the h is given by

$$h = 0.664\,(k/l)\,(\mathrm{Re})^{0.5}\,(\mathrm{Pr})^{0.333}\ \mathrm{W/m^2\,K} \qquad \text{for Re} < 2\times10^5 \tag{5.10}$$

while for high velocity coolant flow, it is given by

$$h = 0.036\,(k/l)\,(\mathrm{Re})^{0.8}\,(\mathrm{Pr})^{0.333}\ \mathrm{W/m^2\,K} \qquad \text{for Re} > 3\times10^5 \tag{5.11}$$

where k is the thermal conductivity, l is the characteristic dimension of the surface, Re is the Reynolds number which is proportional to the fluid velocity and inversely proportional to the characteristic dimension, and Pr is the Prandtl number.

By applying Equations 5.10 and 5.11 to Equation 5.9, we conclude that the external thermal resistances in MCMs are strongly influenced by the coolant velocity and the package dimensions.

To reduce further the external resistance for a fixed value of the convective heat transfer coefficient, a fin structure or a compact heat exchanger is attached to the module case to enlarge the surface area in contact with the coolant. However, the fin structure and an additional bonding layer introduce new thermal resistances. The concept of "fin efficiency" can be used to deal with the conductive resistance of the fin structure. Fin efficiency η is defined as the ratio of the average temperature difference between the fin base and the coolant [10]. The value of fine efficiency ranges from 0 to 1.

Thus, the heat transfer by a fin or fin structure can be expressed as

$$Q = hA[\eta(T_s - T_f)] \tag{5.12}$$

and therefore the external resistance is

$$R_{ex} = \frac{1}{\eta hA} \tag{5.13}$$

5.2.4 Overall Thermal Resistance

According to the above discussion, the overall thermal resistance is

$$R_T = R_{jc} + R_{ex} = \sum_p (\delta x / kA) + 1/\eta hA \tag{5.14}$$

As discussed in the development of the relations for internal and external resistances, Equation 5.14 indicates R_T to be a strong function of both the convective heat transfer coefficient and geometric parameters (thickness and cross-sectional area of each layer). Thus, the introduction of superior coolant, use of thermal enhancement techniques that increase the local heat transfer coefficient, and selection of a heat transfer mode with inherently high heat transfer coefficients will all be reflected in appropriately lower total thermal resistance. Similarly, improvements in the thermal conductivity of materials and reduction in the thickness of the relatively low conductivity bonding materials would also reduce the total thermal resistance. In many designs, an increase in the cross-sectional area reduces the total resistance significantly; however, this results in a larger module.

5.3 MCM THERMAL MANAGEMENT

Since the heat path of an MCM consists of the internal path and the external path, so thermal management technology can be applied to either the internal or external path or to both. In an internal path, if heat is transferred through the substrate, thermal vias can be used to reduce the thermal resistance. If heat is transferred through the chip side, since the thermal resistance of this cooling method is very low, it is sufficient to ensure a high conductivity interface between chip and eat sink. High thermal conductivity solder, a thermal epoxy, or a thermal grease can also be used. There are many alternatives in thermal management in the external path. One means of classification is on the basis of a heat transfer mechanism, such as

1. Free air convection
2. Forced air convection
3. Free liquid convection

4. Forced liquid convection
5. Phase change

These mechanisms are listed in their increasing effectiveness of heat removal. Typically air cooling methods are significantly simpler than liquid convection and immersion methods. However, the heat flux that can be handled by an air convection method is relatively less than with other more exotic methods. Yet another classification of these methods depends on actual equipment or components used. In this section, we introduce some MCM thermal management technologies based on this classification.

5.3.1 Thermal Vias

When heat is transferred through the substrate in an internal path, the substrate becomes the major conductor. But usually the substrates that are being used in MCMs have low conductivities; thus the thermal resistance of MCM substrates is very high. To reduce the thermal resistance, a special heat conductor called a *thermal via* can be used to transfer the heat in a more efficient way, as shown in Figure 5.3. *Thermal vias* are large diameter holes punched into the substrate and are aligned vertically. The resultant via stack is a solid metallic plug, and the vias are filled with paste (or other materials with high thermal conductivities) at each substrate layer.

Using thermal vias, the heat generated from the chips is transferred through the interface between the die and substrate (such as solder-bump in flip-chip) to the thermal vias, and then thermal vias transfer the heat to other heat conductors such as a heat sink. The heat removal is determined by the size and number of vias. The thermal resistance decreases as the via size increases. However, the internal thermal resistance does not decrease when the via size increases beyond a certain limit. At this stage, the interface has become the major impedance to heat transfer. The via distribution also affects the thermal resistance. Thermal vias should be distributed close to the hot spots in dies.

If the size and distribution of thermal vias are designed properly, then the internal path resistance can be reduced significantly. However, thermal vias occupy routing space and may make the routing difficult. using the thermal via

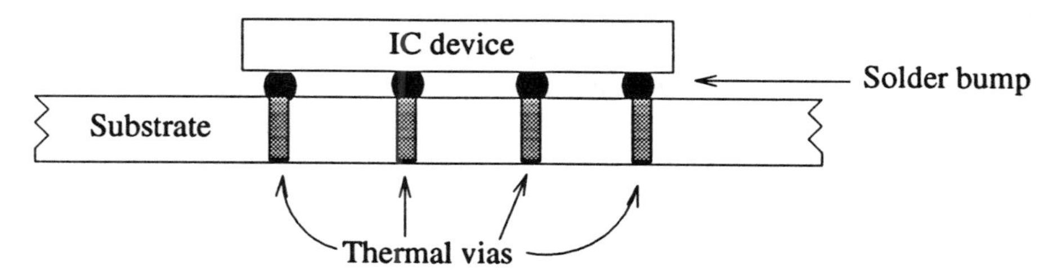

FIG. 5.3. Use of thermal vias in flip-chip MCMs.

cooling method, resistances of 3.0 °C/W for a 1 cm square chip and 10 °C/W for a 0.25 cm^2 chip were achieved [129].

5.3.2 Heat Sinks

According to Equations 5.3, 5.4, and 5.5, the efficiency of heat transfer is strongly influenced by the conductivity and the cross-sectional area of the conductor. MCM chips often have a limited cross-sectional area, and the conductivities of substrate and package materials are low compared with the power dissipated. Thus it is necessary to use some other heat conductor to help transfer heat.

A *heat sink* is a heat removal device with a large fin area made of high conductive materials such as metals. Since a heat sink has good conductivity and a large surface area, it conducts heat away from the MCM and convects it to the ambient. Heat sinks can have several different structures, as shown in Figure 5.4.

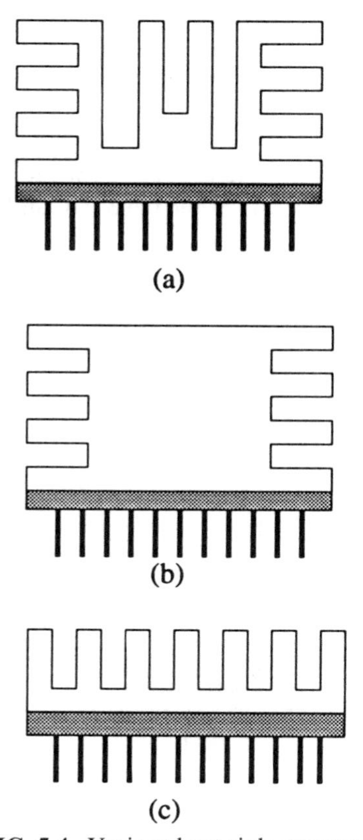

FIG. 5.4. Various heat sink structures.

The heat sink can be mounted on the substrate or on the chip side. When the heat sink is mounted on the substrate, the heat is transferred through the substrate with thermal vias or thermal cutouts to the heat sink. In this method, the primary internal heat path is through the substrate. Since the thermal conductivities of the most common substrate materials are low, their thermal resistances may be high. Substrate thermal resistance can be reduced by using aluminum nitride instead of aluminum or by using more thermal vias. When the heat sink is mounted on the chip backside, the heat dissipation is above the chip. This method leads to very small thermal resistance. The potential problem is that mounting a heat sink on the chip backside may damage the surface of the chips, so only flip techniques can be used.

The sink is a very effective and commonly used heat transfer device. It can improve heat transfer dramatically. To utilize the heat sink efficiently, the interface resistance between the heat sink and the chip or module must be kept small. For this purpose high thermal conductivity greases and high conductivity gases, notably helium and hydrogen, have been used. And, wherever possible, bare interfaces are coated with thermal grease, soft solder, or silicon rubber to enhance interface conductance. Let us discuss some specific examples of *heat sinks.*

The Hitachi SiC RAM module uses a heat sink to transfer heat as shown in Figure 5.5. The module contains six 1 W ECL chips, and each provides 1 kbit of memory. The heat released by each chip is conducted through 77 solder bumps (52 of which are purely thermal in function) to the silicon substrate and then through the low resistance gold eutectic bond to the SiC. The aluminum heat sink is about 8 mm high and 20 × 20 mm at the base. The sink transfers the dissipated heat to the ambient air blown past the RAM module. The theoretical thermal resistance of this module, at an air velocity of 3 m/s, is 34.7 K/W based on the heat

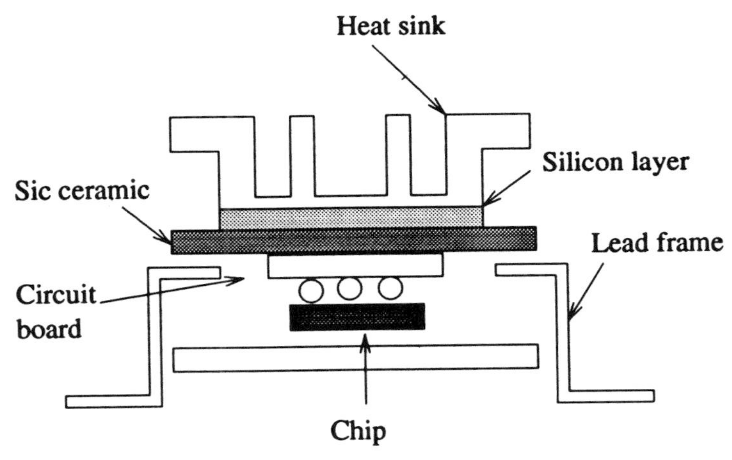

FIG. 5.5. Heat sink of the Hitachi Sic RAM module.

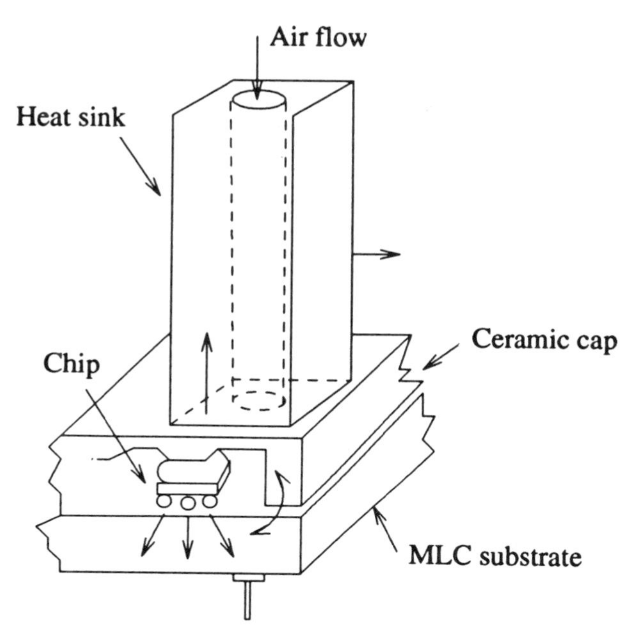

FIG. 5.6. Heat sink used in the IBM 4381.

dissipation of a single chip or nearly 5.8 K/W based on total module dissipation. The measured value of the resistance based on the total module dissipation is 5.5 K/W. Using the stated theoretical thermal resistance values and a module dissipation of 6 W for 25 °C inlet air temperature and an assumed 10 °C rise in the air flowing past the modules, the maximum chip temperature can be expected to approach 70 °C [156].

The IBM 4381 also used a heat sink as its thermal management component. It consists of a single board containing 22 modules. Each of the modules is 64 × 64 mm and approximately 40 mm high and houses up to 36, though typically 31, logic chips of approximately 7,000 elementary components, or 704 circuits, in an area of 4.6 × 4.6 mm [10]. The thermal management structure of the IBM 4381 is shown in Figure 5.6. In this design, heat is transferred through two paths. One is through the substrate, and the other is through the ceramic cap and to the heat sink. It provides an external resistance, based on total module power, of approximately 0.23 K/W. Alternatively, in a module containing 36 identical chips, the chip-to-air thermal resistance, based on each chip, has been found to be 17 K/W, divided nearly equally between the internal (9 K/W) and external (8 K/W). With an air flow of 211 liters/s at 20 mm of water pressure head, the thermal management system transfers up to 3.8W per chip, 90 W per module, and 1.3 kW per board, while keeping all the chips below the maximum specified temperature of 90 °C [10, 22].

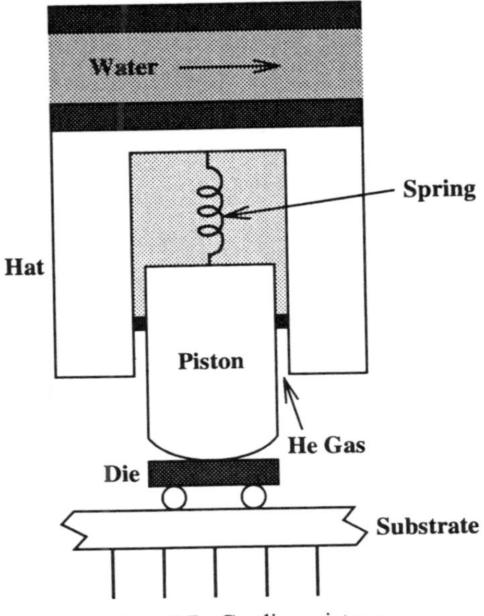

FIG. 5.7. Cooling pistons.

5.3.3 Cooling Pistons

Cooling pistons are another kind of heat removal devices for heat management. A typical heat piston is shown in Figure 5.7 [6]. The spring-loaded pistons press against the backside of the die so that the heat is transferred from the chip to the piston, which in turn conducts it to a cold plate cooled by circulating water. The cavity around the piston is filled with helium, which has higher thermal conductivity than air.

This method was first used by IBM in 1981 in system IBM 3081 [38]. The total resistance of the package is 11°C/W per chip site. The chips can dissipate up to 4 W of power and raise their junction temperatures only 44 °C above the cooling water temperature.

5.3.4 Cooling Channels

Heat transfer in the heat sink method is limited by the interface between the die and heat sink. This interface could be the main impedance to heat transfer in many applications. In the cooling channel method, the heat sink is fabricated directly inside the semiconductor itself. Since the interface is elimainated, the heat transfer can be tremendously improved in cooling channel techniques. The structure of the method is shown in Figure 5.8.

The NTT water-cooled substrate approach is an example of using cooling channels in substrate [114]. The NTT prototype module, a 5×5 array of 8 mm

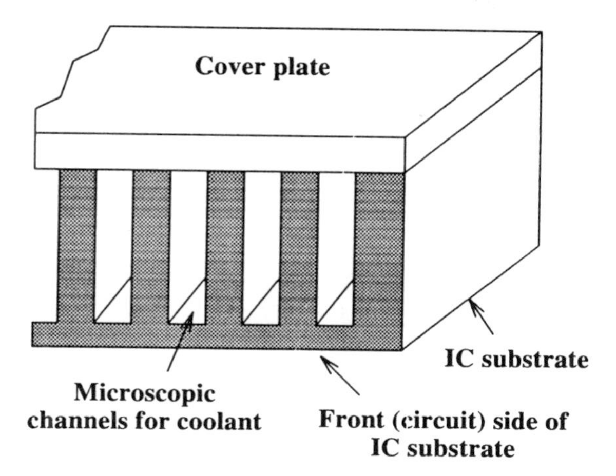

FIG. 5.8. Microchannel cooling.

square VLSI chips, was mounted on a 85 × 105 × 1.2 mm six-conductor layer alumina substrate containing 29 0.8 × 0.4 mm channels and 900 I/Os, both on a 2.54 mm pitch. The coolant distributor and collector were 7 × 77 mm in external dimension. The worst thermal resistance for the prototype module of NTT is 3.3 K/W, and the maximal chip junction temperature is 85 °C. The NTT integrally cooled substrate can thus accommodate 25 identical VLSI devices, each dissipating 15.1 W, with a flow rate of 17 cm^3/s of 35 °C water.

A second approach was developed at Stanford University [200]. It used microscopic channels, 50 μm wide and 400 μm deep cut on the backside of a silicon wafer, that were closed by a cover plate to confine the fluid flow forced through them. With this method, heat fluxes as large as 1 kW/cm^2 can be removed while keeping the junction and ambient temperature differential below 100 °C.

Motorola also has a similar method in which the microchannels are mechanically cut on a 3.8 × 3.8 cm silicon substrate. By forcing water flow of 12 and 63 cm^3/s, thermal resistances of 0.03° and 0.02 °C/W were obtained [133].

5.3.5 Cooling Bellows

Cooling bellow is also a device that can bring the chips in direct contact with the coolant fluid as shown in Figure 5.9. In this method a jet of chilled water flows around the backside of the die directly. A thermal resistance of 2°–3 °C/W for 7.5 × 7.5 mm chips has been achieved, and a chip with 75 W power can maintain the junction temperature only 25 °C above the cooling water temperature [6].

Fujitsu's FACOM M-780 module also uses the cooling bellows method. It has 336 single chip modules mounted on both sides of a 540 × 488 mm PCB. The thermal management system consists of bellows and water jets packaged in it. The tip of the bellows is connected to the chip surface to ensure adequate thermal

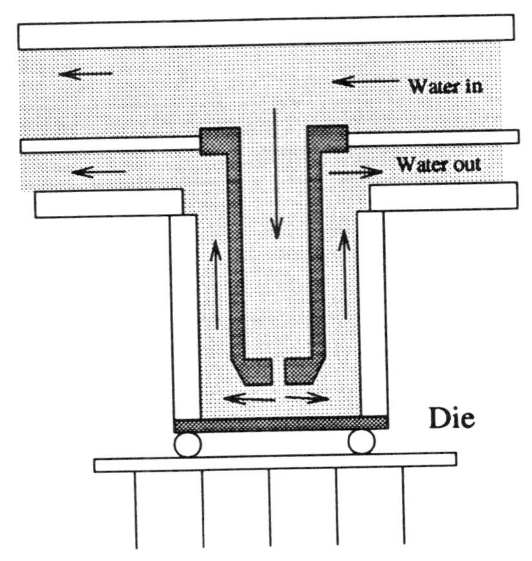

FIG. 5.9. Cooling bellow scheme.

contact. The maximum chip power is 9.5 W, and the total board power dissipation can be as large as 3,000 W [58].

5.3.6 Immersion Cooling

Another cooling method is immersion cooling, which requires immersing of the whole MCM and circuit board into the cooling liquid. One of the main advantage of this method is that it eliminates the interface resistance.

CRAY-2 consists of SCMs mounted on eight PCBs. The total power dissipation is 600–700 W for a heat density of 0.21 W/cm^2 [58]. The coolant used is FC-77 fluorocarbon. Though 0.21 W/cm^2 heat flux can be realized by air cooling techniques, it requires a very high air flow rate, which is impractical.

The immersion method has a low thermal resistance and a large heat flux; thus it can effectively control the temperature of the whole MCM or board. However, immersion methods are costly due to specialized equipment and coolant.

5.4 COMPARISON OF THERMAL MANAGEMENT METHODS

We have introduced several thermal management technologies in the previous section. These thermal management technologies can be categorized into three types: air cooling, water cooling, and immersion cooling. In this section, we compare these three cooling technologies.

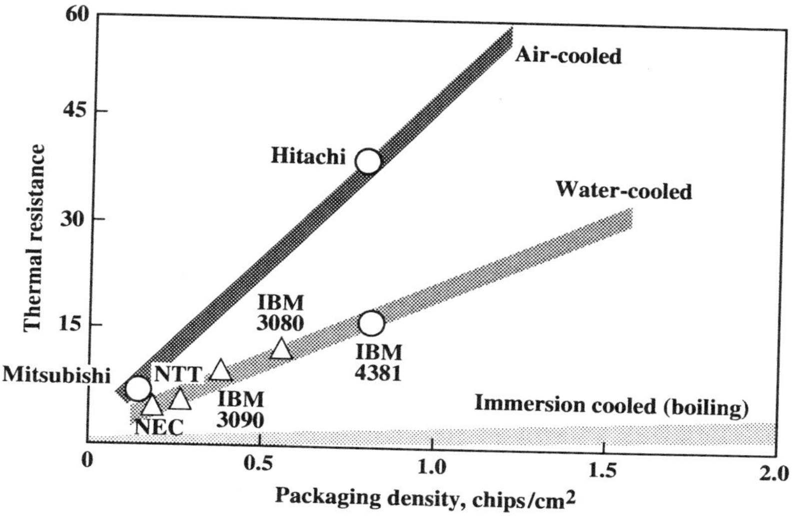

FIG. 5.10. Thermal resistance versus packaging density for MCMs.

1. **Thermal Resistance:** Figure 5.10 [10] shows the thermal resistance versus packaging density for different thermal management technologies. We can see from this diagram that the immersion cooling methods have the lowest thermal resistance; the air cooling methods have the largest thermal resistance, and the water cooling methods have moderate resistance. It also shows that water cooling methods have better heat removal capability than air cooling methods.

   Figure 5.11 [10] shows the internal and external (thus the total) thermal resistance of some practical MCM systems. It suggests that in the air-cooled modules, such as IBM 4381, Mitsubishi's HTCM, and Hitachi's RAM, the external resistance is roughly comparable to or greater than the internal resistance. While in the water-cooled modules, such as IBM 3090, NTT, and NEC, the external resistance is generally less than one-third of the total thermal resistance.

2. **Heat Flux:** Figure 5.12 [10] shows the area versus heat dissipation for different thermal management technologies. With the exception of the Honeywell module, it again indicates that water cooling methods have larger heat flux than air cooling methods. Although it does not give the figure for the immersion method, in fact immersion methods have the largest heat flux among all thermal management methods. The actual thermal parameters of these practical MCMs are shown in Table 5.1 [10].

3. **Cost:** Unfortunately, the costs of the thermal management technologies discussed above are inversely proportional to their performance. The immersion methods have the highest cost since they immerse the whole

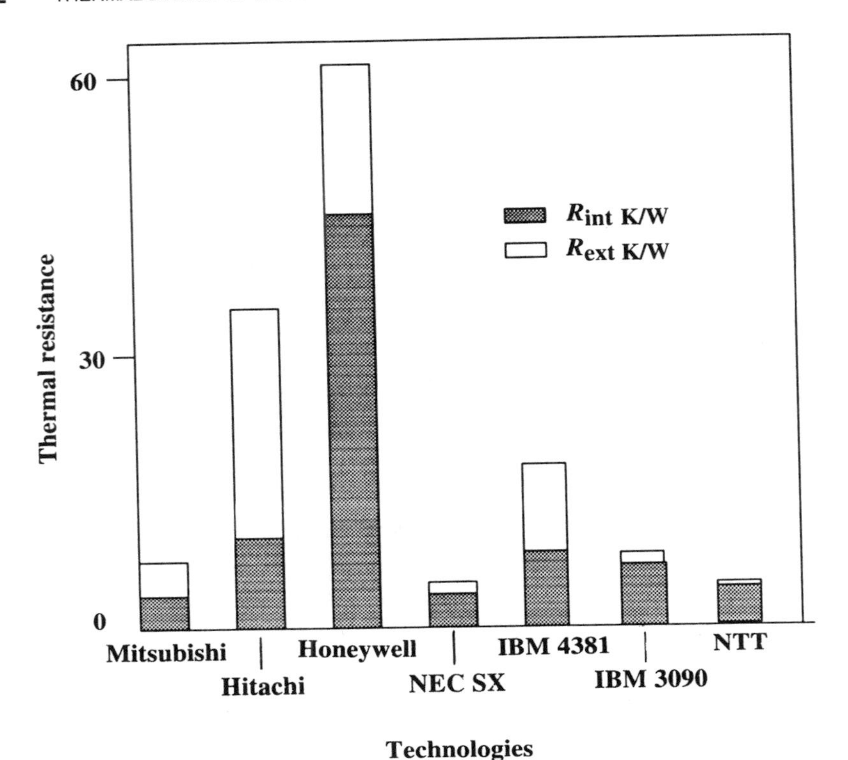

FIG. 5.11. Comparison for thermal resistance.

TABLE 5.1. Multichip Module Thermal Parameters

Technology	Chip Size (mm)	Number of Chips	Dissipation (W)	Heat Flux (W/cm²)	R_T (K/W)
Mitsubishi HTCM	8 × 8	9	36	0.83	7.3
NEC SX	8 × 8	36	250	1.6	5
Hitachi RAM	1.9 × 4	6	6	0.8	34.7
IBM 4381	4.6 × 4.6	36	90	2.2	17.0
IBM 3090	4.85 × 4.85	100	500	2.2	8.7
NTT	8 × 8	25	377	4.2	3.3

module or board into the coolant, and the coolant is much more expensive than air and water. Therefore, the modules have to be protected from moisture. Moreover, in some applications, the module or the board may not have a large total power dissipation, but just a few chips have large heat dissipation. In this case, it is not economical to immerse the whole module or board into the coolant. On the other hand, the air cooling methods have the lowest cost since they use natural air as the coolant and it is still cheaper

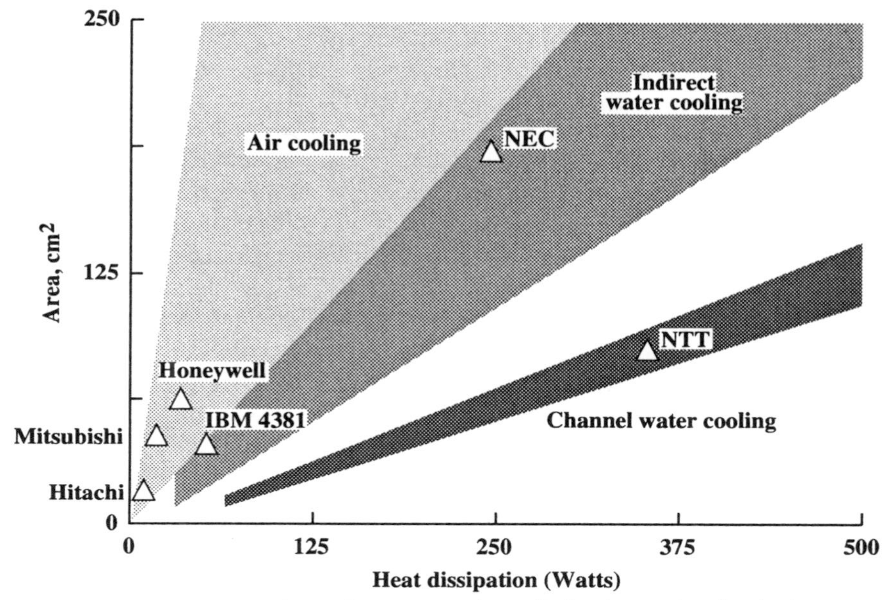

FIG. 5.12. Comparison of area versus heat dissipation and technologies.

to use fans to drive the air. On the other hand, the water cooling methods need exhaust systems to maintain or drive the liquid coolant, so they are more costly than air cooling methods but are still cheaper than immersion cooling methods.

5.5 THERMAL DESIGN CONSIDERATIONS FOR MCMs

In many MCMs, thermal management is the most important constraint. Thermal management constraints may affect the selection of package technology, package size, and physical design. Thermal design considerations should be considered at the beginning of the MCM design. In this section, we discuss some thermal design considerations.

1. **Heat Flux and Thermal Resistance:** Heat flux and thermal resistance are two main parameters of thermal management. Heat flux is determined by total power and total module size. According to Figure 5.12, each cooling method can deal with a certain range of flux. Thus, given a certain amount of heat, we can choose a thermal management method. On the other hand, given a cooling method, we also can determine the minimum package size.

   Thermal resistance affects many thermal management factors. The first is the internal path of MCMs. With chips die bonded to the substrate, heat must flow through the substrate's dielectric layers, which are often polymide or other materials of low thermal conductivity. It becomes necessary to provide thermal vias through the substrate. However, these vias take up space that may be needed for routing. With flip-chip, heat removal through the substrate can be enhanced by providing extra solder bumps not required for electrical connectivity. Though hundreds of extra thermal paths can be obtained this way, each bump would require a thermal via in the substrate, again creating the trade-off between thermal conductivity and wirability. For maximum thermal performance, backside colling can be used. The second factor is the external path. The thermal resistance parameter can determine the heat sink structure, coolant material, coolant temperature and flow rate etc.

2. **Considerations for Applications:** In avionic and military applications, high reliability of the cooling system is required. As a result, immersion cooling and liquid methods may not be suitable for these applications. Highly reliable thermal management methods for avionic and military applications are still being investigated.
3. **Considerations for Cost:** Since immersion methods have a high cost, they are only used in supercomputing systems. For low-end workstation applications, the cost is critical and the air cooling methods can be used. Liquid cooling methods can be used in other applications in which both performance and cost are required. Before choosing thermal technology, manufacturers always first exhaust all lower options, such as air cooling methods, before considering any form of higher level cooling methods, such as liquid or immersion cooling.

5.6 SUMMARY

In this chapter, we have studied the thermal design and thermal management of MCMs. The main goal of thermal management is to reduce the thermal resistance. The internal thermal resistance in MCMs can be reduced by using high conductivity materials, increasing the cross-section area, and choosing a proper mounting technology; the external thermal resistance can be reduced by using a heat sink, enlarging the package dimension, and increasing the coolant velocity. Thermal management techniques using thermal vias, heat sinks, cooling pistons, cooling channels, cooling bellows, and immersion cooling techniques were introduced. Basically, air cooling methods have the lowest cost but relatively low heat flux; immersion methods have the highest heat flux and also the highst cost; and liquid cooling has both moderate thermal performance and manufacturing cost. In MCMs, the thermal management design is an integral part of the design cycle. As electrical constraints, thermal constraints must be considered very early in the design process, in fact well before the design of the MCM itself.

5.7 PROBLEMS

1. What is the difference between MCM thermal management and VLSI chip thermal management?
2. Generate the thermal resistance models for TAB, flip-chip, and wire bond MCMs.
3. Categorize the thermal management methods discussed in this chapter into air cooling, liquid colling, and immersion cooling methods.
4. Generate a thermal resistance model for the IBM 4381 module and indicate the main possible impedance to the heat removal in the whole heat path.
5. Suggest some other methods that can be used to further reduce the thermal resistance in Hitachi Sic RAM module?
6. Why do cooling channel methods have better heat removal?
7. Assume a chip contains 10 output buffers, the power supply is 3 V, and the load capacitance of each buffer is about 70 pF. If the operating frequency is 60 MHz, calculate the total power dissipation of this VLSI chip. If the junction temperature required is under 120 °C and the ambient temperature is 25 °C, what is the thermal resistance of this chip?
8. If an MCM contains 100 chips as described in problem 7, calculate the minimum module sizes for both air cooling and liquid cooling thermal management methods according to Figure 5.10. If the operating frequency is 100 MHz, how are these module sizes?

BIBLIOGRAPHIC NOTES

Many papers and books have been written about thermal management in MCMs. Thermal management technologies in VLSI chips can be found in Bakoglu 6 and Billings 16. The heat path and thermal characteristics of MCMs are discussed by Doane and Franzon [58]. Bar-Chen 10 examines air cooling and liquid cooling thermal management technologies in detail. The thermal management technologies used in eight distinct MCMs also are discussed in that paper. Heat sinks, cooling pistons, and cooling channels are discussed by Bakoglu [6]. Thermal trade-off analysis are covered in Sandborn and Moreno [178], and thermal management is covered in Johnson et al. [108] and in other MCM journals and proceedings.

CHAPTER 6

TESTING AND ASSEMBLY YIELD DESIGN

For an electrical device to function properly, all its components must meet their design specifications and each one must be proved to be free of any defects before assembly. This leads to VLSI IC testing and PCB testing. Moreover, testing is important to ensure high yield and low cost. In fact, testing is now so important for both VLSI IC and PCB designs that more time is spent on testing than any other phase. The requirement for testing is even higher in MCMs, since they are usually used in high performance applications. An MCM can consist of many dies, and it is difficult to repair these dies since they are mounted on a dense substrate. Therefore, it is important to test all the dies before mounting. MCMs suffer from virtually all the problems associated with assembled PCBs. For example, the interconnections on an MCM substrate may have short circuit and open circuit faults due to manufacturing problems or mishandling. In the die placement process, some of the dies may be missing or wrong dies may be selected. Moreover, dies may have functional defects or parametric defects, if the process does not use KGD or there is physical damage to the dies. These problems can cause either functional or parametric defects to MCM systems. To detect these defects, both functional testing and parametric testing are required. Many testing approaches have been developed for functional testing and parametric testing in PCBs [69, 71, 211].

Although the process of MCM testing is similar to that of PCB testing, MCMs present some rather unique difficulties in detecting and diagnosing these problems:

1. MCM substrates often have much higher density pads than PCBs. For example, an MCM with a node count ranging between 64 and 600 can be built on a substrate as small as 35 mm^2, with pad size ranging from 4 to 20

mils. Therefore, the bed-of-nails tester that is used commonly in board testing is not feasible in MCM substrate testing. Moreover, an MCM system can contain up to 100 dies [10] (such as the IBM 3090) and can have hundreds of thousands of internal nodes, while the I/O pins are still limited. Hence, it is extremely difficult to access all the internal nodes through these limited I/O pins in MCM testing.

2. Many of the dies currently available for MCM assembly do not provide the same performance or reliability guarantees as packaged ICs due to the difficulty of unpackaged die testing. Thus, an additional burden is placed on module level testing.
3. MCM testing must be performed at each different fabrication stage, such as before and after die mounting on the substrate, since damage or an improper process at any stage can produce a bad MCM.
4. Most methodologies currently in use for MCM testing are borrowed directly from VLSI testing and PCB testing. These testing methods may not be well-suited to MCM testing. Moreover, test equipment for MCM testing is not readily available.

To ensure high MCM yields, testing must be performed throughout the whole fabrication process to verify the quality of each processing step and component used in the module. During substrate fabrication, the connectivity of the substrate must be verified before any component can be mounted on it; each naked die is tested before it is attached to ensure KGD; after the whole module is assembled, the complete module should also be fully tested to ensure that it meets its design specification. Therefore, MCM testing consists of three steps: substrate testing, die testing, and module testing.

In this chapter, we introduce methods for MCM testing. We first start with the basic concepts of MCM testing. Next, three methods for MCM substrate testing are discussed. In the MCM die testing section, we focus on the methods of handling unpackaged dies. Die burn-in is also addressed. In the subsequent sections, we present module testing methods for improving assembly yield and give basic guidelines for test and yield design for MCMs.

6.1 BASIC CONCEPTS

In this section, we introduce the basic concepts of testing such as fault models, test pattern generation, DFT, and test equipment.

6.1.1 Test Process

A typical IC or a PCB test process is shown in Figure 6.1 where DUT refers to "device under test." It can be either a wafer or a packaged device. The tester actually is a computer that controls the whole test process and analyzes the test results. The interface between the tester and the DUT can be either a probe card

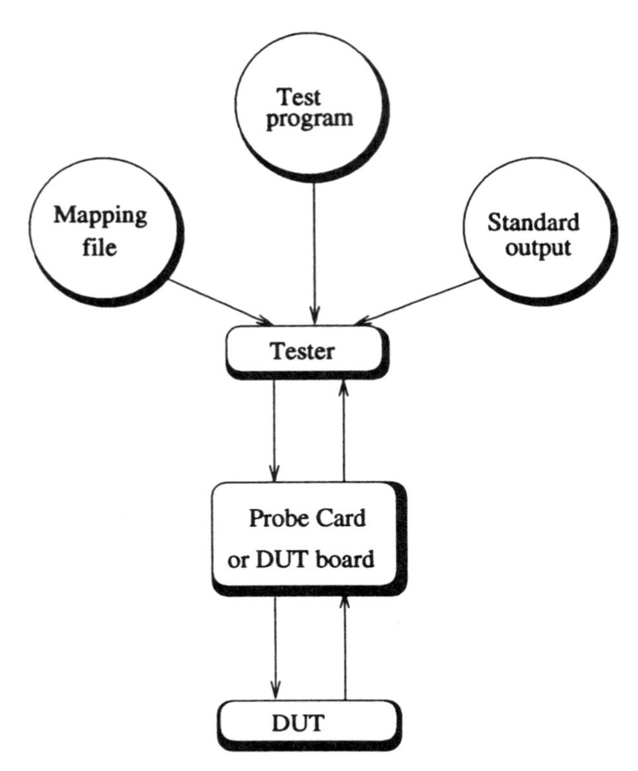

FIG. 6.1. Typical test process.

or a DUT board, depending on whether a wafer or a packaged device is being tested. The test program defines the I/O terminals, input stimulus, and input test vectors. The mapping file contains the interconnecting information of the DUT under a certain test specification. A standard output file provides standard output information under the given test vectors. When testing, the tester first loads the test program file and the mapping file and then controls the probe card or DUT board to connect the DUT to a certain topology according to the connection information contained in the mapping file. Next, the test input vectors are applied at the input of the DUT and the corresponding responses at the output are collected as samples by the tester. Finally, the tester compares the output with the standard output file for fault reporting.

6.1.2 Fault Model

To detect and test the faults in an electrical system, one needs to know the types of faults that one may encounter, that is, one needs a model for how faults occur and impact on circuits. Several fault models have been proposed in [70, 206].

1. **Stuck-at Faults:** This is the most commonly used fault model to represent failures in logic circuits. With stuck-at fault, the output of a gate is stabilized to a state regardless of the input. If the output is stabilized in "1," we say it is "stuck at 1" (S-A-1). Similarly, if it is stabilized at "0," we call it "stuck at 0" (S-A-0).

2. **Shorts and Opens:** A short fault is defined as an electrical short circuit between two nodes that are supposed to be electrically isolated. An open fault represents a failure that causes a line or wire in the circuit to be broken. In MCM manufacturing, the typical short and open faults include [86]

1. Bare-substrate shorts due to incomplete etch
2. Bare-substrate opens due to over etch
3. Shorts due to bridging or lead bent over solder or wire bonds
4. Opens due to dry or unsoldered joints or weak, missing, or broken bonds
5. Shorts of opens due to the damage caused by static, brute strength, or mishandling

Other fault models, such as cross-point faults, memory faults, functional faults, and delay or timing faults, are also used in programmable logic array (PLA) testing, memory testing, and microprocessor testing [70].

6.1.3 Test Pattern and Fault Coverage

To determine whether a given VLSI chip or PCB performs correctly according to specifications, one normally applies a sequence or a set of test input stimuli and samples the responses at the output. If the sampled response differs from the response expected from a good device, then the device is declared faulty. The test stimuli is called the *test pattern* or *test vector*. The test pattern must be so chosen that for every modeled fault at least one input will produce a response that differs from the corresponding good response.

The goodness of a test pattern is measured by the time of testing and fault coverage. The time of testing is the time required to test the whole test pattern. *Fault coverage* is expressed as the ratio of the number of faults detected by input test patterns to the total number of the possible faults the system has. Although it is impossible and not practical to provide 100% fault coverage for most systems, IC functional test fault coverage greater than 99% is necessary for manufacturing high quality ICs.

6.1.4 Test Pattern Generation

The fault coverage is determined by the efficiency and length of the test patterns. To achieve greater than 99% fault coverage, extensive test pattern generation is often required. The exhaustive test method, which uses all 2^n input combinations as the test set, is an excellent test method but is impractical for complex ICs.

Fortunately, methods do exist that allow high quality testing of complex ICs in a cost effective and timely manner. Approaches used to ease test development for complex ICs include manual test pattern generation, automatic test pattern generation, and DFT methods. The first two methods attempt to reduce the number of patterns, development time, and execution time required for a high quality test by intelligent application of the input stimulus test patterns. The last approach attempts to ease the testing problem by modifying the circuit structures. These approaches are often used together to achieve an optimum balance of life cycle costs in meeting time-to-market and quality objectives.

In *manual test pattern generation*, the test patterns are designed by test engineers themselves. The fault coverage of the test patterns depends on a thorough understanding of the system function and structure. After the test patterns are generated, fault simulation is used to determine the fault coverage by using some CAE tools.

Automatic test pattern generation tools are based on algorithms that evaluate the circuit under test and attempt to provide an optimal or near optimal test pattern. Proper partitioning of a complex circuit into more easily tested subcircuits is the key to effective utilization of these tools.

6.1.5 Functional Testing

Functional testing is used to check if the behavior of the device is correct. In functional testing, the device under test is considered as a whole entity in its approximate electrical environment, and the electrical access is usually limited to external I/O pins. Functional testing includes two types of testing: Standard speed functional testing verifies that the device behavior is correct, using static settled-state measurement, and, at-speed functional testing verifies behavior at the rated circuit signal timing.

In functional testing, the fault at each internal node must be propagated to external I/O pins such that it can be detected. Thus the test pattern may be very difficult to generate because the number of internal nodes may be very large and the external I/O pins are limited. For many complex systems, including MCMs, DFT (Section 6.1.7) technology is often used to improve the fault coverage of functional testing.

6.1.6 Parametric Testing

Whereas functional testing tries to verify the function of the device as a whole, another method, *parametric testing*, is used to ensure that the device meets its performance specifications. Parametric testing consists of a dc test and an ac parametric test. The dc parametric test measures the static voltage and current at each pad of the IC; the ac parametric test measures the timing issues of the system, ac parametric testing is especially critical for high speed systems, such as MCMs, due to their high working frequencies, and dc parametric testing is used to test the following parameters:

1. **Shorts and Opens:** Applies current and measures forward voltage drops of diode p–n injections on I/Os
2. **Input Level Tests** (V_{IL}, V_{IH}) : Input switching thresholds
3. **Output Drive Tests** (V_{OL}, V_{OH}, I_{OL}, I_{OH}): Output voltage levels and output current source and sink capability
4. **Static Supply Current** (I_{CC}): Indicator of process problems
5. **Dynamic Supply Current** (I_{DD}): Applicable for CMOS where I_{DD} varies with frequency
6. **Leakage Tests** (tristate, input, pull-up, pull-down): Checks for excessive current flow through internal conduction paths

ac parametric testing is used to test the following parameters:

1. **Propagation Delay:** Time interval from input signal application to output response
2. **Setup/hold Times:** Verifies valid signal before/after assertion of second signal
3. **Clock Frequencies:** Verifies duty cycle, period
4. **Signal Timing:** Verifies signal edge placement
5. **Pulse Width:** Verifies period

6.1.7 Design-For-Testability

With advances in VLSI technology, it is becoming more difficult to test chips through the limited I/O pins. *Design-for-testability* is the deliberate effort to ensure the inherent testability of a circuit. The testability of a system can be described by the following concepts:

1. **Controllability:** The ability to manipulate signal flow within a circuit
2. **Observability:** A measure of the extent to which signal activity can be monitored
3. **Partitioning:** Reducing a complex circuit into a set of minimally interactive subcircuits

There are three major approaches currently being used for DFT namely, ad-hoc approaches, scan based approaches and built-in self-test (BIST). Let us discuss these three approaches.

Ad-hoc approaches are case-specific testing aimed to reduce the combinational explosion of test patterns. Dedicated test access points (bond pads), multiplexers, control gates, test buses, embedded test software, and other case-specific approaches are used to improve an IC's controllability, observability, and partitioning. Ad-hoc testing requires the test engineers to have a thorough understanding of the system structure and testing approaches.

Scan based approaches are also called *structured approaches*. These techniques impose the scan based design rules on the IC design. The scan based design rules require altering sequential storage elements within the device logic and connecting them to act in a serial shift-register fashion. In the normal mode, the system can operate unimpeded; in the test mode, data are synchronously moved through the scan path under the control of a test clock signal. The input test patterns are shifted into the input path, and the state of each storage element is observed by shifting the data to output of the scan path. Complete internal scan design can require from 10% to 20% circuit overhead [86], and the run time may be very slow.

Boundary-scan is a typical scan based technique. Figure 6.2 provides a pictorial view of the boundary-scan concept within an IC and an MCM.

In the boundary-scan technique, the boundary-scan cells are attached to each device I/O pin and connected serially to form a shift-register chain (scan-path) from the test date input (TDI) pin, around the periphery of the device, to the test date output pin (TDO). The boundary-scan control logic unit contains a test access port (TAP), an instruction register, an instruction decode register, and a bypass register. In normal (nontest) operation, the boundary-scan test circuitry is in a reset condition. In this mode, the boundary attached to the chip pins is transparent, although there may be some delay penalty due to the addition of the extra gates. In the test mode, the device under testing is disconnected from the I/O pads. A sequence of bits is then shifted into the boundary such that each register contains the required logic level for each input on the device. The boundary control then connects the internal logic.

BIST is synonymous with BIT (built-in test) and self-test. It can be defined as the capability of a unit such as a bare chip, assembled MCM, or system to test itself with little or no external test equipment or manual intervention required. BIST design techniques are used to create on-chip hardware for input stimulus generation and output response reevaluation. By placing the stimulus and response evaluation hardware (test circuitry) within the same silicon as the IC's normal application circuitry, BIST solves the test access problem. This embedded test circuitry is not prone to the loading effects or signal limitations of an external test system and provides "at-speed" testing of the chip.

6.1.8 Burn-In

Burn-in is a process by which the electrical devices are subject to accelerated aging conditions for many hours. In burn-in, the devices are put inside an environmental chamber, and the ambient temperature in the chamber is elevated to provide accelerated life testing. During the burn-in process, the device can be operated in either a static or a dynamic mode.

The burn-in test is used to stress the electrical interconnection of the device and package and drive any contaminants in the body of the device into the active circuitry, thus causing functional or parametric failures. By conducting burn-in tests, the infant mortality (refers to early failures) from manufacturing defects can be detected.

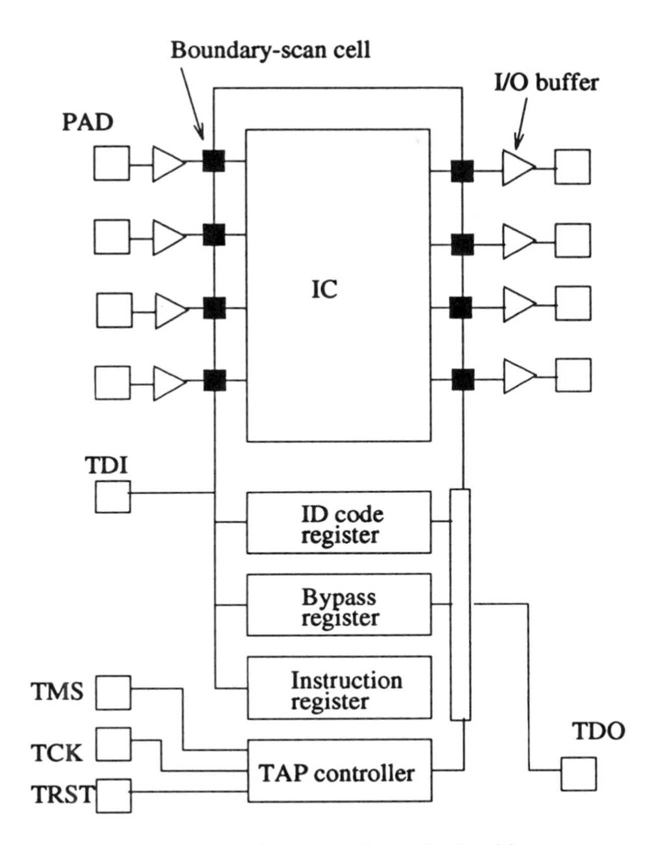

(a) Boundary-scan for a single chip

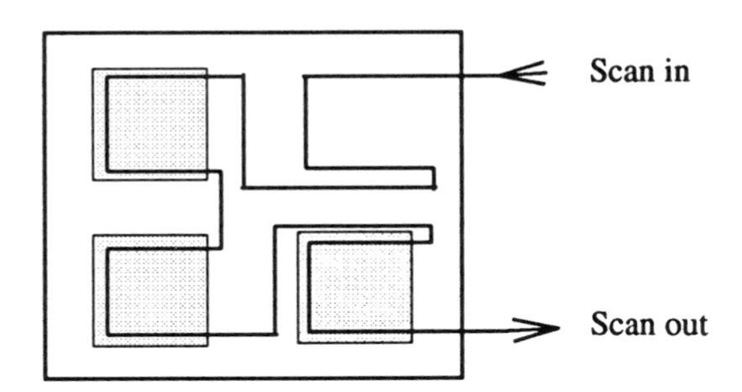

(b) Boundary-scan for an MCM

FIG. 6.2. An IC with boundary-scan.

The burn-in test can be optional. The need for burn-in is related to the maturity of the fabrication process. New IC designs that utilize smaller design rules or new process steps are more likely to fail during a burn-in test. As the process is refined, the failures at burn-in can drop low enough to warrant eliminating the burn-in step. Although burn-in tests are not performed on many commercial ICs, it is required for all high performance devices such as MCMs.

6.1.9 Known Good Devices

The term *known good devices* refers to those devices for which it is known with high probability that they are free of any as-received defects and that they will remain fully functional and defect free [86] under the following conditions.

1. After assembly onto the next level board or substrate
2. Over the design temperature range and operating speeds
3. After next level burn-in or environmental stress screening
4. After having been fielded for a time approaching at least the minimum guaranteed field life

To ensure a known good IC, full functional tests must be performed by socketing the single-chip packages and running them at full operating frequency and over the full temperature range. Burn-in tests can also be performed to weed out infant mortality failures. After this extensive testing, the single-chip package ICs can have a KGD probability of 0.99999 or even higher [86].

6.1.10 Bed-of-Nails

Bed-of-nails is a kind of probe card. It uses a head that contains many probes to contact all of the pads on the board or substrate at the same time. All the probes are connected to a switching network (electrical multiplexer) such that each pad can be connected or disconnected to any other pads on the substrate or board. By measuring the resistance of any two pads, the short and open faults between pins can be detected. If the resistance of any two pads within the same net is high, then an open or a high resistance fault is indicated; on the other hand, if the resistance of two pads in two different nets is very close to zero, a short fault is declared.

A bed-of-nails contacts all the pads at the same time, and the testing is finished by switching the multiplexer. Thus it has a fast testing speed. A bed-of-nails is commonly used to test printed wiring board (PWB), but it is not suitable for MCMs since MCMs have high density pads and contain naked dies, with no pads on a grid.

6.1.11 Flying Probe Tester

The *flying probe tester* is another kind of probe used to test MCM substrates. It has two electrodes. One electrode moves under computer control. The other is stationary and is in electrical contact with the backside of the substrate. The capacitance of each net is measured and recorded. The tester software then compares the measured capacitance with a table of prestored expected capacitance value. If the difference is larger than the tolerance, the net is flagged. Since the substrate database is available from the layout system, each pad location can be fed to the tester with the expected capacitance based on the layout and substrate process information. This way the layout information can be used for the substrate.

6.2 SUBSTRATE TESTING

A substrate in an MCM can have an open or a short fault and timing errors due to fabrication failure. Thus, substrate testing is necessary to verify the connectivity of the substrate and to monitor the fabrication process for quality control. Typically, substrate testing measures and tests the capacitance and resistance of interconnection nets on MCM substrates.

The substrate testing problem in MCM is similar to the PCB testing problem. The difference between the testing of a substrate and a PCB is in the test machinery used. PCB is often tested by bed-of-nail probes. Due to the small dimensions and high density of the MCM substrates, they are usually tested by a flying probe tester.

In this section, we briefly describe the substrate testing methods. A detailed discussion of substrate testing has been published elsewhere [58].

6.2.1 Single Probe Testing

In *single probe testing,* the tester uses the capacitance meter as the measure unit. One side of the meter is connected to the ground plane, and the other side is connected to a single probe. The probe in turn is moved to contact the testing pad, and thus the capacitance of each pad can be measured. Then the open and short faults can be determined by comparing the measured capacitance with a reference capacitance. The reference capacitance can be obtained from its design, or from measurement of its capacitance on a known good part.

To show how single probe testing works, let us consider the short and open cases shown in Figure 6.3. Five nets with nominal capacitances of 20, 15, 17, 16 and 2 pF, respectively, are shown.

Nets N3 and N4 are bridged due to the small separation between them or any manufacture failure. When using a single probe to test the pads within nets N3 and N4, an approximate 16 + 17 = 33 pF, will be measured; thus a short fault between N3 and N4 is reported. Moreover, this fault can be located, since both N3 and N4 have similar measured capacitances.

Whereas a short fault is indicated by an increase in capacitance, the open fault within a net is indicated by a reduction in the capacitance. Net N1 has a nominal capacitance of 20 pF. If a open fault occurs at point B as shown in Figure 6.3, the

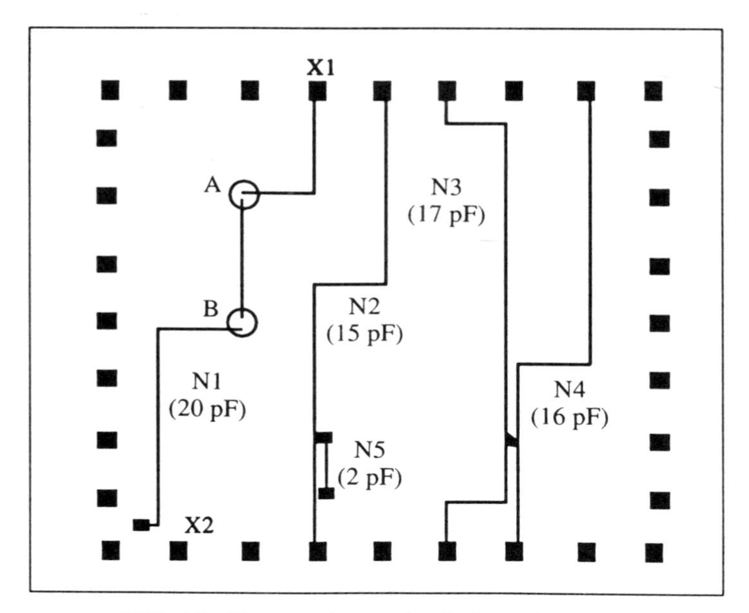

FIG. 6.3. Shorts and opens in single probe testing.

measured capacitance at pad X1 and X2 would be approximately 10 pF since B is near the middle of N1 net. But if the break occurs at point A, the measure capacitance at pad X1 will be less than 10 pF, and the fault can be detected since it is much less than 20 pF. But the measure capacitance at pad X2 will be less than 20 pF but larger than 10 pF, and thus it will be difficult to detect since the difference between measured capacitance and nominal capacitance is minor. To detect all possible open faults, the capacitance measurement is performed for each pad within the net.

However, there are some faults that capacitance measurement is unable to locate. For example, nets N2 and N5 in Figure 6.3 are shorted. N5 has a small capacitance value (2 pF) compared with N2 (15 pF); thus the net N2 may not be indicated as a fault since the measured capacitance of 17 pF at the pads of N2 is very close to its nominal value. Although these shorts can be detected by testing pads of net N5, they cannot be located. This kind of problem can only be overcome by resistance measurement.

Single probe testing can be used to provide open and short testing on a wide variety of MCM substrates. The drawback of this method is that it is unable to locate some short faults due to the lack of resistance information of the nets.

6.2.2 Two Probe Testing

To test the resistance of the nets, an extra probe is required. The two probe testing approach is basically the same as the single probe approach except that another probe is used to test a pair of pads at the same time. The two pads can either be within the same net and thus the resistance information between the two pads can be obtained or are in different nets such that shorts between nets can be detected.

Measuring the resistance of each pair of pads may require a long time. Hence, two probe testing usually uses the capacitance measurement ability as in single probe testing to detect and locate the short and open faultsfirst. Whenever a short fault is unable to be located, that is, a net with excessive capacitance and all other nets with similar capacitance as their nominal values as the case of N2 and N5 in Figure 6.3, a resistance measure is performed to determine which pair of nets is shorted.

The short, open, and high resistance faults can be detected and located by using a two probe testing approach. Other measurements can also be performed in two probe testing by connecting other external measuring devices to the probes. This approach is feasible for most MCM substrate testing. The disadvantage of this approach is the time required due to the mechanical motion of the probes, especially for high density MCM substrates.

6.2.3 Electron Beam Testing

Both bed-of-nails testing and flying probe testing use a physical probe to contact the pads that can cause physical damage to the substrate. Moreover, in some cases such as thin film MCMs, they are built up one layer at a time and interconnect pads may not exist until the final layer is applied. Thus the mechanical probe testing methods are not feasible.

To solve this problem, *electron beam testing* is used. Electron beam testing works in the same way as a scanning electron microscope. While testing a net, an electron beam is ejected toward a pad of the net, and thus the pad will be charged to a preset voltage. Then, another electron energy analyzer is used to detect if other pads on the substrate are also charged. An open fault is indicated if any pad of the same net has not charged. Similarly, a short fault can be detected if any pad in other nets has been charged to the same voltage. Electron beam testing uses an electron beam flood gun to charge and discharge the pad on the substrate. Unfortunately, electron beam testing is expensive since it requires that the substrate be placed in a high vacuum chamber, which takes some time to pump down.

6.3 DIE TESTING

By definition, an MCM contains two or more bare dies mounted on a common substrate. The quality of the die used in an MCM plays a key role in the MCM yield. Consider an MCM with 10 dies. Even if there is no yield loss from the substrate and assembly process, but 5% of the dies fail during module burn-in, then the MCM yield will be $(0.95)^{10}$, or only 60%. This means that 40% of the

MCMs will require some sort of rework to obtain a fully functional module. Generally, the use of unpackaged dies with KGD probabilities much below 0.999 will lead to a very low first-pass module yield, massive rework required to repair MCMs having faulty ICs, and very expensive resulting MCMs. Therefore, it is clear that all the dies to be used on an MCM should be as free of defects as possible prior to installation on the substrate or to be KGDs. To ensure the KGDs, die testing is a must.

While VLSI chip testing is a mature technology, MCM die testing presents some unique challenges. Most ICs today are not ac tested and burned-in until they have been mounted into their packages since it is much easier to perform the final testing of the chip in its packaged form. Dies that are used in MCMs must be tested fully as bare dies since it may be difficult to perform a complete test once the dies are mounted to the MCM substrate. Moreover, when dies are handled in an unpackaged format they are much more difficult to test thoroughly than after they have been sealed into single-chip packages. As a result, the unpackaged dies do not have the same 0.9999+ probability of being good that packaged ICs typically have since they do not have the same benefit of individualized testing in sockets.

Like VLSI chip testing, MCM die testing also consists of functional and parametric tests. Since ac testing of a die improves the final test yields and the final test yields greatly influence the MCM cost, performing ac testing on die prior to packaging into an MCM is advantageous. For high end MCMs, ac testing at the die level will be necessary. Basically, the strategies of functional and parametric testing in MCM die testing are similar to those in VLSI testing. The only difference is how to handle the unpackaged dies in MCM die testing. A variety of methods are in use for pre-testing individual unpackaged dies prior to assembly into MCMs. They fall into two categories: employing pressure contacts to make connections with the die pads and employing metallurgical connections to the die pads. All of these methods must address the issues of producing adequately low resistance connections to the die and avoiding damage to the die pads that later will be assembled into modules.

Besides die testing, die burn-in is another concern for MCM die testing. In many MCM designs, the final module testing will verify only that it has been assembled correctly and operates at its rated speed. It is unlikely that a complete test of each die will be performed after MCM assembly since the access to each device I/O is limited. Since the burn-in process stresses the internal logic of the dies, testing after burn-in can detect many manufacturing defects that cannot be detected before burn-in. Therefore, each die must be burned-in before installation on the module.

In the following section, we discuss the pressure contact and metallurgical connection tests for dies. We then discuss die burn-in.

6.3.1 Pressure Contact Methods

The following are some commonly used pressure contact testing methods:

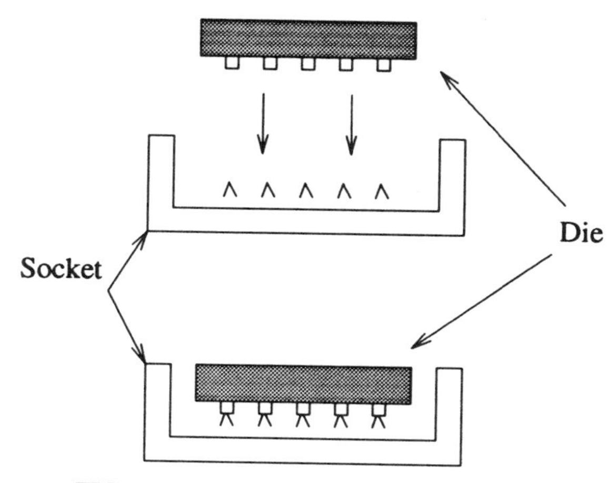

FIG. 6.4. Socket approach for die testing.

1. **Probe Cards:** In this method, the tester contains fixture equipment and many probe needles. During testing, the die singulated from the wafer is held by the fixture equipment inside a heated or cooled environment. Probe needles contact the die pads and provide an interconnect between the load board and the die under test. The probe needles are built to protrude from the probe card on an average for about 1.4 inches, depending on the number of die I/Os. They are designed to avoid inducing the undesired parasitic inductance and resistance, and thus a correct ac testing result can be obtained. Since the die is tested at the wafer level, probe needles may damage the die pads and testing cannot be performed at high frequencies. The testing cost is also high. Therefore, this method has limited applications [17, 86].

2. **Sockets:** Sockets have been widely used in hybrid applications for many years as shown in Figure 6.4. They can also be used in MCM die testing. The naked die is mounted on a socket and thus the full tests are performed through the socket in the same way as the testing of a packaged IC. However, it is very difficult to ensure the alignment of the contacts and the planarity of the contacting sufrace since the die pads are usually depressed in wells below the surface of the passivation layer. In other words, making a good pressure contact between the socket and the die under test is a major challenge with this method. Moreover, contacting the naked die also has disadvantages of pad damage, limited high frequency, fragility, and cost [86].

3. **Anisotropic Conductive Compliant Films:** Both probe card and socket methods may damage the naked die pads since they contact the pads to test the board directly. To avoid direct contacts, anisotropic conductive compliant film (ACCF) can be used to attach the die to the test board. The ACCF is loaded with vertical conductor elements deposited in a certain pattern. In this structure, the whole film plane is isolated, whereas the vertical conductor

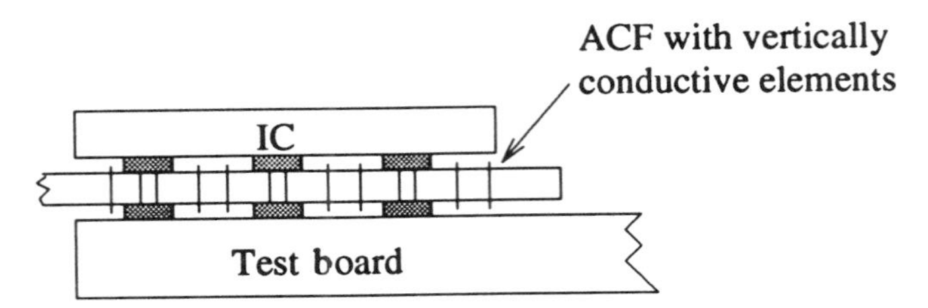

FIG. 6.5. ACCF method for die testing.

elements provide current paths in vertical direction such that the die pads can be connected to the test board. ACCF does not cause physical damage to the die pads, and the die can easily be detached and cleaned after testing. However, as the pitch of the vertical conductors cannot be made very small, the ACCF method does have an alignment problem.

4. **Anisotropic Conductive Adhesive Films:** To solve the alignment problem with ACCF, the anisotropic conductive adhesive film (ACAF) method can be used. Similar to ACCF, ACAF is used to attach the die to the test board and to maintain the isolator characteristics along the XY plane (film plane). But ACAF is embedded with metallic particles instead of vertical elements. When the ACAF is compressed by the die and test board, these metallic particles will form Z-axis connections between die pads and the test board. Typically, the metallic particles are made of Ag, Ni, or Au. By controlling the size of metallic particles, ACAFs can successfully connect dies with 4 *mil* I/O pads [58]. It is also easy to make a good contact pressure and achieve alignment with the small sized metallic particles. The die can also be detached from the test board after testing. Moreover, ACAF does not need to be patterned and thus can be used to test a variety of MCM dies. The main problem of this method is the thermal expansion and corrosion mechanisms of ACAF materials. This method is shown in Figure 6.6.

5. **Flip-Bump Adhesive:** In this method, protruding metal bumps are used on the die, or the substrate or both, and an adhesive is used to maintain the die and substrate together in pressure contact. This method is shown in Figure 6.7. Gold, copper, indium, and tin-lead have been used to form the bumps in various applications. This method also allows easy die removal and cleanup after testing. The problem with this method is the cost.

6.3.2 Metallurgical Connection Methods

In this category of die testing strategies, a fixed metallurgical connection, instead of a pressure contact, is made to the die pads. The configurations and metallizations used are as follows.

1. **Tab Lead Frame Bonded to Die:** TAB technology is well suited for die testing and burn-in. In testing, a TAB frame carrier is used to support the tape

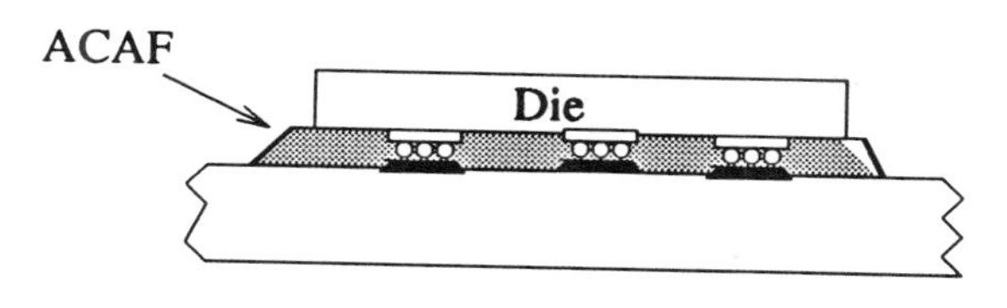

(a) IC to circuit board ASAF

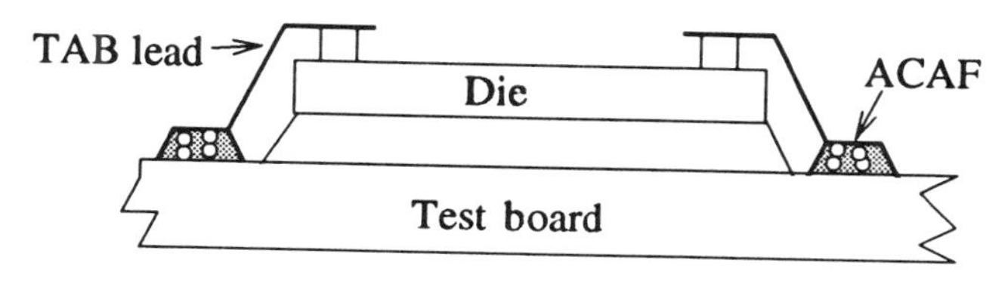

(b) TAB to circuit board ACAF

FIG. 6.6. ACAF method for die testing.

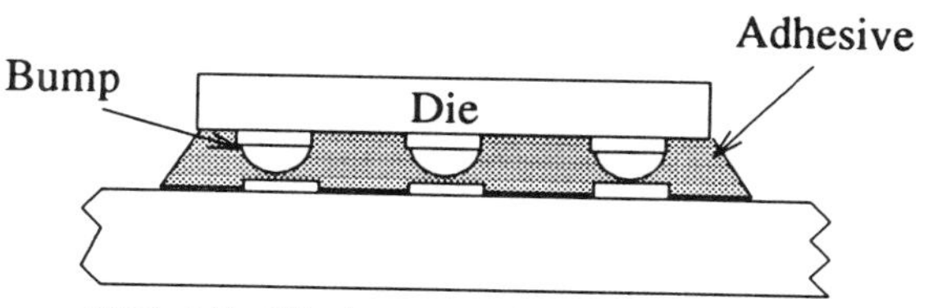

FIG. 6.7. Flip-bump method for die testing.

lead and the test probes are directly in contact with the tape leads as shown in Figure 6.8. The end of the tape leads connecting to the pads are typically on a fine pitch and are fanned out to a larger pitch at the other end to mate the test probes. This method has the advantages of miniaturization, automation, and capability for very high I/O fine pitch dies and provides a format well suited for die testing and burn-in of MCM dies. However, TAB technology and its industry infrastructure are still expensive and immature [86]. Thus, this method has limited applications.

2. **Flip-Chip IC Bonded to Sacrificial Test Carriers:** The die is mounted to the carrier by a flip-chip solder attachment, and the carrier can be designed to permit easy mounting and removal of the die with no damage. The known die testing methods for flip-chip dies in MCM include probing of the solder bumps at the wafer level prior to dicing [116] and burn-in of dies temporarily flip-chip soldered to underside pads on burn-in substrates. The largest problem of this method is the limited availability of flip-chip dies.

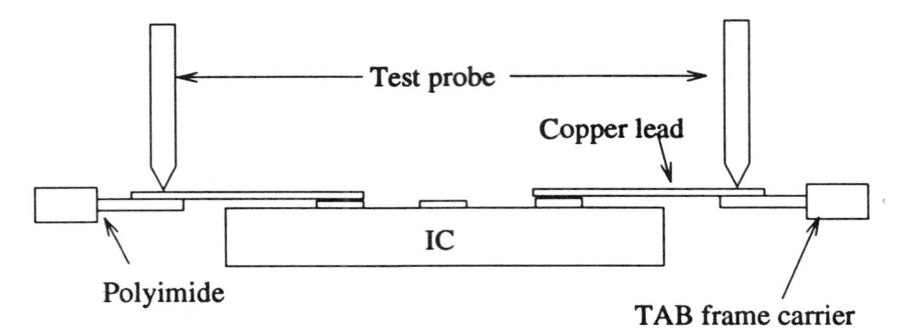

FIG. 6.8. TAB lead frame bonded method for die testing.

3. **ICs Bonded to a Miniature Device (Microchip) Carrier:** In this method, a miniature device carrier (MDC) or a microchip carrier (MCC) is used as a testability aid, and the die under testing is mounted on it. Die testing is performed through the carrier. The interconnect between the die and the carrier can be a wire bond, a TAB, or a flip-chip. The interconnect from the carrier to the substrate or test board can also accommodate several different technologies: wire bond, TAB, flip-chip, or leads/solder joints. The known MDC and MCC materials include glass, epoxy, ceramic, flex circuit, silicon, and aluminum nitride. Testing with MCC technology can ensure that the die is known good, and it can provide good mechanical protection prior to assembly into MCMs. Moreover, it can also allow easy die/MCC replacement.

4. **Chip-on-Tape (IC Wire Bonded to a TAB Frame):** In this approach, the die is wire bonded to a TAB, which in turn is placed in a standard TAB carrier frame for burn-in and testing. Since wire bonding is easily programmable to different bond pad geometries, many different ICs can be used with a standard TAB design, thus eliminating the significant tooling cost penalties associated with unique inner lead bond designs for each different IC. By designing a tape with large extra I/O lines, a "universal" TAB design can accommodate hundreds of different IC designs. Each TAB carrier consists of a lead fan-out, which is connected electrically to the pads on the die. Once the die has been tested and burned-in, the TAB lead frame is trimmed to its final size and the die is mounted directly to the package.

5. **HDI Removable Overlay Fabricated Over IC Pads:** In this method, an overlay is used either to apply larger bond pads to bond the IC temporarily into a single-chip package for die testing or to apply a TAB-like frame for die testing in TAB fixtures. After testing, the overlay is peeled off the die, leaving the original wire-bond sites available for assembly of the MCM.

6.3.3 Design-For-Testability

Testing of dies suitable for MCM assembly depends on a sufficient electical test access to the die under test. The pressure contact methods and metallurgical techniques presented earlier provide an overview of these approaches and the associated trade offs. However, with the growing complexity of ICs external test access methods alone are not adequate. The circuits must therefore be deliberately designed to ease testing. On-chip circuitry must be allocated to improve internal test access or even provide testability to the extent of allowing the ICs to be largely self-testing. The incorporation of emerging DFT approaches such as boundary-scan and BIST can greatly simplify die testing.

To be able to perform boundary-scan and BIST, it will be necessary to design each die such that it has the ability to test itself. This is normally done by partitioning each die into specific functional blocks (such as memory, control, I/O, etc.) and they by architecting boundary-scan and BIT for each of these functional blocks. As indicated elsewhere [86], more than 68% high-reliability ASIC vendors now have offered either boundary-scan or BIST testability techniques in their products, and more and more dies used in MCMs are expected to have boundary-scan and/or BIST testability in future.

6.3.4 Die Burn-In

An MCM may contain many dies fabricated by different processes, such as CMOS, TTL, and ECL. Different dies may have different burn-in requirements. Burn-in of the full module may not meet the needs of all of the dies on the module. Therefore, bare die burn-in is necessary for high yield.

Bare die burn-in can be performed through wafer level reliability and wafer level burn-in procedures [58]. In wafer level reliability procedures, dies are stressed with the elevated temperature, voltage, and output currents. Increasing the temperature and current will cause shorts or opens in improperly formed structures. High voltage will cause electromigration and breakdown in gate oxides and junctions. In wafer level burn-in, the individual dies in the wafer are connected together using additional layers of metallization. The wafer is then connected using a socket or a probe card and placed in the burn-in system. After burn-in, the metallization is either stripped off or disconnected in some way.

Methods are also available for die burn-in using TAB [199]. TAB leads are bonded to the die prior to mounting the die on the module. A test fixture can be made that connects the TAB leads. Test fixtures similar to cantilever probe technology can be made to burn-in wire bond dies. Test fixtures for flip-chip MCMs could also be designed and implemented using a bed-of-nails testing philosophy.

6.4 MODULE TESTING

After testing the substrate and the dies, and mounting the dies onto the MCM substrate, the complete module must be fully tested to ensure that it meets the design specifications. This is necessary since MCM assembly manufacturing process can damage the substrate interconnections or the bare die or the bonds, or mismount dies, or use wrong components. Moreover, the performance of the whole MCM can only be tested after the whole MCM is assembled.

Module testing in MCMs also consists of a functional test and a parametric test and uses a strategy similar to ASIC testing and PWB testing. Due to the complexity and small size of MCMs, it is very difficult to access the internal components of MCMs. Thus the DFT is the key to cost effective MCM testing.

Because of the high cost of MCMs, the assembly yields of MCMs are also a critical issue in MCM design. All substrate testing, die testing, and module testing are used to ensure high assembly yields of MCMs. In addition, some other methods including rework can also be used to improve the assembly yields.

In this section, some basic module testing issues are addressed. We do not focus on the traditional ASIC testing methods, since that material can be found elsewhere [69, 71, 211]. The methods used to improve assembly yields and test design considerations are also introduced in this section.

6.4.1 Traditional Test Approaches

Traditional functional and parametric testing approaches can be used in module testing; that is, a test stimulus file and tester and DUT specific information are used to generate a test vector along with the I/O location and I/O specific information for the IC tester.

Depending on the system structure, the cost, and the time required for testing, full functional testing or limited functional testing can be performed. From a fault coverage standpoint, the ideal module test would combine all of the tests for individual components into one large test pattern. Running this test pattern will guarantee the functionality and performance of all of the internal logic in the module. Full functional testing requires the generation of a test pattern that can fully exercise all of the internal logic from the limited I/O pins. Generation of this test pattern is extremely difficult and time consuming. In many applications, it is even impossible. Limited functional testing just covers a limited scope of defects. If the assumption is made that the primary defects that show up during module testing are assembly errors (KGD and substrate are ensured) then limited functional testing is enough. Limited functional testing requires much less effort to create. In some cases, adequate coverage can be obtained without considering the internal behavior or gate level logic inside ICs.

To secure a good functional test for an MCM, one must have

1. High quality models for the devices used on the module
2. Timing information for each of the nodes
3. A good simulation system
4. An input vector sequence that exercises the module very thoroughly
5. A high-speed test system capable of performing fault diagnosis

A good model must be capable of duplicating the actual device's outputs for any input sequence. An inaccurate model, even with slight inaccuracy, can cause immense difficulties. However, accurate models for some devices, especially complex devices, are very difficult to acquire. Timing information for a device is somewhat more easily obtainable. Most good simulation systems are capable of providing this data. This data is needed so that a device that does not meet its propagation-delay specifications can be detected. The simulation system chosen must be capable of generating test data that is compatible with the chosen tester and the selected diagnostic technique. The input vector sequence to be used should be exhaustive and reflect the expected use of the module. In the case of a complex module, this might be many thousands, even hundreds of thousands, of vectors. This again shows the importance of good device models.

Finally, a high-speed tester capable of very accurate edge placement and receiver strobing is required. Generally, for module testing, modified chip testers are utilized. These systems have the driver/receiver accuracies needed and can generally be interfaced to a microprobe for fault isolation. Additionally, experiments in the use of contactless probing techniques are underway in many test-system research and development laboratories.

6.4.2 Testing With DFT

Because of the high complexity and the limited I/O pins of MCMs, the testing and accurate diagnosis of MCM defects will prove to be a formidable task until proper DFT procedures are adopted in the module and ASIC design process. Using the traditional test approaches, all the problems in MCM testing cannot be solved. Even the ad-hoc DFT approaches may not be sufficient for complex MCMs.

In MCM applications, boundary-scan increases the fault coverage of internal interconnects that cannot be distributed throughout the I/O connector. The IEEE 1149.1-1990 (JTAG) standard partitions scan channel rings at each I/O signal buffer in order to provide control and access through a scan base tester. If memory devices are included on an MCM, a BIST system can be implemented that would provide a self-test of the memory devices with only a few test vectors. Figure 6.9 shows a preferred approach for MCM design: standard test bus, scan techniques, and BIST [86].

A standard bus is used to provide a defined common interface between the testability circuitry incorporated in an electronic component or assembly (MCM) and the test equipment (built in or external). This allows multivendor interoperability in order to lower cost and cycle time for test generation, logic simulation, fault detection, fault isolation, and repair.

The whole test process of this test strategy can be divided into five steps [86]:

1. Verify the test circuitry itself. This will ensure that the test circuitry is reset and does not affect the normal MCM function and test the path integrity. This can be done by the internal mode of the boundary-scan circuit.

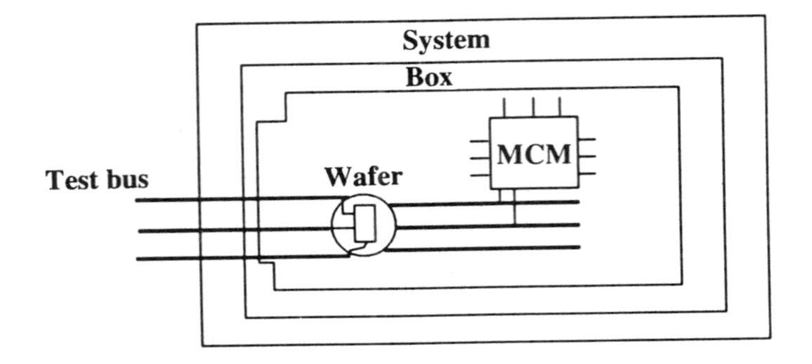

(a) Standard test bus

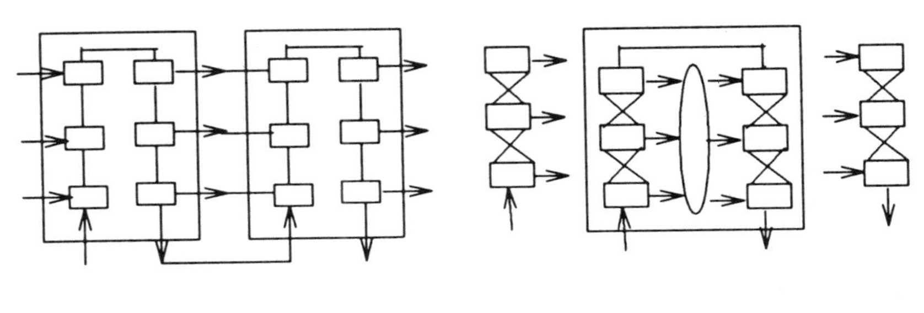

(b) Scan technology (c) Built-in-self-test

FIG. 6.9. A hierarchical approach to testing.

2. Test the interconnects using external test mode. This will verify the interconnects between dies.
3. Test microcircuit core logic. This will verify individual components inside the chips. This is done by using an internal test mode, internal scan path, or BIST.
4. Use the available boundary-scan access path to test remaining logic clusters.
5. Execute an overall performance test. Dynamic functional testing, emulation, or BIST software functional testing techniques can be used.

6.5 IMPROVING ASSEMBLY YIELD

Testing is the key policy to ensure the assembly yields for MCMs. However, there are some other methods that can be used to improve the assembly yields of MCMs. In this section, we introduce some of these approaches, including MCM rework.

6.5.1 Automated Component Placement

The substrate of an MCM has a limited area but very dense interconnections. Thus the alignment of component placement can be a problem. To place the dies and other components correctly and automatically, "pick-and-place" equipment can be used. Pick-and-place systems are available with video camera/pattern recognition features that verify that the correct part is about to be picked up by comparing the component surface pattern to a picture previously stored in memory and also verify that the IC is rotated in the current orientation for the substrate pattern. Assembly yields in such automated facilities are high, typically above 99.9%.

6.5.2 Interconnection Technologies

The electrical connection between the die and the substrate can be made by different techniques, such as flip-chip, TAB, wire bond, and ACAF interconnections. Each technique has advantages and disadvantages as discussed in Chapter 2.

Basically, wire bond MCMs need special probing equipment, and hence it is difficult to test wire bond MCMs; the flip-chip and microdevice carrier technologies can use a sacrificial carrier and temporary socket, respectively, to perform testing, and thus they are not as hard as wire bond MCMs to test. TAB MCMs are the easiest to test because they use sockets.

From the viewpoint of rework, flip-chip MCMs are the easiest to be repaired, but TAB MCMs are difficult. MCMs using wire bond and microdevice carriers are moderately difficult to rework.

6.5.3 MCM Packaging

Low cost, low IC count MCMs are frequently packaged by attaching a lead frame to the MCM substrate and molding a plastic body around the substrate assembly. There are many research efforts to develop adequate reliability with thin encapsulant coatings for MCM substrate assemblies, which would approach the reliability of hermetically packaged parts. "Glob-top" epoxy coatings have been used successfully for years in COB assemblies for consumer products, but do not have sufficient durability under thermal cycling to survive environments such as avionics or military applications. For those environments a hermetic package is required until encapsulant technology improves.

6.5.4 MCM Reworking

Many MCMs may consist of ASICs costing as much as \$400 to \$600 (or more) each. When a defect is detected in an MCM, it is unlikely that manufacturers will be willing to discard this MCM (which may contain multiple sets of these expensive devices). Some rework can be done to save the MCMs with faulty components; that is, a faulty chip is removed from the substrate and replaced with a new chip.

The cost of reworking must be less than the cost of a finished module; otherwise, rework does not make economic sense. MCMs costing thousands of dollars may receive cost-justified repair operations, while it may be justifiable to throw away a low cost MCM rather than repair it.

Module level rework begins with fault detection by either visual or electrical tests, followed by accessing the faulty chip. Rework methods at the lower assembly levels may simply involve removal and replacement of a faulty IC using a carefully planned rework strategy that does not degrade the rest of the subassembly. The chip is demounted, normally, by applying heat to the chip connection points. At the higher assembly levels it may involve unsoldering the MCM from the circuit board and removing the package lid in order to replace a failed IC. Rework at this level requires MCM packages designed to allow removal and de-lidding operations.

The choice of materials and manufacturing technologies determines whether a module can or should be reworked. An inherent rework philosophy is available for flip-chip MCMs. Once a bad die is located and removed ultrasonically or through a thermal probe by directing heat only to the specific die being reworked, the solder balls connected to the substrate can be cleaned and a new die can be attached using a thermal probe. Reworking with TAB or wire bonded chips is still a labor-intensive procedure that can quickly destroy any economies of MCM use.

High-value MCMs need to be designed to be repairable, that is, to allow replacement of defective components at every assembly level. This implies that the test methods need to be able to isolate faults at each assembly level tested to identify the particular fault component.

6.6 TEST DESIGN CONSIDERATIONS FOR MCMs

As more dies are integrated on MCM substrates, MCM testing becomes extremely difficult, even with the use of DFT technologies and the advanced test equipment. For example, an MCM with as many as 1,000 I/O pins, which is becoming common now, has exceeded the capabilities of most commercially available automated test equipment. To ease the task of testing, the system testability and test strategy must be carefully considered both in MCM system design and in MCM test design phases. In this section, some basic guidelines for system designers and test designers are listed [86].

1. **System Design:** To allow ease of test and repair, the system designer should incorporate as many testability features as possible while still providing the required performance. A list of design specifications for test features are as follows.
   a. Use a top-down approach to design the system with modularity. This can facilitate developing modular test programs and isolating faults.

b. All chips must be independently isolatable. This requires all the chips to have the tristate output such that each chip can be isolated from all other chips by simply turning off the other ICs. When a chip is isolated with other chips, it can be tested with its standard IC test vectors if its I/Os are accessible.

c. Isolate Faults for a single electrical net. This needs to bring as many internal chip I/Os to MCM I/O pins as possible.

d. Incorporate boundary-scan and BIST in system design. DFT technology is the key to test MCMs, and it is impossible to test complex MCMs successfully without using DFT technology. Boundary-scan and BIST need to be designed in testing blocks inside the MCM.

e. Design MCMs with standard sizes and footprints.

f. Use ICs equipped with boundary-scan and BIST.

2. **MCM Testing:** To ensure high quality testing, MCM testing should also have the following criteria:

a. Ensure 100% MCM I/O stuck-at fault coverage.

b. Ensure 100% MCM I/O adjacent short fault coverage.

c. Ensure 100% MCM chip I/O stuck-at fault coverage.

d. Ensure >85% MCM chip internal stuck-at fault coverage.

e. Ensure 100% MCM interconnect short fault coverage.

f. Ensure 100% MCM interconnect open fault coverage.

3. **High Performance MCMs:** The following additional tests must also be performed:

a. MCM I/O ac and dc parametric tests.

b. MCM dynamic functional test with 100% MCM chip I/O stuck-at fault coverage.

6.7 SUMMARY

In this chapter, MCM testing issues were addressed. MCM substrates have a limited area with very high density. To verify the connectivity of MCM substrates, the flying probe tester and electron beam methods can be used. The bed-of-nails method can also be used for low density MCM substrates. To ensure that KGD are used in MCMs, each naked die must be fully tested and burned-in. The methods employing pressure contacts to die pads and methods employing metallurgical connections with the die pads can be used to handle the unpackaged dies. After an MCM has been assembled, it also needs to be completely tested. The traditional functional test and parametric test approaches can be used in module testing. But, since one MCM can contain many ICs, it is extremely difficult to access every internal node inside the chips through the limited module I/O pins. So, DFT technology must be used to perform the complete module test.

The testing of MCMs presents many challenges. DFT plays a key role in the success of bringing up a new design and in isolating the faults to a single component. It is expected that boundary-scan and BIST will be essential components in every MCM design in the future since they provide the ability to probe internal nodes without extra test pads or pins.

6.8 PROBLEMS

1. Describe the differences between PCB testing and MCM testing.
2. What are the key differences between VLSI IC testing and MCM testing?
3. Compare the measurements that are required to test an MCM substrate with n nets when using single probe testing, two probe testing and electron beam testing.
4. Review Chapter 2 and compare the testability of wire bonded, TAB, and flip-chip MCMs.
5. Show how boundary-scan and BIST can improve MCM testability.

BIBLIOGRAPHIC NOTES

VLSI IC testing and PCB testing have been well developed. The functional testing and parametric testing in VLSI ICs and PCBs can be found in Giacomo [71] and Zobrist [211]. The IEEE standards board has issued a standard for boundary-scan testing in IEEE Std. 1149.1–1990, "Standard Test Access Port and Boundary-Scan Architecture." Other discussions for substrate testing have also been published [58, 74, 94]. Hagge and Wagner [86] wrote an excellent paper about die tests. Die testing and MCM rework are discussed by Trent [199]. Module testing is also discussed elsewhere [58, 86, 199].

CHAPTER 7

RELIABILITY ISSUES IN MCMs

The reliability of an electronic system is one of the most important concerns in many applications, such as aviation and military applications. For example, avionic equipments usually operate in hostile environments. They must operate reliably over extremes of temperature, shock, humidity, and pressure. Therefore, avionics systems must meet rigid reliability requirements. The electronic systems designed for military applications must also be able to satisfy very strict reliability, environmental, and performance specifications. For example, the equipment developed for aiming of tank guns must be able to perform reliably and accurately in severe polar weather as well as in desert conditions.

System reliability is also an important concern in other applications as well, since the maintenance cost of an electronic system is directly related to its reliability. The reliability degradation can reduce product life, which in turn can lead to an increase in maintenance cost. The reliability improves if possible failure mechanisms are eliminated or reduced. Reliability also improves when the heat generated can be removed efficiently so that the system is operated below a specified temperature. The improved reliability in turn will reduce the maintenance cost.

One of the most important advantages of using MCMs over traditional packaging technologies in an electronic system is the improvement in the system reliability. The MCM technology eliminates a whole package level with the accompanying thermal resistance, lead, and solder joint problems and thus improves the reliability of the system (Fig. 7.1). The reduction of package levels results in a dramatic decrease in the number of fatigue-prone solder joints, hermetic seals, that are prone to failure. Further reliability improvements can be achieved by using packaging materials of high thermal conductivity in MCMs, thereby improving heat removal paths and allowing MCMs to run at lower

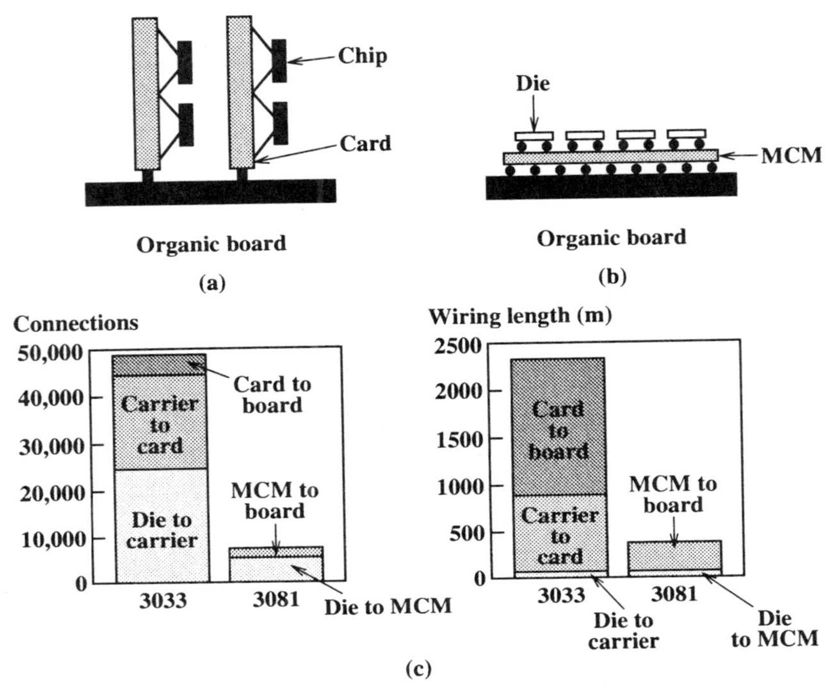

FIG. 7.1. Improved reliability accompanies reduced numbers of interconnects in MCMs.

junction temperatures. Lower junction temperatures prolong the lifetimes of devices. Reliability is a major concern in any MCM package design because an MCM carries many expensive leading edge chips, making it prohibitively expensive to discard because of field failure.

In this chapter, we first discuss the basic reliability concepts and failure mechanisms in general. Then the reliability concerns of individual components are discussed. Finally, methods of improving the reliability of an electronic system are presented.

7.1 RELIABILITY CONCEPTS AND FAILURE MECHANISMS

In this section, some important parameters that are used to measure reliability are discussed. In addition, failure mechanisms are presented.

The reliability of an electronic component is measured statistically since electronic components are rarely worn out. The important metrics of reliability include, MTTF, MTBS, and MTTR. *Mean time to failure* (MTTF) is the average life time of an operational electronic device before failing. *Mean time between service* (MTBS) (or *mean time between repair* [MTBR]) is the average time

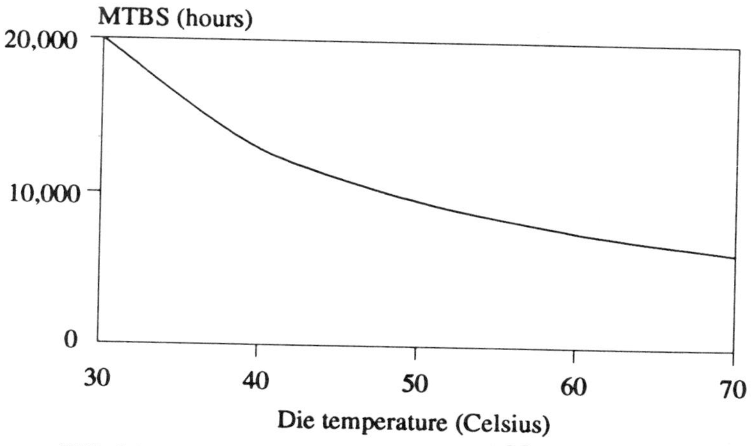

FIG. 7.2. The effects of the operating temperature on MTBS.

between repairs. *Mean time to repair* (MTTR) is the average time needed to repair a failure. The relationship among MTTF, MTBS, and MTTR is

$$\text{MTBS} = \text{MTTF} + \text{MTTR}$$

The system reliability of MCMs depend on the operating temperature of IC dies. Each die generates a certain amount of heat. In MCMs, many bare dies can be placed in a small area, which results in the generation of a large amount of heat in that small area. If the heat is not removed efficiently, the temperature, of the dies can become too high. The operating temperature of a die must be kept low to achieve high reliability. Figure 7.2 shows the effect of temperature on the MTBS for mainframes. To achieve high reliability, the heat generated by the dies in an MCM must be removed efficiently. One of the methods to remove the heat efficiently is to choose materials of high conductivity during the material selection process.

In addition to the concerns of removing heat efficiently, *thermal stability* and *glass transition temperature* are also important concerns in choosing the proper materials for MCMs to acheive high reliability. *Thermal stability* measures the weight loss rate at a given temperature or a given heating rate. *Glass transition temperature* (T_g) is the temperature at which glassy polymers soften [58]. Weight loss for many materials becomes much more severe at higher temperatures. As a result, thermal stability at the temperature in question is important because the polymer properties must remain unchanged during the high temperature process, such as the curing of subsequent layers or chip joining. Thermal stability means that the weight does not change at a specified high temperature. Otherwise, thermal degradation can destroy the multilayer structure by causing delamination [58]. Glass transition temperature is important because several properties change above this temperature. For example, stresses are released, the CTE increases, defects might be healed, adhesion might increase, and dimension might change due to polymer flow. As a result, the glass transition temperature of a material should be measured accurately to predict the behavior of the material under different temperatures. Generally, a lower T_g is beneficial for the adhesion between polymer and other layers. But a lower T_g can result in a larger dimensional change.

Opposite to stability, failure is a phenomenon in which certain properties of MCM change. To achieve high reliability, we should understand the failure mechanisms so as to eliminate or reduce failure mechanisms. *Thermal failure*, which is defined as the thermally induced total loss of electronic function of a component, can occur as a result of melting and vaporization of a part of the component, thermal fracture of a support, or separation between leads [159]. In addition, prolonged use of components at high temperatures can result in a gradual change in the material properties, creep in the bonding material, parasitic chemical reactions in connectors and switches, and diffusion in solid state devices. These results may also lead to thermal failure.

In addition to thermal failure, mechanical failure of an MCM can also occur. The mechanical failure of an MCM is caused by the excessive stress in the MCM [58]. Excessive stress can lead to the release of stress. The stress release can occur due to delamination, cracking, crazing, or the formation of deformation zone. *Delamination* refers to the process in which a layer is completely detached from the substrate. *Cracking* is a phenomenon in which two new surfaces are generated and are not connected across the crack, whereas *crazing* refers to a pre-crack phenomenon in which the material is filled with fibrils separated by voids. *Deformation zones* are pre-crack zones where shearing of the material has occurred but crazing has not occurred. The propagation of a crack through film can result in catastrophic failure. To evade mechanical failure, excessive stress must be avoided.

Excessive stress can be caused by the mismatch of CTEs among the principle materials used in an MCM. The MCM substrate is formed by laying conductive material on top of dielectric material. When the temperature increases, any material will expand. However, if the expansion rates of two materials are different, the connections between the two different materials may be broken or weakened when temperature changes. The larger the difference in their expansion rate, the more likely is the failure. To increase the reliability of an MCM substrate, the CTE of the dielectric material that measures the expansion rate of a material when temperature changes should closely match the CTE of the conductive material used in the MCM. Furthermore, silicon dies are attached to the MCM substrate. As a result, the CTE of the dielectric material should also closely match the CTE of silicon. In case that support substrate is present, the CTEs of the dielectric material and the base material should also match each other. However, if the mismatch of the CTEs between two materials is severe, the lifetime of MCMs is significantly shortened. In the following, an example of the effect of CTE mismatch is discussed.

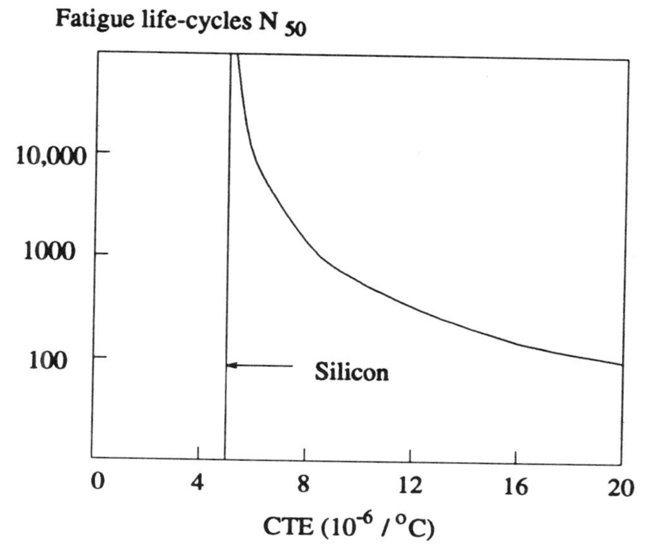

FIG. 7.3. Relationship between fatigue life and thermal expansion coefficient (CTE)

The fatigue life of the solder connecting the substrate and the chip as a function of CTE of the substrate dielectric material is shown in Figure 7.3. It is assumed that silicon dies are solder bonded to the MCM substrate. Note that the CTE of the silicon is about 6. It can be seen from Figure 7.3 that the fatigue life is reduced sharply when the CTE value of the dielectric material moves away from the CTE value of silicon. For example, the fatigue life of the solder connecting the substrate and the chip is over 10,000 when the CTE of the dielectric material is around 6, whereas the fatigue life of the solder is less than 1,000 when the CTE of the dielectric material is 8. Therefore, to achieve high reliability, the CTEs of the principle materials used in an MCM should be closely matched.

7.2 RELIABILITY CONCERNS OF INDIVIDUAL COMPONENTS

The overall reliability of an electronics system also depends on the reliability of individual components, the hierarchy of failure dependence (fault tree analysis), and additional safeguards taken to minimize system failure caused by separate parts [80]. The primary concerns of individual components in an electronic system are three critical components, the die carrier, the test socket, and the impact on the bare dies [80] as well as MCM interconnections and chip attachment. In this section, we discuss the reliability issues in dies, die carriers, test sockets, interconnections, and die attachment.

7.2.1 Reliability of Die, Die Carrier, and Test Socket

Concerns regarding reliability include

1. The carrier
   a. Substrate integrity and durability.
   b. Carrier integrity and durability, including the lid, latch, and the overall support mechanism.
2. The socket
   a. Socket component integrity.
   b. Socket-to-carrier contact integrity.
   c. Socket-to-board contact integrity.
3. Bare dies
   a. Environmental contamination of the bare dies.
   b. Damage to the bond pad surface.
   c. Damage to the structures under the pad.
   d. Bond pad degradation due to the outside environment.

Evaluation of the reliability of the entire system requires evaluation of the reliability of carrier, socket, and bare dies. The criteria for evaluating carrier reliability include the contact resistance, which is monitored over operating temperature excursions, and the surface contamination due to the foreign materials that can cause damage to the carrier or the IC. In addition, the life tests for the degradation during service and the physical damage that could prevent the carrier from operating properly are also included in the evaluation criteria.

The socket lifetime is evaluated as part of the overall system reliability. Acceptance tests for bare dies, after they have been tested and burned-in, include wire bond pull, die shear, inspection of bond pad and underlying structure for damage, and passivation intergrity testing [80].

7.2.2 Reliability of MCM Interconnections

Reliability issues of MCM interconnections concern opens in wires and shorts between neighboring wires. For example, when aluminum is used as the material of interconnection, there can be reliability issues such as electromigration and step coverage [6]. Electromigration is caused by the transport of the metal atoms when electric currents flow through the interconnections. It is the result of the interaction between the aluminum atoms and the electron current. The metal atoms are transported mainly by grain boundary diffusion as the atoms collide with the drifting electrons. The displacement of the metal atoms may eventually cause undesirable opens in the wires or shorts between two neighboring wires, if material accumulated as a result of electromigration forms a bridge between the neighboring lines (Fig. 7.4).

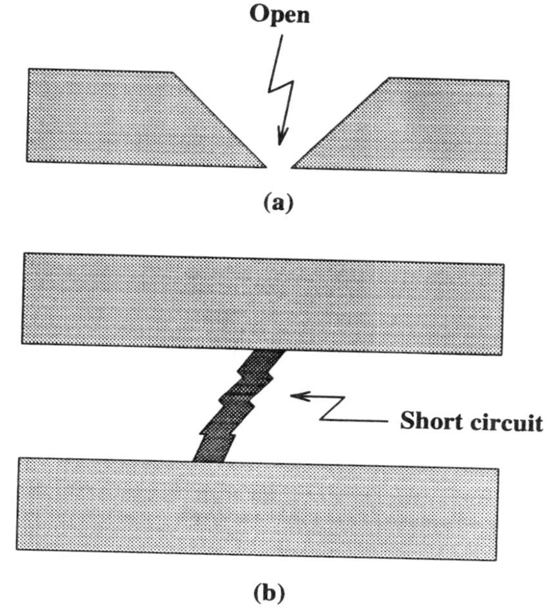

FIG. 7.4. Electromigration.

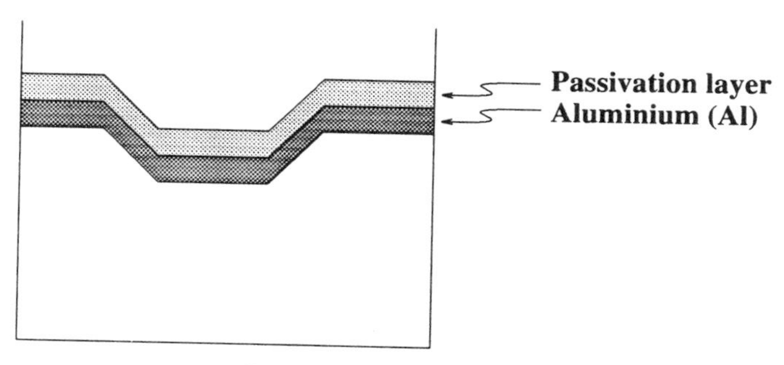

FIG. 7.5. Step coverage.

Step coverage is caused by the shadowing effect of the surface topography during the metal deposition process. The thinning of the line at the steps can create opens and cracks and generate electromigration-prone and high-resistance points due to reduced cross-sectional area (see Fig. 7.5).

7.2.3 Reliability in Chip Attachment

In addition to MCM interconnections, chip attachment may also cause reliability problems. In the following, we discuss the reliability issues in each of the three chip bonding techniques.

1. **Wire Bonding:** One of the major reliability issues in wire bonding is the intermetallic reaction related to time and temperature [58]. Intermetallic diffusion may happen that may result in an increased joint electrical resistance or low pull strength due to eventual voiding at the surface. For the gold ball bond on an aluminum alloy pad, the thickness of the aluminum is the critical factor. If the thickness of aluminum is not sufficient, the pad can be consumed by an intermetallic formation, eventually causing voiding. For the aluminum wedge bond on a gold pad, the ratio between the bond heel thickness and the gold thickness on the pad becomes critical.

Reliability is also affected by the contamination from wafer processing, plastic package outgassing or epoxy die attach outgassing, and bleedout [58]. Contaminants accelerate failures by reducing the onset temperature or the time to failure due to intermetallic growth.

2. **TAB:** The critical areas in TAB for reliability are TAB tape, inter lead bonding (ILB) joints, and the outer lead bonding (OLB) joints. TAB tape is the main source of concern for reliability. The failures of the TAB lead and the OLB joint are mechanically induced by temperature cycles. Two types of stresses can occur. Transient stresses are the result of rapid temperature excursions, whereas steady-state stresses come from the inherent differences in the CTEs among the principal materials. It has been revealed through research studies that TAB leads are apparently the weak link in the mechanical assembly due to the relatively smaller area of the cross section of the TAB lead compared with that of surface mount leads. Other concerns of the reliability of TAB tape include metal to dielectric delamination, dielectric expansion, chemically caused failures due to the moisture absorption of the dielectric or planting inconsistencies.

The gradual degradation of the ILB joint is also a major reliability concern in the TAB technique. The causes of ILB failure includes separation between the bump and die pad due to insufficient cleaning of the aluminium prior to the deposition of the bump metallization and the failure of the under-bump metallization to prevent interdiffusion between Au and Al. In addition, ILB failures can also be caused by the separation between the bump and TAB lead and the separation between the TAB lead and bump interface. The separation between the bump and TAB lead is due to contamination at the bump and lead interface. The separation between the TAB lead and bump interface is due to intermetallic formation, particularly in reflow bonded joints. Assembly process issues such as restrictive process windows and materials issues such as poor lead frame design and material selection can also lead to ILB failures. Another reliability concern is the chemical corrosion induced by moisture penetration through the encapsulate or passivation. The moisture penetration is largely due to the presence of halides, remnants of the encapsulation, or OLB (flux) processes. Moisture is absorbed by the dielectric material during tape manufacturing. Planting issues such as contaminants, variable thickness, and shelf life also contribute to ILB failures.

The metallurgies of OLB joints are mostly solder or Au to Au. The quality of the solder joint is the major issue in the reliability of OLB joints. Gang reflowed solder joints show nonuniformity based on thermode temperature variations or planarity and flatness problems. Incomplete reflowed solder joints may then be produced, resulting in mechanical failures in the poor quality solder joints. The quality of solder joints also depends on the choice of tape plating. Tin-plated tape is subject to shelf life and oxidation concerns and can result in poor solder joints if used beyond its expected life or is not stored properly, whereas gold-plated tapes used with tin-lead solder systems can form brittle intermetallics in the solder joint. The migration of metal between adjacent leads under an applied potential can cause low resistance leakage pats or even complete electrical short by bridging. It has been found that failures occur profusely between leads with exposed copper over a polyimide surface and also occur in tin plated leads. Needle-like whiskers can grow from the tin plating as a result of stresses imposed during the plating process. This phenomenon, known as tin whiskering, can also cause potential reliability problems.

3. **Flip-Chip:** The CTE mismatch between silicon dies and the MCM substrate can cause the displacement of flip-chip solder bump joints. The amount of strain in a solder joint depends on the distance from the zero displacement point. The concern for premature failure due to the CTE mismatch is a critical issue for the flip-chip solder bump technology. The joint is sensitive to thermal fatigue failure as it is the pliable member separating die and substrate, which are relatively rigid elements. Therefore, it is crucial for the solder joints to withstand strain to avoid premature failure and loss of functionality.

7.3 IMPROVING SYSTEM RELIABILITY

Reliability of multichip packages must be dealt with at the following three levels to improve the overall reliability of the entire system [201]:

1. Design for reliability up front.
2. Build in reliability by process controls during manufacturing.
3. Accelerate testing of final products to guarantee the design and the process.

To improve the system reliability, it is very important to choose the proper fabrication technology and the principle materials. MCM-C is the technology of choice if reliability is a major concern. MCM-C provides high temperature stability and reliability better than any other MCM technology [91]. The high reliability of MCM-Cs are partly due to the materials in them that are completely imprevious to moisture. In addition, the dielectric constants of ceramics do not change with humidity even without hermetic packaging. MCM-C substrates have superior strength and rigidity compared with the alternative interconnect materials. They are resistant to elevated temperature assembly and operating conditions.

The reliability of an electronic system improves when the number of package levels is reduced. Therefore more components should be packaged at a package level to increase the reliability. In addition, the CTEs of principle materials in an

MCM should be closely matched, and the operating temperature of components should be reduced. Reliability can also be improved by process control of all critical parameters such as green-sheet thickness, metal width, and via hole and by accelerated testing at high temperature, voltage, and humidity conditions [201].

Reliability tests are used to predict the limits of temperature, thermal gradients, and vibrating environments necessary to induce excessive stress in MCMs and to predict stresses, strains, and temperatures occurring in the respective MCMs [65].

7.3.1 Test Vehicles To Enhance System Reliability

Various tests can be made to analyze the failure mechanisms and reliabitiy of MCMs. The analytical techniques for characterizing materials in their cured and uncured states, as well as the environmental and mechanical stress tests, are discussed by Doane and Franzon [58]. By performing these tests, comparisons among different materials can be made. The failure modes and failure rates can be predicted by using the information obtained in stress testing. To improve the overall reliability of an MCM, extensive coverage of failure modes should be carried out. Mechanical failure modes such as delamination should be avoided through material selection that reduces high temperature exposures during cure. Excessive thermally induced stress must be avoided. In the following, several analytical and environmental tests and their purposes are listed [58].

1. **Differential Scanning Calorimetry** (DSC): Determines % cure, T_g thermal stability, reactivity, phase change versus temperature
2. **Thermal Gravimetric Analysis** (TGA): Determines % cure, thermal stability, reactivity, and weight loss versus temperature
3. **Thermal Mechanical Analysis** (TMA): Determines, T_g and mechanical performance versus temperature
4. **Residual Gas Analysis** (RGA): Identifies impurity and gas composition in hermetic modules and breakdown products of organic adhesives
5. **Scanning Electron Microscopy** (SEM): Determines failure mode and microstructure of die attach adhesives
6. **Thermal Shock:** Identifies poor adhesion properties and brittle materials and accelerates fatigue and crack generation and propagation
7. **Temperature Cycle:** Identifies poor adhesion properties and brittle materials and accelerates fatigue crack generation and propagation
8. **High Temperature Steady State** (HTSS): Determines temperature related electrical sensitivity and corrosion sensitivity due to residual contanminates within the MCM
9. **Highly Accelerate Stress Test** (HAST): Determines temperature related electrical sensitivity and corrosion sensitivity due to moisture entering the MCM

10. **Burn-in:** Determines temperature related electrical sensitivity and corrosion sensitivity due to residual contaminates within the MCM
11. **Mechanical Shock:** Determines mechanical stability and identifies poor adhesion
12. **Mechanical Vibration:** Determines mechanical stability and strength and identifies poor adhesion
13. **Solvent Resistance:** Identifies process vulnerability to other process steps and solvents and determines chemical nature of adhesives

7.3.2 Post-Test Analysis

The analytical and environmental tests should provide a detailed understanding of the failure mechanisms for the technology being tested. The long-term reliability aspects being predicted in the tests include [65]:

1. Technology strengths and weaknesses
2. Acceptable field use environments
3. Process/material variation affects
4. Realistic failure mechanisms
5. Fielded hardware expected lifetimes

In addition, the reliability analysis should also include

1. A description of the MCM screen tests/procedures to ensure the quality of materials, processing methods, and performance integrity
2. Determination of the MCM failure sites and elements and type of failures (catastrophic or degradation)
3. Establishment of activation energies associated with identified failure mechanisms and MTBS extrapolations versus temperature to confidence levels commensurate with sample size
4. ac performance versus lifetime and temperature data
5. Electrical, mechanical, and environmental limitations of the industry process
6. Establishment of operating behavior and characteristics under dc or ac electrically applied excitation, determination of equivalent stress levels (dc vs. ac) and using appropriate factors, and prediction of reliabilty at use temperatures
7. Documentation of the necessary corrective actions for the MCM reliability improvement

To ensure the design for high thermal reliability, the operating temperature of the components should be precisely measured or predicted. There exist numerical thermal analysis programs to analyze thermal effects in a two-dimensional or three-dimensional environment. Typically, finite element or finite difference techniques are applied to solve a nodal representation of the circuit layout. In the thermal analysis, the cooling mechanism must be taken into account. Usually electronics can be cooled by three methods: conduction, covection, or radiation. In the conduction method, material of high conductivity and/or external heat sinks are used to dissipate heat from hot components. This type of cooling is expensive. Conduction cooling is often modeled as a two-dimensional resistive network of interacting components. In convection cooling, the flow of fluid such as air or water is used to cool hot components. The heat flow may be analyzed in terms of a resistive network where component temperature primarily depends on only those components that are upstream in the fluid flow. Usually the heat removed by radiation is much less significant than that removed by the other two methods.

After the cooling mechanism is modeled, a thermal analysis that includes the interaction of components can be carried out. As a result, a more accurate prediction of thermal reliability can be obtained. Trade off analysis pertaining to the cooling mechanism, reliability, maintainability, performance, and maintenance cost can be carried out.

7.4 SUMMARY

Reliability is an important issue in MCM design and testing. In some applications such as aviation and military applications, the reliability is critical to the success of using MCMs, while in other applications the reliability affects the maintenance costs of the MCMs. The reliability of MCMs is affected by the operating temperature of the dies. Prolonged use of dies at high temperature reduces MTBS which is an important parameter affecting reliability. To remove the heat efficiently, materials of high conductivity should be chosen. Excessive stress can cause delamination, cracking, and crazing. Mismatch of the CTEs among the principle materials in an MCM is a major source of excessive stress. Therefore, it is important to choose materials with matching CTEs during the material selection process. In addition, the thermal stability of a material, which measures the weight loss of the mateiral during a high temperature process, should also be considered during the material selection process. Materials with the minimum weight loss should be chosen to avoid the change of properties during a high temperature process such as the curing of the MCM substrate or chip joining. Various test and analysis techniques can be used to study the failure mechanisms and to predict the operating temperatures of the components, the stress and strain in the system, and the expected lifetime of the system. Results can then be used in the trade off analysis and to eliminate or reduce failure mechanisms in MCMs, thereby improving system reliability.

7.5 PROBLEMS

1. Is the following statement true? Why? The reliability issue is easier to handle and less critical in MCMs than in traditional packaging technologies since the number of package levels and the number of fatigue prone solder joints and hermetic seals are significantly reduced in MCMs.
2. What are the properties that need to be considered in choosing principle materials in an MCM in order to improve the reliability of the MCM?
3. Describe the potential causes of thermal failures, mechanical failures, and chemical failures, respectively. How can these failures be prevented?
4. Explain the purposes of reliability tests and anlysis. What tests are best suited for determining thermal stability, T_g, and mechanical stability and strength, respectively?
5. Why is it useful to model the cooling mechanism in the thermal analysis? In the thermal analysis, how can a forced air cooling method be modeled? How can the radiation method be modeled?

BIBLIOGRAPHIC NOTES

Doane and Franzon [58] present the basic reliability concepts and failure mechanisms and discuss reliability issues in different chip bonding techniques and several analytical and environmental tests. The reliability issues in individual dies, die carriers, and by Gucciardi and Green test sockets are addressed [80]. Some accelerated and enviromental tests, as well as reliability modeling and post-test anlysis, are introduced by Fayetle [65].

CHAPTER 8

PHYSICAL DESIGN FOR MCMs

The size and complexity of MCMs makes them very difficult if not impossible to complete manually even one aspect of MCM design. Therefore, it is necessary to automate the MCM design process as much as possible to improve performance, reduce cost, and increase reliability. The design of an MCM is carried out in several steps as discussed in Chapter 3. MCM physical design is one of the important steps in the MCM design cycle. The physical design cycle in MCMs is the process of determining an exact physical layout for a given circuit diagram. The steps in the MCM physical design cycle are similar to IC or the PCB physical design. In spite of this similarity, the tools available for the physical design automation of ICs and PCBs cannot be used for MCMs. This is mainly due to the design constraints, which are significantly different from those of ICs or PCBs. Furthermore, the tools used in the design of PCBs and ICs cannot handle the high complexity of MCMs due to the multitude of electrical, thermal, and testing considerations. In an attempt to meet the CAD needs of MCMs, several commercial CAD tools are now available; however, these tools are essentially extensions of PCB or IC design tools and do not satisfy all the requirements of MCM physical design. As a result, the design and development of CAD tools for MCMs is of paramount importance and is considered critical to further growth of MCMs.

MCM physical design automation is essentially the study of algorithms and data structures related to the physical design process. The objective is to study optimal arrangements of devices (chips, routing wires, etc.) in a three-dimensional space (or plane) and efficient interconnection schemes between these devices to obtain the desired functionality. In addition, algorithms must use the space very efficiently to lower costs and improve performance and yield. In addition, the relative arrangement of chips plays a key role in determining the electrical and thermal performance of an MCM. Algorithms for physical design must also ensure that all the rules required by the fabrication process are observed and that the layout is within the tolerance limits of the fabrication process. Finally, algorithms must be very efficient in memory usage and computer time and should be able to handle very large designs. Efficient algorithms not only lead to fast turn-around time, but also permit designers to improve the layouts iteratively.

Most of the objects in MCM physical design are very simple geometric objects, such as rectangles and lines. However, a pure geometric point of view ignores the electrical (both digital and analog) and thermal aspects of MCMs. In an MCM, rectangles and lines have interrelated electrical and thermal properties that exhibit a very complex behavior and depend on constraints imposed by fabrication processes and performance requirements, among others. Therefore, it is necessary to keep the electrical and thermal aspects of the geometric objects in perspective while developing algorithms for MCM physical design automation.

In this chapter, we discuss in detail the different steps of the MCM physical design. The basic concepts and related algorithms for MCM partitioning, MCM placement, and MCM routing phases are aso discussed.

8.1 MCM PHYSICAL DESIGN CYCLE

The physical design cycle in MCMs is the process of converting an MCM circuit diagram into a layout (circuit layout). Due to the high complexity and a host of constraints involved in converting the circuit diagram into an MCM layout in one step, this cycle has been divided into a number of steps. The physical design cycle of MCMs is shown in Figure 8.1. As shown, the physical design cycle of MCMs consists of MCM partitioning, MCM placement, MCM routing, and design verification steps. Also notice from Figure 8.1 that it is necessary to perform the physical design for each chip at the end of the MCM partitioning step. The physical design cycle shown can be considered conceptual because in reality the whole design process starts at the MCM placement phase as the integrated circuits are already available to be placed. However, for establishments with the capability of completing in-house designs of MCMs, the entire design cycle shown is applicable. In this section, we briefly discuss these steps. Each of these physical design steps are discussed in detail in subsequent sections.

1. **MCM Partitioning:** The size of the circuit assigned to an MCM is significantly large. Hence it is not possible to fabricate the whole circut as a single unit (chip). This necessitates the decomposition of the given circuit into a set of smaller subcircuits. MCM partitioning is the process of dividing the given MCM circuit into subcircuits such that each subcircuit can be fabricated on a single chip and the number of subcircuits is less than or equal to the number of chips that the MCM can carry. MCM designs require a performance driven approach. This requirement necessitates that timing, noise, and thermal constraints are to be

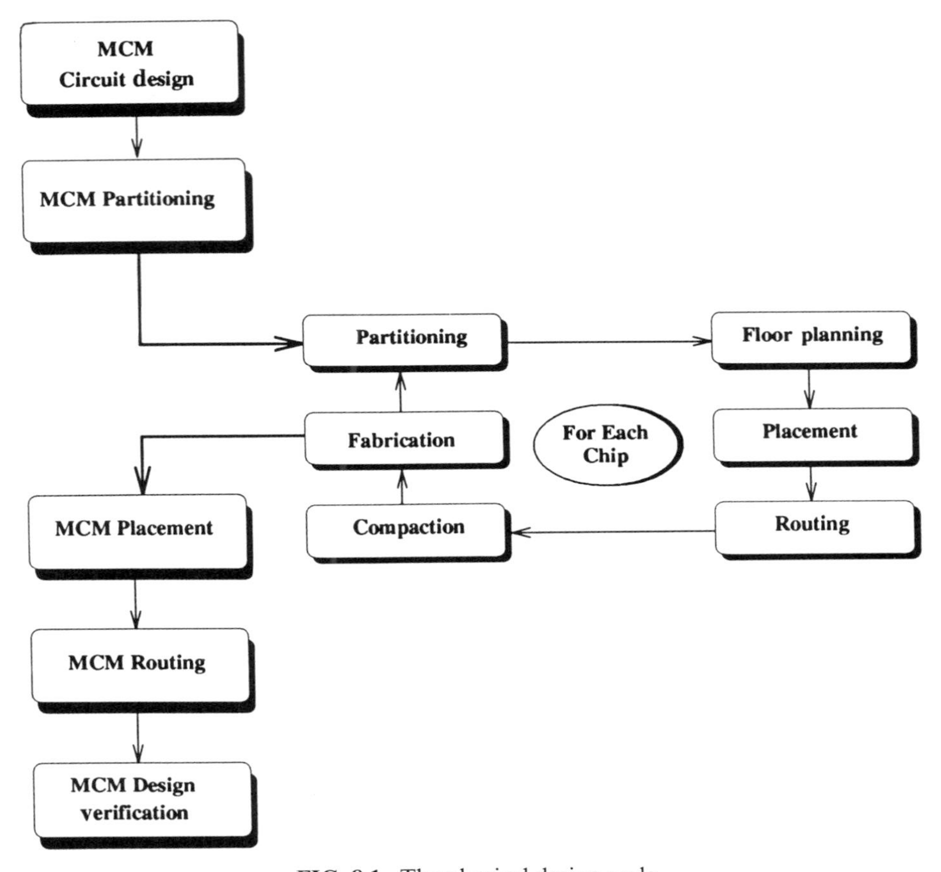

FIG. 8.1. The physical design cycle.

considered in the MCM partitioning step, in addition to the I/O and area constraints.

MCM partitioning should minimize the number of times a critical path crosses a partition so as to meet the timing constraints. Since both chip delay and interchip (interconnect) delay contribute to the total delay in a critical net, it is imperative to reduce the number of times a critical net crosses a chip boundary.

Thermal constraint is another important factor that has to be satisfied during the MCM partitioning phase. This is primarily dependent on the type of cooling process used. In general, the amount of heat that can be efficiently removed from an MCM is determined by the thermal management system. Depending on the type of thermal management system, partitioning has different objective functions. In certain cases a combined effort of partitioning and placement is required to satisfy the thermal constraints.

After an MCM has been partitioned into a set of subcircuits, each subcircuit is fabricated as a chip. Figure 8.1 depicts this process. The process of MCM partitioning can lead to MCMs containing well over 100 chips. This phase is discussed in detail in Section 8.2.

2. **MCM Placement:** After the completion of the MCM partitioning phase, chips are available as bare dies. The process of assigning the bare dies to their exact locations while satisfying the timing, thermal, and routability constraints is MCM placement.

MCM placement is based on the type of design style used. In the case of chip array style, the placement step is concerned with mapping the chips to the chip sites on the MCM substrate. When the chip sites are prefabricated, the MCM placement problem lends itself to a gate array based approach. On the other hand, in full custom style the objective of the placement phase is to place the chips on the MCM substrate such that the total area used is minimized. The maximum operating frequency as well as the correct operation of an MCM based system depend critically on the timing constraints. For example, the chips that lie on the critical path must be placed such that the longest critical path must be minimized. This minimization process must be performed for all the critical paths.

Thermal considerations in MCM placement are critical because bare chips are placed closer and generate a significant amount of heat. To ensure proper operation of design, the heat generated must be dissipated. The total heat generated in an MCM is dependent on the placement of the chips. In fact, thermal constraints are critical and have to be considered during the placement phase, when area wide cooling is adopted so as to avoid "hot spots." However, as discussed previously, thermal constraints have to be considered during both partitioning and placement phase to satisfy them.

A "bad" placement not only takes up large areas and degrades performance but also makes it difficult or sometimes impossible to complete the routing task. Routability in MCMs is based on two factors; number of layers used and type of MCM such as thick or thin film. Depending on the number of layers used, the process of cooling may introduce some thermal vias that may block the nets and, hence, effect the routability. This phase is discussed in detail in Section 8.3.

3. **MCM Routing:** The MCM placement phase guarantees that chips are not placed too far such that the timing constraints are violated. In comparison, MCM routing phase ensures that the timing constraints are satisfied in addition to the noise and fabrication constraints while connecting the chips as specified by the netlist. One of the main goals of the MCM physical design is to optimize performance, especially in the case of high-end systems. The objective of the routing phase is to meet the timing and noise constraints specified by the circuit design and optimize the number of routing layers.

The long interconnect wires used in MCM design are subject to delay, crosstalk, and skin effect, all of which must be considered to meet the specification of timing constraint. In particular, in MCM-D, skin effects of the interconnect become more severe. The parasitic effects also degrade the performance if not accounted for in MCM routing. Hence MCM routing algorithms have to consider these factors during the MCM routing phase. Also, crosstalk is another

TABLE 8.1. Factors To Be Considered in MCM Physical Design

Factors	Partitioning	Placement	Routing
Delay	Critical	Critical	Critical
Noise	Not critical	Not critical	Critical
Thermal effects	Critical	Critical	Not critical
Yield	Critical	Critical	Critical
Size	Critical	Critical	Critical

factor that needs to be considered by the MCM routing algorithms. Crosstalk may lead to unnecessary logic transitions resulting in unreliable systems.

The MCM routing algorithms need to optimize the number of routing layers used. Number of routing layers used is not a constraint for thick film MCMs. Additional layers are added to complete the routing. The number of routing layers is fixed in thin film MCMs. If thin film MCMs are used, the number of layers used is limited and, hence, becomes a critical factor which must be considered by the routing algorithms.

The layout size and the number of nets to be routed in an MCM is much higher than that in the VLSI domain. Hence, the routing phase requires an enormous amount of memory and computing resources to complete the interconnections on the MCM. This phase is discussed in detail in Section 8.4

Table 8.1 summarizes critical factors that must be considered during different steps of the MCM physical design so as to guarantee performance. There may exist several ways to overcome these factors. For example, one of the probable ways to reduce the effect of crosstalk is to ensure a minimum separability constraint for the critical nets. Let us discuss each phase of the MCM design cycle in detail.

8.2 MCM PARTITIONING

The first step in designing any system is dividing it into smaller components or subsystems such that each subsystem can be designed efficiently. This reduces the complexity involved in the design of the system. After the division, each subsystem can be designed independently and simultaneously to speed up the design process. A system must be divided carefully so that the original functionality of the system is maintained. During the division, an interface specification is generated that is used to connect all the subsystems. Usually, the basic aim of system division is to minimize the interface interconnections between any two subsystems while satisfying a host of constraints. At the same time, it helps to reduce a complex problem into much smaller problems that can be solved efficiently. The process should be simple and efficient so that the time required is a small fraction of the total design time. The process of division is called *partitioning*.

Partitioning plays a key role in the design of a computer system in general and an MCM in particular. A computer system consists of tens of millions of transistors. To simplify the design process, it is partitioned into several PCBs. The partitioning at this level is based on functionality. Each PCB is then divided into a set of MCMs (modules). The circuit assigned to an MCM may still be too large to be fabricated as a single chip. Hence, each module/ MCM is further partitioned into a set of subcircuits such that each subcircuit can be fabricated as a chip. Partitioning at the module level is called *MCM partitioning.* The partitioning process of a system into PCBs, PCBs into MCMs and MCMs into a set of chips is physical in nature. That is, partitioning is required by the limitations of the fabrication process. The MCM partitioning depends on the design style. There are two design styles used in MCMs.

1. **Chip Array:** If a gate array based design style is used, the MCM partitioning problem is similar to the gate array partitioning except that each "gate" refers to a chip. We refer to this as the *chip array* approach.
2. **Full Custom:** In this style, each chip can have a different size, and the MCM partitioning problem is analogous to the full custom partitioning problem of a chip.

Several CAD tools for partitioning at the chip level are available and have been in existence for some time. These tools have been assisting the designers in partitioning specifications onto multiple chips while satisfying a host of constraints [123] such as chip areas, pin counts, system performance, system delay, and so on. The SPARTA (System PARTitioning Aid) [170] tool developed at GTE does not generate partitionings, but evaluates users's partitionings. SPARTA estimates area, power dissipation, and number of I/O.

In this section, we discuss the basic concepts of partitioning, connection topologies, and some of the known algorithms for MCMs.

8.2.1 Basic Concepts

Partitioning is done at every level of the system. The levels of partitioning can be categorized as system level, board level, module level, and chip level. In this section, we discuss the basic types of partitioning and connection topologies.

8.2.1.1 Types of Partitioning Hierarchically, a system can be partitioned into different levels. This is analogous to the well known algorithm technique of *divide and conquer*. The objectives of partitioning at different levels are discussed below.

1. **System Level:** The partitioning of a system into a group of boards is called *system level* partitioning. At the system level, the given circuit is partitioned into many subcircuits such that each subcircuit can be designed as a PCB.

2. **Board Level:** Partitioning the circuit assigned to a board into a set of MCMs is referred to as *board level* partitioning. The objective of partitioning at this level is to reduce the number of MCMs and to satisfy thermal and timing constraints.
3. **Module Level Partitioning:** The partitioning of an MCM into a set of chips is called *module level* partitioning. We refer to module level partitioning as MCM partitioning. MCM partitioning is characterized by high performance and high density design. The constraints and objectives of module level partitioning differ from those at the board level. Instead of satisfying the area and the terminal constraints of each partition, as in the case of board level partitioning, the objective is to satisfy timing and thermal constraints and to minimize the area of each chip. In addition, the number of chips used for each MCM must be minimized to increase the reliability. Since the off-chip delay is critical at the module level, minimization of the number of chips also helps in enhancing the performance of a module.
4. **Chip Level Partitioning:** The partitioning of a chip into smaller subcircuits is called *chip level* partitioning. Partitioning at this level is not physically necessary since the circuit assigned to a chip can be fabricated as a single unit. However, it is usually done to manage complexity.

8.2.1.2 Connection Topologies *Connection topology* is defined as the physical orientation of one partition to another [178]. This criterion is normally not considered while partitioning the system as a whole based on functionality to different modules or chips. There are two topologies that can be applied to inter module of interboard connections:

1. **Plane-in-plane:** Pin grid array (PGA) single chip packages attached to boards represent plant-in-plane topologies. Figure 8.2b shows a plane-in-plane topology.
2. **Edge-to-plane:** Edge connector technologies represent folded or double folded edge-to-plane topologies, that is, a DIP attached to a board is a folded edge-to-plane topology, while a quad package attached to a board is a double folded edge in plane topology. Figure 8.2a shows an edge-to-plane topology.

The connection topologies determine the number of interconnect crossovers. A crossover occurs when, in a selected plane, one interconnection between partitions crosses over to another one. Crossovers tend to penalize the system size, cost, reliability, and electrical performance. The physical size of a system and the cost associated with it grow linearly while the number of crossovers increases quadratically. The number of connection crossovers can be minimized by appropriately partitioning the problem. Parallel bussed connections result in large numbers of crossovers if they are completely implemented within a plane. Bussed

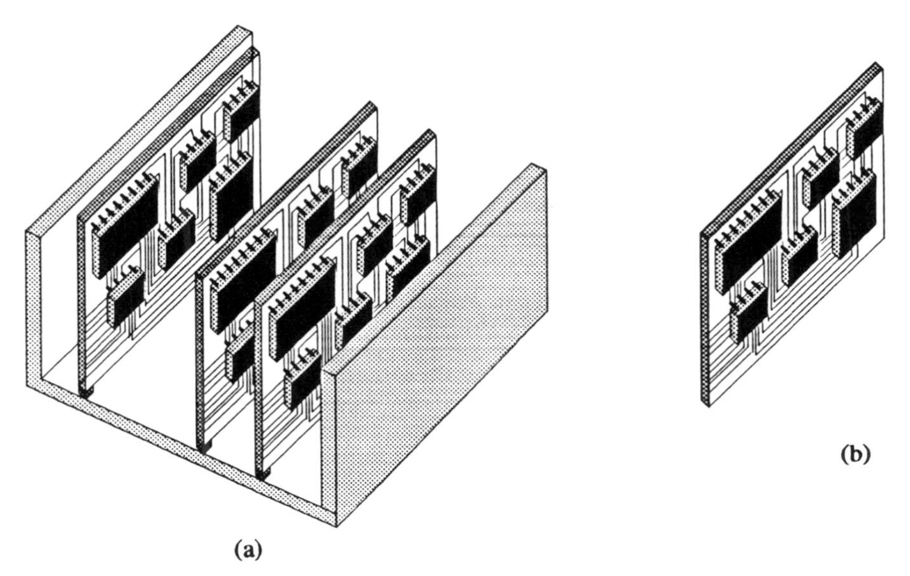

FIG. 8.2. Different types of topologies: **(a)** edge-to-plane, **(b)** plane-in-plane.

connection are those that have equivalent access points in the network. However, if an edge-to-plane topology is used, all the crossovers can be eliminated. For pseudorandom connection distributions, it is not possible to eliminate the crossovers by selecting the correct topology. However, it can be minimized.

8.2.2 Problem Formulation

As discussed above, several factors have to be considered during the MCM partitioning phase. The constraints and the objective functions for the partitioning algorithms vary with different design styles used. This makes it very difficult to state a general partitioning problem that is applicable to all design styles. As a result, in this section we list all the constraints and the objective functions that are applicable. Partitioning problems concern one or more of the following parameters.

1. **Interconnections Between Partitions:** The number of interconnections between partitions has to be minimized. Reducing the interconnections not only reduces the delay but also reduces the interface between the partitions, making it easier for independent design and fabrication.
2. **Delay:** The partitioning of a circuit might cause a critical path to go in between partitions a number of times. As the delay between partitions is significantly larger than the delay within a partition, this is an important factor that has to be considered while partitioning MCMs.

3. **Number of Terminals:** Partitioning algorithms must partition the circuit so that the number of nets required to connect a subcircuit to other subcircuits does not exceed the terminal count of the subcircuit.
4. **Thermal Effects:** The total heat generated by a partition must be less than or equal to the thermal capacity of that partition.
5. **Number of Partitions:** The number of partitions is a constraint at all levels of partitioning. A large number of partitions may reduce the design complexity of individual partitions. However, this may also increase the cost of fabrication and the number of interconnections between the partitions, resulting in violating a host of performance constraints. On the other hand, if the number of partitions is small, it may not be possible to handle the design efficiently. As a result, it is necessary to have an optimum number of partitions.

The MCM partitioning problem requires that the area constraint, interconnection constraint, and constraint on the number of partitions be satisfied. In addition, any performance driven partitioning problem must consider the timing and thermal constraints. The constraints and objective functions for the partitioning algorithms depend on the design style used. For example, while using chip array based designs, algorithms used for the gate array style designs can be extended by making some modifications. Similarly, full custom partitioning algorithms can be extended for the design of full custom MCMs.

8.2.3 Comparison of IC and MCM Partitionings

Partitionings of ICs and MCMs are performed at the chip level and the module level, respectively. Partitioning at the module level (MCM) is a physical process of partitioning the circuit assigned to the module into a set of subcircuits such that each subcircuit can be mapped to an individual chip. Partitioning at the chip level is a logical partitioning that is performed to reduce the complexity involved in the design of a chip. In this section, we list several differences between IC and MCM partitionings. Table 8.2 summarizes the differences.

1. **Timing:** Timing is a critical factor in both MCMs and high speed large area chips. In the case of MCMs, the interconnects behave like a transmission line.

TABLE 8.2. Comparison of IC and MCM Partitionings

Factors	IC	MCM
Timing	Not critical	Critical
Thermal effects	Not critical	Critical
I/O	Not critical	Critical
Area	Minimize	Minimize

2. **Thermal Effects:** Heat generated from a module (MCM) can be removed in two different ways: from each chip individually or simultaneously from the whole module. If the former method is used, then the partitioning algorithms must consider that the amount of circuit being assigned to a chip should be such that the heat generated by the chip is less than or equal to the heat that can be removed from the chip based on the cooling method. In the latter case, partitioning plays a supporting role along with the placement. As a result, the total heat that can be dissipated from the substrate needs to be considered. On the other hand, the thermal effect is not critical in the design of ICs.
3. **I/O:** In MCMs, the number of I/O terminals is based on the type of MCM used. For example, in the case of periphery terminal MCMs, terminals are only located at the periphery of the dies. However, in the case of area terminal MCMs, terminals are distributed over its area and, hence, are more in number. In ICs, the number of terminals per block is determined based on the type of routing strategy used. For example, if channel routing algorithms are used, then the number of terminals is dependent on the aspect ratio of the blocks. However, if over the block or over the cell routing schemes are used, then the number of terminals per block increases.
4. **Area:** In the case of MCM partitioning, it is important to minimize the area based on the fabrication constraints. At the chip level, the size of each partition is not so important as long as the partitions are balanced. Partitioning at the chip level is based on functionality.

As can be seen from the above discussion, while there are many common concerns between IC and MCM partitioning there are certain unique aspects to MCMs. In particular, the timing and the thermal constraints are critical and must be considered during the MCM partitioning phase. As a result, the IC partitioning algorithms cannot be used directly in MCM partitioning.

8.2.4 Basic Algorithms

Many MCM partitioning methods are based on the basic partitioning algorithms. However, it is necessary to modify the existing IC and PCB partitioning algorithms so as to handle the thermal and timing constraints before being used for MCM partitioning. To extend the IC and PCB partitioning algorithms to handle MCM partitioning, it is desirable to understand the basic partitioning algorithms in detail. There are two ways in which the partitioning algorithms can be classified based on the process used [184].

1. **Group Migration Algorithms:** The *group migration* algorithms belong to a class of iterative improvement algorithms. These algorithms start with some initial partitions, formed by using a constructive algorithm. Local changes are then applied to the partitions to reduce the cut size. This process is repeated until no further improvement is possible. Kernighan and Lin [110] proposed a graph

bisectioning algorithm for a graph that starts with a random initial partition and then uses pairwise swapping of vertices between partitions until no improvement is possible. Schweikert and Kernighan [181] proposed the use of a net model so that the algorithm can handle hypergraphs. Fiduccia and Mattheyses [66] reduced time complexity of the Kernighan and Lin algorithm to $O(t)$, where t is the number of terminals. An algorithm using a vertex-replication technique to reduce the number of nets that cross the partitions was represented by Kring and Newton [119]. Goldberg and Burstein [73] proposed a ratio-cut model in which the sizes of the partitions do not need to be specified. The algorithms based on group migration are used extensively in partitioning VLSI circuits and by making suitable modifications can be used in the partitioning of MCMs. Some of these algorithms are discussed in detail in the following sections.

2. **Simulated Annealing and Evolution Based Algorithms:** The *simulated annealing/evolution* [35, 78, 113, 173] algorithms carry out the partitioning process by using a cost function, which classifies any feasible solution, and a set of moves, which allows movement from solution to solution. Unlike deterministic algorithms, these algorithms accept moves that may adversely affect the solution. The algorithm starts with a random solution, and, as it progresses, the proportion of adverse moves decreases. These degenerate moves act as a safeguard against entrapment in local minima. These algorithms are computationally intensive compared with group migration and other methods. *Simulated evolution* is a class of iterative probabilistic methods for combinatorial optimization that exploits an analogy between biological evolution and combinatorial optimization.

Partitioning is a computationally hard problem. As a result, several heuristics have been proposed. The group migration and simulated annealing/evolution methods for partitioning are the most widely used. Extensive research has been carried out in these two methods. We now present some of these well known algorithms.

8.2.4.1 The Kernighan–Lin Algorithm

The Kernighan–Lin algorithm is a heuristic bisectioning algorithm. It starts by initially partitioning the graph $G = (V, E)$ into two subsets of equal size. Vertex pairs are exchanged across the bisection if the exchange improves the cutsize. The procedure is carried out iteratively until no further improvement can be achieved.

Figure 8.24 (see below) presents the formal description of the Kernighan–Lin algorithm. The procedure INITIALIZE finds initial bisections and initializes the parameters in the algorithm. The procedure IMPROVE tests if any improvement has been made during the last iteration, while the procedure UNLOCK checks if any vertex is unlocked. Each vertex has a status of either *locked* or *unlocked*. Only those vertices whose status is *unlocked* are candidates for the next tentative exchanges. The procedure TENT-EXCHGE tentatively exchanges a pair of vertices. The procedure LOCK locks the vertex pair, while the procedure LOG stores the log table. The procedure ACTUAL-EXCHGE determines the maximum partial sum of $g(i)$, selects the vertex pairs to be exchanged, and fulfills the actual exchange of these vertex pairs.

The time complexity of the Kernighan–Lin algorithm is $O(n^3)$, where n is the number of vertices. The Kernighan–Lin algorithm is, however, quite robust. It can accommodate additional constraints, such as requiring a group of vertices to be in a specified partition. This feature is very important in layout because some blocks of the circuit are to be kept together due to functionality. For example, it is important to keep all components of an adder together. However, there are several disadvantages of this algorithm. For example, it is not applicable for hypergraphs, it cannot handle arbitrarily weighted graphs, and the partition sizes have to be specified before partitioning. Finally, the complexity of the algorithm is considered too high even for moderate sized problems.

Fiduccia and Mattheyses [66] developed a modified version of the Kernighan–Lin algorithm. The first modification is that only a single vertex is moved across the cut in a single move. This permits the handling of unbalanced partitions and non-uniform vertex weights. The other modification is the extension of the concept of cutsize to hypergraphs. Finally, the vertices to be moved across the cut are selected in such a way that the algorithm runs much faster. A's in the Kernighan–Lin algorithm, a vertex is locked when it is tentatively moved. When no moves are possible, only those moves that give the best cutsize are actually carried out.

The Fiduccia–Matheyses algorithm is much faster than the Kernighan–Lin algorithm. A significant weakness of the Fiduccia-Matheyses algorithm is that the gain models the effect of a vertex move upon the size of the net cutsize, but not upon the gain of the neighboring vertices. Thus, the gain does not differentiate between moves that may increase the probability of finding a better partition by improving the gains of other vertices and moves that reduce the gains of neighboring vertices. Krishnamurthy [120] has proposed an extension to the Fiduccia–Matheyses algorithm that accounts for high-order gains to get better results and a lower dependence on the initial partition.

8.2.4.2 Simulated Annealing *Simulated annealing* is a special class of randomized local search algorithms. The optimization of a circuit partitioning with a very large number of components is analogous to the process of annealing, in which a material is melted and cooled down so that it will crystallize into a highly ordered state. The energy within the material corresponds to the partitioning score. In an annealing process, the solid-state material is heated to a high temperature until it reaches an amorphous liquid state. It is then cooled very slowly according to a specific schedule. If the initial temperature is high enough to ensure a sufficiently random state, and if the cooling is slow enough to ensure that thermal equilibrium is reached at each temperature, then the atoms will arrange themselves in a pattern that closely resembles the global energy minimum of the perfect crystal.

Early work on simulated annealing used the Metropolis algorithm [146]. Since then, much work has been done in this field [35, 78, 113, 173]. The simulated annealing process starts with a random initial partitioning. An altered

```
Algorithm SA(G)

Input:   Graph, G = (V, E)
Output:  Graph decomposed into two partitions with minimum number
         of crossings.
begin
        t = t0;
        cur_part = ini_part;
        cur_score = SCORE(cur_part);
        repeat
            repeat
                comp1 = SELECT(part1);
                comp2 = SELECT(part2);
                trial_part = EXCHANGE(comp1, comp2, cur_part);
                trial_score = SCORE(trial_part);
                δs = trial_score − cur_score;
                if (δs < 0) then
                    cur_score = trial_score;
                    cur_part = MOVE(comp1, comp2);
                else
                    r = RANDOM(0, 1);
                    if (r < e^(−δs/t)) then
                        cur_score = trial_score;
                        cur_part = MOVE(comp1, comp2 );
            until (equilibrium at t is reached)
            t = αt     (* 0 < α <1 *)
        until (freezing point is reached)
end.
```

FIG. 8.3. Algorithm *SA*.

partitioning is generated by exchanging some elements between partitions. The resulting change in score, δs, is calculated. If $\delta s < 0$ (representing lower energy), then the move is accepted. If $\delta s \geq 0$, then the move is accepted with probability $e^{-\delta s/t}$. The probability of accepting an increased score decreases with the increase in temperature t. This allows the simulated annealing algorithm to climb out of a local optimum in search for a global minimum. This idea is presented as a formal algorithm in Figure 8.3.

The SELECT function is used to select two random components, one from each partition. These components are considered for exchange between the two partitions. The EXCHANGE function is used to generate a trial partitioning and does not actually move the components. The SCORE function calculates the cost for the new partitioning generated. If the cost is reduced, this move is accepted and the components are actually moved using the MOVE function. The cost evaluated by the SCORE function can be either the cutsize or a combination of cutsize and other factors that need to be optimized. If the cost is greater than the cost for the partitioning before the component was considered for the move, the probability to accept this move is calculated using the RANDOM function. If the move is accepted, the MOVE function is used actually to move the components in between the partitions.

Simulated annealing is an important algorithm in the class of iterative, probabilistic algorithms. The quality of the solution generated by the simulated annealing algorithm depends on the initial value of temperature used and the cooling schedule. Temperature decrement, defined above as αt, is a geometric progression where α is typically 0.95. Performance can be improved by using the temperature decrement function, $t = te^{-0.7t}$. However, initial temperature and cooling schedules are parameters that are experimentally determined. The higher the initial temperature and the slower the cooling schedule, the better the result, but the time required to generate this solution is proportional to the steps in which the temperature is decreased.

8.2.5 Partitioning Algorithms for MCMs

MCM partitioning is defined as an optimum mapping of the design to a set of chips. However, as the performance considerations enter the design, the MCM partitioning process must consider other constraints as well. Thus, for high performance system designs, MCM partitioning can be defined as a partition of the design to a set of chips that minimizes the interchip wire crossings subject to timing, area, thermal, and I/O pin count constraints.

Therefore, any performance driven partitioning algorithm may be applied by taking into consideration the thermal constraints. For the gate array based approach, where each chip slot is of equal size, any gate array partitioning algorithm may be applied with appropriate modifications. Similarly, any high performance full custom partitioning algorithm may be applied for the generalized full custom based MCMs.

Few results have been reported for MCM partitioning problems. Shin et al. [185] present an integrated partitioning and placement algorithm. This algorithm is discussed in Section 8.3. Saucier et al. [179] present a new approach for circuit partitioning with a constraint on the number of I/O pins and containment of the critical path within the partitions. This method is based on the use of cone structures. Shih et al. [185] also present an integer programming approach for the MCM system partitioning under timing and capacity constraints. Beardslee and Sangiovanni-Vincentelli [13] present a new approach for improving a topological partition of a logic circuit. We now discuss the different MCM partitioning algorithms in detail.

8.2.5.1 *Functional Partitioning With Cone Structures* Saucier et al. [179] present a new approach for circuit partitioning with a constraint on number of I/O pins and containment of the critical path within the partitions. This method is based on the use of cone structures. Cone structures are the minimum cut partitioning structures for netlists with high fanout. A cone structure can be defined as a block containing all the nodes between an output and all the inputs that lead to this single output. The method of cone partitioning selects and merges combinational groups within cones that exist because of low fanouts. From the definition of cone structures, it is relatively easy to include the critical path within a cone. As cones refer to partitions, this approach can be adapted to contain the critical paths within the partitions to a maximum. Furthermore, the definition of a cone can also be viewed as functional, and each cone can be mapped to a functional unit.

In the first step of the algorithm, a root list *ROOT* of combinational node children for all primary output nodes and sequential element nodes is created. This step determines the number of cones in the circuit that is equivalent to the size or the number of elements in the *ROOT*. In the second step, the combinational nodes of the circuit from each node of the *ROOT* list are scanned to their primary inputs. While scanning, a label number referring the cone number from the *ROOT* list is added to each combinational node. Hence, each combinational node has a label list (*LL*).

$$LL(Node) = \bigcup_{i=1}^{\#P(Node)} LL\left[P(Node)_i\right]$$

where $\#P(Node)$ is the number of parents $P(Node)_i$ of *Node*. If there does not exist a parent with an equal label list, then combinational node *NODE* is added to the *ROOT* list. These new *ROOT* list entries have label sizes greater than 1. The new size of the list *ROOT* is the number of clusters in the circuit. In the third step, the overlapping cone sections are clustered by grouping combinational nodes with common label lists. If the cluster size of a circuit is greater than the number of cones, then this indicates high cone overlap and fanout. In the fourth step, sequential elements are added to the clusters to reduce the size and I/O ratio. In the fifth step, clusters are merged that have combinational elements on the same critical path. If the clusters cannot be merged due to constraints, then replicate sequential elements of the critical path that exist between the two clusters end. Finally, two clusters *i* and *j* are merged if the SIZE and I/O constraints are not violated. The time complexity of this algorithm is $O(nN)$, where n is the number of cones and N is the number of nodes.

This algorithm is illustrated on a combinational network in Figure 8.4a. This network has a total node size of 12 and an I/O of 8. Figure 8.4b shows the nodes scanned of each cone during the labeling step. Figure 8.5a shows the overlapping cone sections (shaded regions). In Figure 8.5b, all combinational gate nodes with the same set of labels are assigned to the same cluster. The process of merging can begin now with the largest cost cluster and evaluate it with all the other clusters. The cluster cost is defined as the ratio of the number of nodes in a cluster to the

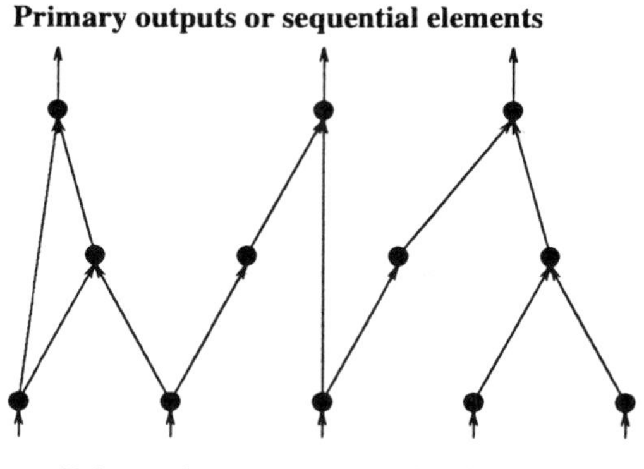

(a)

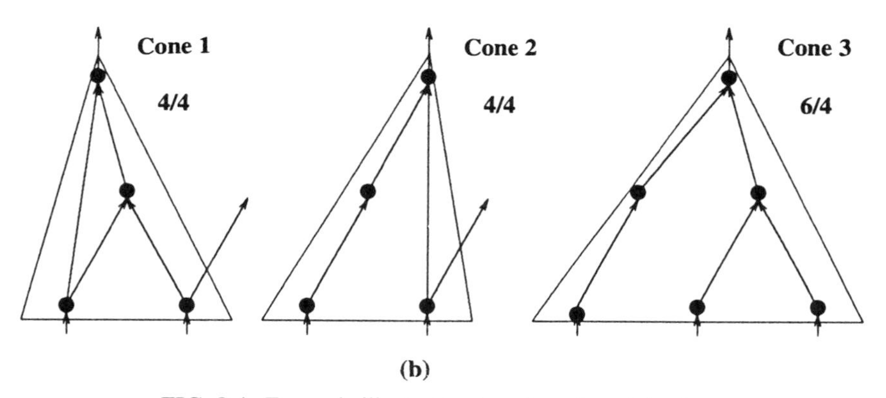

(b)

FIG. 8.4. Example illustrating functional partitioning.

sum of inputs and outputs of the cluster. In the example, cluster 5 is chosen, as it has the largest cluster cost. Next, the clusters are merged based on the merge costs. Merge costs can be defined as the sum of the largest cost cluster and the cluster that will produce a new cluster with the largest cost. For our example, the merge cost of clusters 1 and 5 is the maximum of 8/7. Clusters 1 and 5 are merged if this meets the package constraint. The partitions are thus filled sequentially. The algorithm stops when no further clusters can be merged.

The worst case time complexity for clustering is the time to perform clustering and is equal to $O(nN)$, where n is the number of cones and N is the number of combinational nodes. The worst case labeling occurs when each cone overlaps all other cones completely except for one node at the cone root. However, even when this worst case labeling exists, cluster merging is the most expensive step. The proposed algorithm has been run on real industrial benchmarks, and good partitioning solutions were obtained. Detailed experimental results and evaluations can be found in Savcier et al. [179].

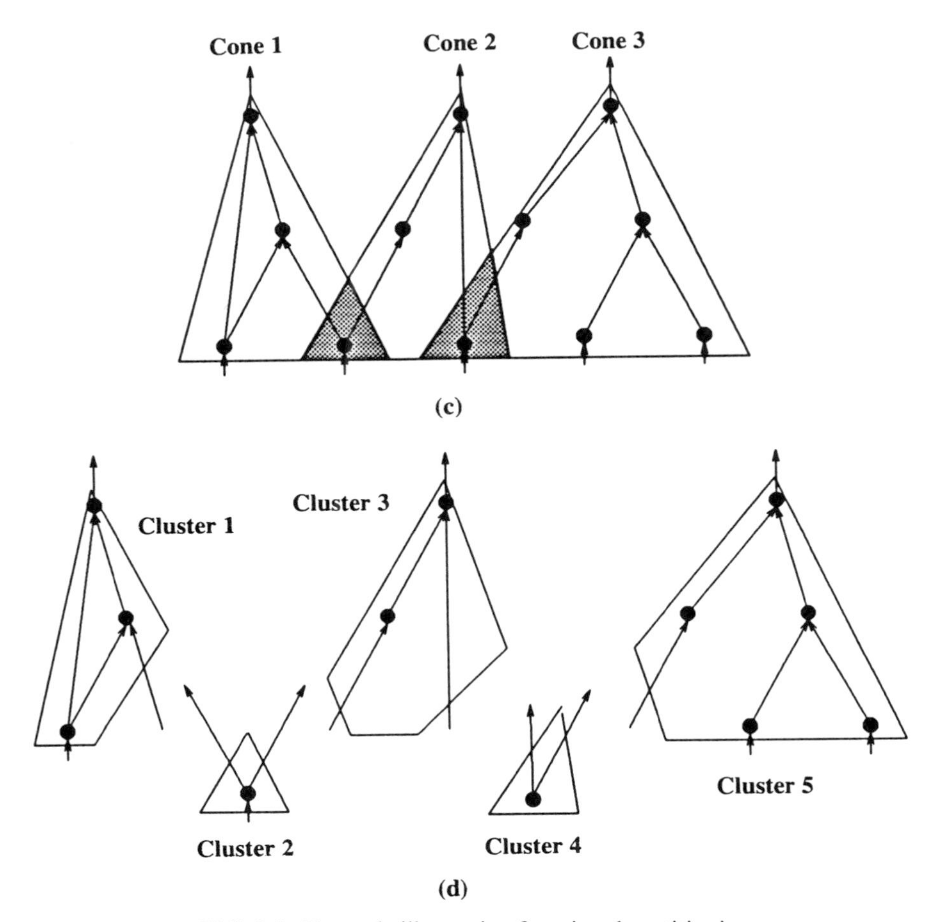

FIG. 8.5. Example illustrating functional partitioning.

8.2.5.2 Partitioning Using Integer Programming

Shih et al. [185] have presented an integer programming approach for the MCM system partitioning satisfying the timing and capacity constraints. This approach uses the fact that the number of elements in a chip interconnect are small and, hence, an exhaustive search technique such as integer programming is feasible. In addition, the authors also propose a new approach known as constraints decoupling, which divides the dual-constraint problems into two independent single constraint problems. In this section, we briefly discuss the problem formulation and integer programming formulation.

Given I set of M partitions, J, a set of N modules; s_j, the size of module j; c_i, the capacity of partition i; P, an $M \times N$ matrix where p_{ij} is the cost of assigning module j to partition i; D_p, an $M \times M$ matrix where $D_p(i_1, i_2)$ is the routing distance between partitions i_1 and i_2; and D_c, an $N \times N$ matrix where $D_c(j_1, j_2)$ is the maximum routing distance allowed between module j_1 and j_2: Find and map F: $J \rightarrow I$ such that the following constraints ($G1$, $G2$) are satisfied:

$G1$: *Capacity*

$$\sum_{\forall j,\, F(j)=i} s_i \le c_i \quad \forall i \in I$$

$G2$: *Timing*

$$D_p\left[F(j_1),\ F(j_2)\right] \le D_C(j_1, j_2), \quad \forall j_1, j_2 \in J$$

The objective is to

$$\textit{minimize} \sum_{\forall i,\, j,\, F(j)=i} p_{ij}, \qquad \textit{subject to } G1 \textit{ and } G2.$$

The above problem formulation is transformed into an *integer programming problem* by defining an $M \times N$ matrix X of binary variables, which defines the mapping. $x_{ij} = 1$ if module J is assigned to partition i; otherwise, it is set to zero. The matrix of binary variables introduced to define mapping is also defined as a constraint. The integer programming problem is defined as follows;

$G1$: *Capacity*

$$\sum_{j=1}^{N} s_j x_{ij} \le c_i \quad \forall i \in I$$

$G2$: *Timing*

$$D_p(i_1, i_2) \le D_C(j_1, j_2), \quad \forall\ x_{i1j1} = 1 = x_{i2j2}$$

$G3$:

$$\sum_{i=1}^{M} x_{ij} = 1, \quad \forall j \in J$$

The objective is to

$$\textit{minimize} \sum_{i=1}^{M} \sum_{j=1}^{N} p_{ij} x_{ij}, \qquad \textit{subject to } G1,\ G2,\ G3.$$

As stated above, the algorithm is exhaustive in nature. However, as the input size is relatively small, the approach is feasible. Experimental results show a converging pattern while satisfying the timing constraints. Detailed results can be found in Shih et al. [185].

8.2.5.3 Improving Predetermined Partitions Beardslee and Sangiovanni-Vincentelli [13] have presented a new approach for improving a topological partition of a logic circuit. Their attempt is to reduce the number of I/O pins necessary for communication between the modules and in turn reduce the size of

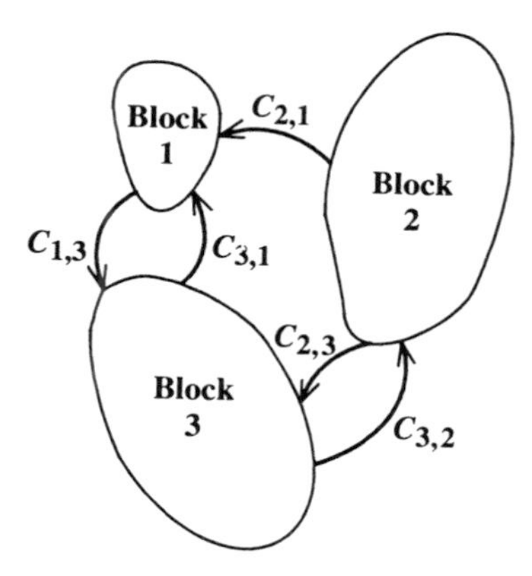

FIG. 8.6. Network partitioning.

necessary for communication between the modules and in turn reduce the size of the modules or chips. Interpartition communication is organized into many unidirectional channels connecting the blocks. The number of lines necessary for implementing a communication channel is reduced by minimizing the amount of information that the channel must transmit and by encoding the information that is transmitted. The reduction in the channel size is usually accompanied by an increase in the amount of logic in the partition modules. Hence, this approach is appropriate for designs that are *pin-limited.*

This approach first finds a topological partition derived from one of the traditional methods and then improves the given partition by optimizing the communication among the modules. Figure 8.6 shows three blocks and five communication channels that have been obtained as a result of the topological partitioning. The channels are denoted by $C_{1,3}$, $C_{2,1}$, $C_{2,3}$, $C_{3,1}$, and $C_{3,2}$. Each channel is a collection of nets that have driving gates in one block and receiving gates in a destination block. Each channel transmits information unidirectionally from the driving block to the receiving block. The proposed methodology has been applied to digital networks for minimizing the number of communication channels.

8.3 MCM PLACEMENT

After partitioning, individual chips are fabricated and are available as bare dies. These bare dies are ready for placement. As a result, the input to the placement problem is a set of bare dies and their interconnection netlist. The placement problem in MCM is to assign exact locations to chip sites on the substrates subject to constraints. MCM placement is also dependent on the type of design style. For example, the objective of placement algorithms or chip array based designs is to map a chip to a suitable site on the MCM. On the other hand, for full custom designs, the objective of placement algorithms is to find a suitable location on the substrate while minimizing the substrate area. In addition, the MCM placement has to meet the timing, thermal, and routability constraints. If these constraints are not satisfied, then it may be impossible to complete the design.

The thermal and timing considerations in the MCM placement problem make it significantly different than the IC placement problem. With the increase in the density of the individual chips, the thermal requirements have also gone up. High speed VLSI chips may generate as much as 40–100 Ws. To ensure proper operation, such a large amount of heat must be dissipated efficiently. The heat dissipation of an MCM depends directly on the thermal management system and on how the chips are placed. In addition, the timing constraints for the design must also be satisfied. These timing constraints are responsible for the proper operation of the module at high frequencies. In this section, we discuss the basic concepts of placement and some of the known placement techniques.

8.3.1 Basic Concepts

Placement of MCMs is a computationally hard problem. Very little work has been done in this area. Most of the placement methods used for MCMs are extensions of well known methods from the VLSI domain or the PCB area. One of the major challenges concerns the thermal effects. In addition, placement also affects routing efficiency. To understand the problems involved in MCM placement and the various methods proposed for MCM placement, it is desirable to learn the basic placement techniques.

8.3.2 Types of Placement

Placement is an important step in the physical design cycle. As mentioned above, a bad placement may use larger areas and, hence, affect the performance. Further more, a bad placement may also lead to inefficient thermal management and routing, thus reducing the performance, reliability, and yield of the system. Placement can be considered at the following levels;

1. **System Level Placement:** The placement problem at the system level is to place all the MCMs together so that the area occupied is minimal. At the same time, the heat generated by each of the MCMs should be dissipated properly so that the reliability of the system increases and the system functions properly.

2. **Board Level Placement:** The objective of board level placement is to optimize the location of MCMs on a PCB such that the timing and thermal constraints are satisfied. It is desirable to place MCMs on a PCB such that the total lengths of the critical paths between the MCMs are minimized. This is essential to reduce the delay associated with the wires laid between the MCMs. Each MCM generates a certain amount of heat. Hence, another objective

function is to position the MCMs such that the heat generated per unit area of the board is uniformly distributed or is generated in a fashion that can be efficiently dissipated by the thermal management scheme.

3. **Module/MCM Level Placement:** At the MCM level, all the chips on a module have to be placed within a fixed area of the module. All chips are fixed and rectangular in shape. In addition, some chips may be pre-placed. There is essentially no restriction on the number of routing layers in thick film MCMs, whereas thin film MCMs do have a restriction on the number of layers. Therefore, generally speaking, no matter how badly the components are placed, the nets can always be routed in case of thick film MCMs, while in the case of thin film MCMs chips must be placed properly so that a suitable area router can be used to complete the routing. The objective of module-level placement algorithms is to minimize the number of routing layers and satisfy system performance requirements. For MCMs, the critical nets should have lengths that area less than a specified value and, hence, the placement algorithms should place the critical components closer together. Another key placement problem is the temperature profile of the module. The heat dissipation on a module should be uniform, that is, the chips that generate maximum heat should not be placed too close to each other.

4. **Chip Level Placement:** At chip level, the problem can be either placement or floorplanning, along with pin assignment. The blocks are either flexible or fixed, and some of them may be preplaced. The key difference between the board level placement problem and the chip level placement is the limited number of layers that can be used for routing in a chip. In addition, the circuit is fabricated only on one side of the substrate. This implies that some "bad" placements maybe unroutable. However, the fact that a given placement is unroutable will not be discovered until routing is attempted. This leads to very costly delays in completion of the design. Therefore, it is very important to determine accurately the routing areas in the chip-level placement problems. Generally, two or three layers are used for routing; however, chips with three layer routings are more expensive to fabricate. The objective of a chip-level placement or floorplanning algorithm is to find a minimum area routable placement of the blocks. In some cases, a mixture of macro blocks and standard cells may have to be placed together.

8.3.3 Problem Formulation

The input to the MCM placement phase is a set of chips or dies and a netlist. As a result, an MCM placement problem can be defined as follows: Given a set of dies or chips and a netlist defining the interconnection pattern among the dies, the MCM placement problem is to construct a layout indicating the position of each die such that the thermal, timing, and routability issues are satisfied. In addition, it is necessary to minimize the total delay by minimizing the lengths of the critical paths.

The MCM placement problem is also dependent on the design style used. For example, for chip array based MCMs, the objective of placement is to map a given set of chips to a predefined set of rectangular areas on the MCM. On the other hand, if the design style used is full custom, then the objective is to place the given set of chips on the MCM substrate such that the overall area used is minimized. In this section, we discuss the MCM placement problem formulation for full custom based designs.

Several factors have to be considered for achieving a high quality placement solution:

1. Layout area
2. Routability
3. Circuit performance
4. Thermal effects
5. Number of layers used

The layout area and the routability of the layout are usually approximated by the topological congestion, known as a *rat's nest*, of interconnecting wires.

Let us now formally state the MCM placement problem for k layers. Let $C_1, C_2, \ldots, C_n$ be the chips to be placed on the MCM. Each C_i, $1 \leq i \leq n$, has associated with it a height h_i and a width w_i. Let $\mathcal{N} = \{N_1, N_2, N_3, \ldots, N_m\}$ be the set of nets representing the interconnection between different chips. Let L_i denote the estimated length of net N_i, $1 \leq i < m$. The placement problem is to find a layout of placing the chips C_i such that the following criteria are satisfied:

1. No two chips or dies overlap, that is, $C_i \cap C_j = \phi$, $1 \leq i, j \leq n$.
2. Placement is routable in k layers.
3. The number of layer (k) is minimized.
4. The total substrate area is minimized.
5. The total wire length is minimized, that is, $\sum_{i=1}^{m} L_i$ is minimized.
6. The length of the longest net, max $\{L_i | \, i = 1, \ldots, m\}$ is minimized. This is a criterion for all high performance systems.
7. The timing constraint is satisfied.
8. The thermal constraints are satisfied.

The actual wiring paths are not known at the time of placement; however, a placement algorithm needs to model the topology of the interconnection nets. An interconnection graph structure that interconnects each net is used for this purpose. The interconnection structure for two terminal trees is simply an edge between the two vertices corresponding to the terminals. To model a net with more than two terminals, rectilinear steiner trees are used as shown in Figure 8.7a to estimate optimal wiring paths for a net. This method is usually not used by routers because of the NP-completeness of steiner tree problems. As a result, minimum spanning tree representations are the most commonly used structures to connect a net in the placement phase. Minimum spanning tree connections

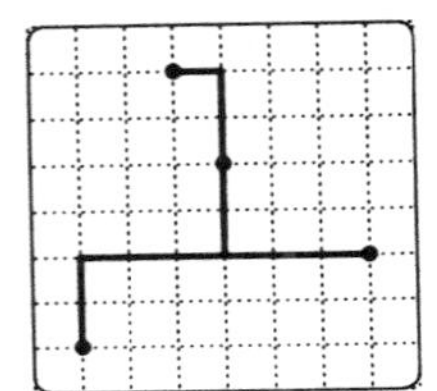

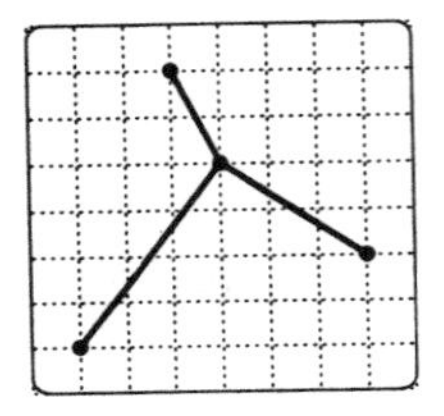

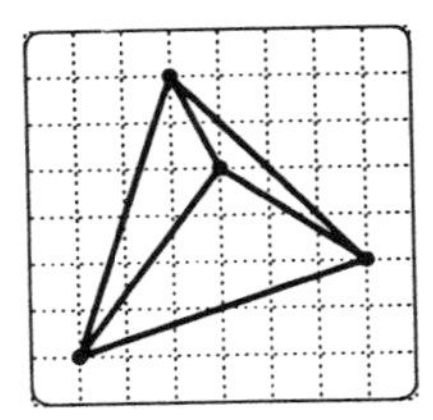

FIG. 8.7. Interconnection topologies: **(a)** Steiner tree (length = 13), **(b)** minimum spanning tree (length = 15), **(c)** complete graph (length = 32).

(shown in Fig. 8.7b) allow branching only at the pin locations. Hence, the pins are connected in the form of a minimum spanning tree graph. A *complete graph* interconnection is shown in Figure 8.7c. It is easy to implement such structures. However, this method causes many redundant interconnections and results in longer wire length.

The large number of objective functions can be classified into two categories; net metrics and congestion metrics. The net metrics deal with the assumption that all the nets can be routed without interfering with other nets or with the components. Usually, the length of a net is important, as the interconnection delays depend on the length of the wire. The net metrics only quantify the amount of wiring and do not account for the actual location of these wires. The examples of this kind of objective functions are the total length of all nets and the length of the longest net. Congestion metrics are used to avoid the buildup of many nets in a particular area, leading to congestion.

8.3.4 Comparison of IC and MCM Placements

As discussed previously, placements of ICs and MCMs are performed at the chip level and module level, respectively. In this section, we list several differences between IC and MCM placement phases of the physical design cycle. Table 8.3 summarizes the differences in MCM and IC placement. The IC placement phase has been discussed in detail [184].

1. **Input Size:** The input to the MCM placement phase is a set of chips, while the input to an IC is a circuit diagram with a set of logical blocks.

TABLE 8.3. Comparison of IC and MCM Placements

Factors	IC	MCM
Input size	Large	Small
Timing	Critical	Critical
Thermal effects	Less critical	Critical
Routability	Critical	Not critical
Area	Minimize	Minimize

2. **Timing:** In MCMs, the transmission line behavior has to be considered in satisfying the timing constraint. However, in the case of ICs, transmission line behavior is less critical and, hence, is not a major factor in satisfying the timing constraint.
3. **Thermal Effects:** Thermal effect is an important consideration during the placement phase depending on the type of cooling method used. For example, if the cooling method used is to dissipate heat globally from the MCM substrate, then thermal effects are critical and have to be considered during MCM placement. As a result, the chips have to be placed such that the heat generated is uniformly distributed over the substrate. On the other hand, thermal effects are far less critical for ICs and, hence, are not considered during the IC placement phase.
4. **Routability:** For thick film MCMs, routability is not a constraint, since routing layers can be added if needed to complete the routing. However, for ICs and thin film MCMs, routability is a constraint, as additional layers cannot be added to complete the routing. As a result, placement algorithms must estimate the area required for routing and then place the blocks in such a manner that there is sufficient area so as to complete the routing using the available layers.
5. **Area:** MCM placement and IC placement are similar for *chip array* and *gate array* based approaches. In the case of *full custom*, the objective for both MCM and IC placement algorithms is to minimize the area of the substrate and the wafer. Hence, minimizing area is critical for both MCMs and ICs.

It can be seen from this discussion that there are some similarities between MCM and IC placements. However, there are significant differences also. As a result, it is a good practice to extend the IC placement algorithms to better suit the MCM placement. In the next section, we present a few of the well known placement algorithms that can be extended to handle MCM placement by making appropriate changes.

8.3.5 Basic Algorithms

The algorithms for placement can be classified in several ways [184]. One such classification is based on the process used by the algorithms. There are two important classes:

1. **Simulation Based Algorithms:** This type of algorithm simulates some natural phenomenon. Simulated annealing is one of the most well developed placement methods available [9, 78, 79, 87, 102, 124, 155, 172, 173, 182, 183]. The simulated annealing technique has been used successfully in many phases of VLSI physical design, for example, circuit partitioning. The detailed description of the application of simulated annealing is used in placement as an iterative improvement algorithm. Given a placement configuration, a change to that configuration is made by moving a component or interchanging locations of two components. In case of the simple pairwise interchange algorithm, it is possible that a configuration achieved has a cost higher than that of the optimum, but no interchange can cause a further cost reduction. In such a situation the algorithm is trapped at a local optimum and cannot proceed further. Actually, this happens quite often when this algorithm is used on real life examples. Simulated annealing avoids getting stuck at a local optimum by occasionally accepting moves that result in a cost increase. There are many problems in the natural world that resemble placement and packaging problems. Molecules and atoms arrange themselves in crystals such that the crystals have a minimum size and no residual strain. Herds of animals move around until each herd has enough space and it can maintain its predator—prey relationships with other animals in other herds. The simulation based placement algorithms simulate such natural processes or phenomena. There are three major algorithms in this class: simulated annealing, simulated evolution, and force directed placement. The simulated annealing algorithm simulates the annealing process that is used to temper metals. Simulated evolution simulates the biological process of evolution, while the force directed placement simulates a system of bodies attached by springs. These algorithms are characterized as simulation based algorithms.
2. **Partition Based Algorithms:** This type of algorithm uses partitioning for generating the placement. This an important class of algorithms in which the given circuit is repeatedly partitioned into two subcircuits. At the same time, at each level of partitioning, the available layout area is partitioned into horizontal and vertical subsections alternately. Each of the subcircuits thus partitioned is assigned to a subsection. This process is carried out until each subcircuit consists of a single gate and has a unique place on the layout area. During partitioning, the number of nets that are cut by the partition is usually minimized. In this case, the group migration method can be used.

In the following sections we discuss some of the well known simulated based algorithms.

8.3.5.1 Simulated Annealing As stated above simulated annealing is one of the most well developed placement technique [9, 78, 79, 87, 102, 124, 155, 172, 173, 182, 183]. In simulated annealing, given an initial placement, moves are made such that all moves that result in a decrease in cost are accepted. Moves that result in an increase in cost are accepted with a probability that decreases over the iterations. In an analogy to the actual annealing process, a parameter called *temperature* T controls the probability of accepting moves that result in an increased cost. Moves that increase cost are accepted more readily at higher values of temperature than at lower values. The acceptance probability is usually expressed by the relation $e^{-\Delta C/T}$, where ΔC is the increase in cost. The algorithm starts with a very high value of temperature that gradually decreases so that moves that increase cost have a lower probability of being accepted. Finally, the temperature reduces to a very low value that causes only moves that reduce cost to be accepted. In this way, the algorithm converges to an optimal or near optimal configuration.

In each stage, the configuration is shuffled randomly to get a new configuration. This random shuffling could be achieved by displacing a block to a random location, an interchange of two blocks, or any other move that can change the wire length. After the shuffle, the change in cost is evaluated. if there is a decrease in cost, the configuration is accepted; otherwise, the new configuration is accepted with a probability that depends on the temperature. The temperature is then lowered using some function that, for example, could be exponential in nature. The process is stopped when the temperature has dropped to a certain level. The simulated annealing algorithm is shown in Figure 8.8.

The parameters and functions used in a simulated annealing algorithm determine the quality of the placement produced. These parameters and functions include the cooling schedule, consisting of initial temperature (*init-temp*), final temperature (*final-temp*), and the function used for changing the temperature (SCHEDULE), *inner-loop-criterion*, which is the number of trials at each temperature, the process used for shuffling a configuration (PERTURB), acceptance probability (F), and the cost function (COST). A good choice of these parameters and functions can result in a good placement in a relatively short time.

Sechen and Sangiovanni-Vincentelli [183] developed Timber Wolf 3.2, which is a standard cell placement algorithm based on simulated annealing. Timber Wolf is one of the most successful placement algorithms. In this algorithm, the parameters and functions are set as follows: For the cooling schedule, *init-temp* = 4,000,000, *final-temp* = 0.1, and SCHEDULE $(T) = \alpha(T) \times T$ where $\alpha(T)$ is a cooling rate depending on the current temperature T. $\alpha(T)$ is relatively low when T is high, for example, $\alpha(T) = 0.8$ when the cooling process starts, which means the temperature is decremented rapidly. Then, in the medium range of temperature, $\alpha(T)$ is increased to 0.95, which means that the temperature changes more slowly. When the temperature is in a low range, $\alpha(T)$ is again reduced to 0.8, thereby speeding up the cooling procedure to go fast again. In total, there are 117 temperature steps. The graph for the cooling schedule is shown in Figure 8.9. The value of *inner-loop-criterion* depends on the size of the

Algorithm *SIMULATED-ANNEALING(G,FINAL-TEMP)*

Input: Graph *G=(V,E) and FINAL-TEMP*
Output: Partitioned Graph with minimum number of crossings.

```
begin
        temp = INIT-TEMP;
        place = INIT-PLACEMENT;
        while (temp > FINAL-TEMP) do
            while (inner_loop_criterion = FALSE) do
                new_place = PERTURB(place);
                ΔC = COST(new_place) - COST(place);
                if (ΔC < 0) then
                    place = new_place;
                else if (RANDOM(0,1) > e^(ΔC/T)) then
                    place = new_place;
            temp = SCHEDULE(temp);
end.
```

FIG. 8.8. The simulated annealing algorithm.

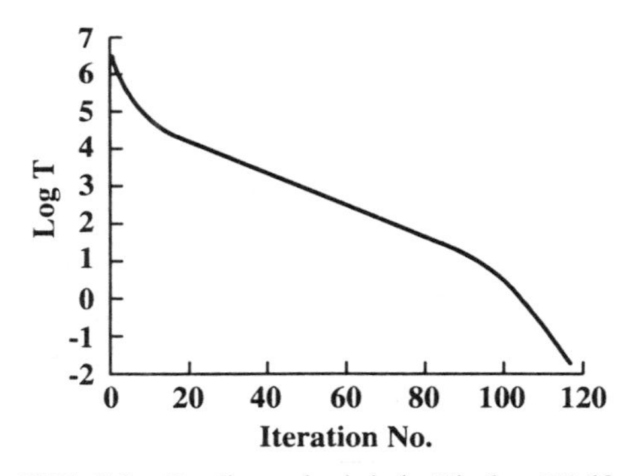

FIG. 8.9. Cooling schedule in Timber Wolf.

circuit, for example 100 moves per cell for a 200 cell circuit and 700 moves per cell for a 3,000 cell circuit are recommended in [183]. The new configuration is generated by making a weighted random selection from one of the following:

1. The displacement of a block to a new location
2. The interchange of locations between two blocks
3. An orientation change for a block

The third alternative is used only when the new configuration generated by using alternative 1 is rejected. The ratio r of single block displacement to pairwise interchange should be carefully chosen to give a best overall result. An orientation change of a block is simply a mirror image of that block's x-coordinate. The cost function depends on the weighted sum of estimate length of all nets, the penalty cost for overlapping, and the penalty cost for uneven length among standard cell rows.

Simulated annealing is one of the most established algorithms for placement problems. It produces good quality placement. However, simulated annealing is computationally expensive and can lead to longer run times. Therefore, it is only suitable for small to medium sized circuits.

8.3.5.2 Force Directed Placement Force directed placement explores the similarities between placement problems and classical mechanics problems of a system of bodies attached to springs. Various results have been reported based on the force directed approach [5, 75, 89, 90, 155, 166, 167].

In this method, the blocks connected to each other by nets are supposed to exert attractive forces on each other. The magnitude of this force is directly proportional to the distance between the blocks. According to Hook's law, the force exerted due to stretching of the springs is proportional to the distance between the bodies connected to the spring. If the bodies were allowed to move freely, they would move in the direction of the force until the system achieved equilibrium. The same idea is used for placing the blocks. The final configuration of the placement of blocks is the one in which the system achieves equilibrium.

8.3.6 Placement Algorithms for MCMs

Chip level placement determines the relative positions of a large number of blocks on an IC as well as organizes the routing area into channels. As opposed to IC placement, MCM placement involves fewer components per MCM compared to the large number of cells per IC. The sizes and shapes of ICs on an MCM are less variable than the general cells within the IC. MCM placement is more complex because many interrelated factors determine layout quality. Wide buses are very prevalent, and propagation delays and an efficient thermal management are much more important. The timing, thermal, and routability considerations in the MCM placement problem render it significantly different and more complex than IC placement.

The timing constraints are important as they determine the proper operation of the MCM at high frequencies. During the MCM partitioning phase the number of times a critical path crosses a partition is minimized so as to reduce the delay. However, if the MCM placement phase locates the chips at positions on the MCM substrate such that the length of the critical path outside the chip is not optimized, then the delay associated with the system increases. Hence, one of the objectives of the MCM placement phase is to minimize the total length of the critical path outside the chip.

High speed chips may generate heat from 40 to 100 Ws. In high performance systems such as MCMs, a large number of high speed bare chips are packed in a small area. This results in the chips being placed close to each other. As a result, a large amount of heat is generated from the MCM substrate. The heat generated has to be dissipated from the substrate. It is desirable to distribute heat uniformly over the substrate so as to increase performance and yield in an MCM. The distribution of heat generated from an MCM substrate depends directly on the relative positions of the chips. Hence, it is very critical for the MCM placement algorithms to consider the thermal constraints. In Chapter 3 we discussed several ways to dissipate heat from an MCM substrate.

Routability is another important consideration in MCMs. In the case of thin film MCMs, the number of layers used for routing is small but the density of nets to be routed is relatively large. As a result, for a given placement it may be impossible to route a circuit especially with a host of constraints as discussed previously. On the other hand, for thick film MCMs, number of layers used for routing is relatively large, and, hence, the cooling method used may cause obstructions for the routing in addition to several other problems. For example, if thermal vias are being used, then the routing algorithm may face blockades while routing critical nets and, hence, increase the delay. Furthermore, thermal vias may affect the completion of routing.

There are a very few results available for MCM placement. Shin et al. [186] presented a performance driven integrated partitioning and placement technique for MCMs. They have considered timing and area constraints in this approach.

In this section, we discuss the chip array based placement, full custom style placement, and the integrated partitioning and placement technique presented by Shin et al. [186].

8.3.6.1 Chip Array Based Approach The MCM placement approach, when the chip sites are symmetric, becomes very similar to the conventional gate array approach. In the chip array approach, the MCM placement problem is the assignment of the chips to predefined chip sites. We may consider the problem as follows: Given a set of chips and a set of slots, assign each chip to one of the slots such that all the constraints specified in Section 8.3.2 (Problem Formulation) are satisfied. The key difference between the IC placement and the MCM placement problem is the type of constraints involved. The two approaches to MCM placement problem have been discussed by La Potin [126] as part of the early design analysis, packaging, and technology trade offs.

8.3.6.2 Full Custom Approach In full custom style, chips of varying sizes and shapes are placed on a substrate. As a result, each chip can be optimally fabricated using the best available technology suited for that chip. This kind of placement style allows optimal fabrication of each chip, reduces cost of fabrication, and improves the overall performance and yield. However, this may lead to chips generating variable amounts of heat and different I/O counts. One of the important features of the full custom style is that it allows the integration of mixed technologies.

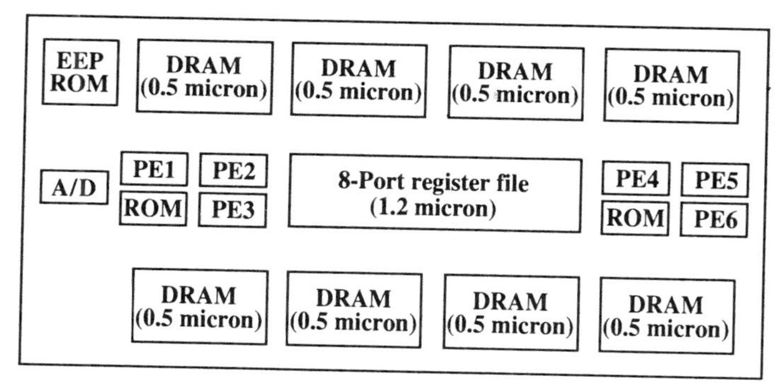

FIG. 8.10. 2.5-D integration placement in MCM.

Figure 8.10 shows an arrangement depicting a concept of the *2.5-D integration scheme* derived from ideas postulated by McDonald [142] and Tewksbury [196] for wafer scale integration. This concept can be viewed as an advanced version of the existing MCMs. It is envisioned that this hypothetical system will respond directly to the cost limitations of VLSI technologies. The system could be assembled on a large-area active substrate. The technology of such a substrate could be optimized for yield, thermal effects, and speed of the interconnect. This substrate could dissipate a large percentage of the total heat generated and could be cost effective if fabricated with relaxed design rules in stepper-free, interconnect-oriented technology. The performance-critical system components could be fabricated separately on fabrication lines oriented toward high volume and high performance. They could be attached to the active substrate with rapidly maturing flip-chip technology. This way only those system elements that really require ULSI technology (e.g., data path) would be fabricated with the most expensive technologies. It is obvious that a placement problem in the 2.5-D integration scheme is that of full custom approach. In addition to the usual area constraints, the placement tool of this type must be able to complete the task of placement, subject to the thermal and timing constraints.

8.3.6.3 Integrated Partitioning and Placement for MCMs Shin et al. [186] presented a performance driven integrated partitioning and placement technique for MCMs. They only considered timing and area constraints. Two different delay models have been considered: (1) constant delay model and (2) linear delay model. For constant delay model, their approach is essentially a partitioning algorithm, which is briefly described below.

Given a system graph, assume that delay time for the signal traveling between a combinational block and a register that are grouped into the same partition is negligible [186]. In addition, the delay time for the signal traveling between a combinational block and a register that are partitioned into different groups is a constant. Each group is called a *super node* and corresponds to a chip.

For each combinational block c_i, the algorithm finds the two registers r_j and r_k that are adjacent to the block in the system graph. The procedure of constructing

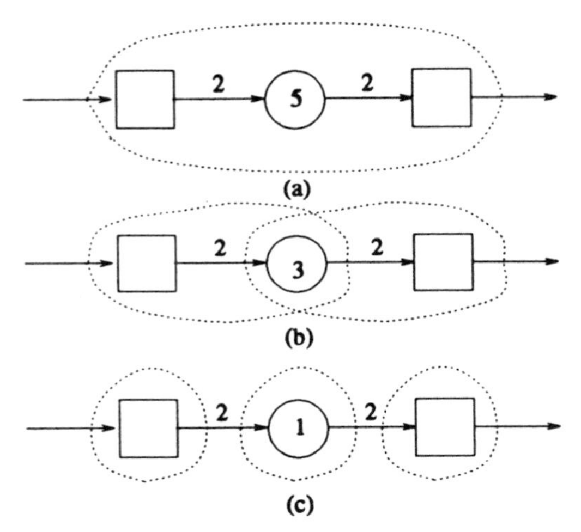

FIG. 8.11. Super node construction.

super nodes is shown by an example in Figure 8.11. In this example, we assume that the system cycle time is $T_{cycle} = 6$, which requires that the maximum delay time between any two registers be less than or equal to 6. The delay time between a combinational block and a register is assumed as $D = 2$. The super nodes are constructed according to the following three cases.

1. **Both Registers Must Be Combined:** In this case, the condition $d_i + D > T_{cycle}$ must be satisfied. Consider the example shown in Figure 8.11a, with $d_i = 5$. If one of r_j and r_k is assigned to a different partition than c_i, the time delay will be at least $2 + 5 = 7$, thus violating the timing constraint. So, all these vertices have to be included in the same super node.
2. **At Least One Register Must Be Combined:** In this case, the conditions $d_i + D < T_{cycle}$ and $d_i + 2 \times D > T_{cycle}$ must be satisfied. Consider the example shown in Figure 8.11b, with $d_i = 3$. If both registers are assigned to different partitions other than c_i, the time delay will be 7, thus violating the timing constraint. In this situation, the super node consists of a combinational block and either one of the registers.
3. **No Registers Need To Be Combined:** In this case, the condition $d_i + 2 \times D < T_{cycle}$ must be satisfied. Consider the example shown in Figure 8.11c, with $d_i = 1$; the register can be assigned to any partitions without violating the timing requirement. Thus, each super node consists of only one vertex.

The algorithm repeats until no nodes can be combined. At this stage, the number of super nodes is equal to the number of chips required in the MCM.

8.4 MCM ROUTING

After the completion of the MCM placement phase, the next phase is to connect the chips specified by the netlist. The main goal of the MCM routing phase is to optimize performance. Hence, the objective of the routing phase is to meet the timing constraints specified by the circuit design and optimize the number of routing layers.

The long interconnect wires used in MCM design are subject to delay, crosstalk, and skin effect, all of which must be considered so as to satisfy the timing constraints set by the circuit design.

Delay due to interconnecting wires must be considered for losses due to transmission. This contributes significantly toward an increase in delay. The effects of transmission line losses are discussed in Chapter 3.

In the design of high speed systems, crosstalk is a primary concern. Excessive crosstalk compromises noise margins, possibly resulting in false receiver switching. The crosstalk between the lines can be minimized by ensuring that no two lines are laid out in parallel or next to each other for longer than the maximum length specified by the circuit design.

In addition to crosstalk, the skin effect is also a major consideration in MCM routing. Skin effect is defined as the characteristic of current distribution in a conductor at high frequencies by virtue of which the current density is greater near the surface of the conductor than in its interior. As the rise time of digital pulses is reduced to the subnanosecond range, the skin effect becomes an important issue in high speed digital systems. As the frequency, conductivity, and permeability of the conductor are increased, the current concentration is also increased. This results in increasing resistance and decreasing internal inductance at frequencies for which this effect is significant. These effects must be taken into account while routing long lines. In particular, in MCM-D, crosstalk, skin, and parasitic effects of the interconnect become more critical.

In MCMs the number of layers used for routing must be optimized because the cost of an MCM is directly proportional to the number of layers used. As a result, the MCM routing phase must also consider this objective function.

In this section we discuss basic concepts in routing such as different graph models and graphs, global routing, detailed routing, some basic and well known routing algorithms, and currently available MCM routing methods.

8.4.1 Basic Concepts

The process of determining the geometric layouts for all the nets is called *routing*. The process of finding these geometric layouts is undertaken in the routing regions. Routing is traditionally divided into two phases. The first phase is the *global routing phase*, and the second one is the *detailed routing phase*.

8.4.1.1 Global Routing *Global routing* refers to generating a "loose" route for each net. It assigns a list of routing regions to each net without specifying the actual geometric layout of wires. The global routing problem can be studied as a graph problem. The routing regions and their relationships and capacities can be modeled as graphs. The primary goal of global routing is to distribute the routing density uniformly over the routing medium. This is particularly useful as the nets are distributed with a global perspective. This enables the detailed router to run fast and efficiently. Global routing can be associated with a set of objective functions such as optimizing performance or minimizing routing area among others. The global routing problem of two terminal nets is to find a path for each net in the routing graph such that the desired objective function is optimized. For a net with more than two terminals, the solution is to construct a steiner tree such that the desired function is optimized.

8.4.1.2 Detailed Routing *Detailed routing* refers to finding the actual geometric layout of each net within the assigned routing regions. Unlike global routing, which considers the entire layout, a detailed routing considers just one region at a time. The exact layout is produced for each wire segment assigned to a region, and vias are inserted to complete the layout. The ordering of the regions is determined by several factors, including the criticality of routing certain nets and the total number of nets passing through a region. The area of the routing region can be determined exactly only after the routing is completed. If this area is different from the area estimated by the placement algorithm, the placement has to be adjusted to account for this difference. Detailed routing includes channel routing and switchbox routing.

8.4.2 Problem Formulation

The MCM routing problem is specified by a set of nets to be routed and by a multilayer routing substrate (*routing space*). The substrate consists of multiple signal routing layers. A three-dimensional grid is superimposed upon the routing layers in an MCM where the spacing between the grid lines is based on the *design rules* for a given technology such that signal wires can only be laid out along the grid lines.

The formal statement of the routing problem is as follows: Given a netlist $\mathcal{N} = \{N_1, N_2, \ldots, N_n\}$, where each $N_i = (t_1, t_2, \ldots, t_k)$, where $t_1, t_2, t_3, \ldots$ and t_k are terminals, and the routing graph $G = (V, E)$, find a three-dimensional steiner tree T_i for each net N_i, $1 \leq i < n$, such that the objective functions listed below are optimized while satisfying timing, capacity, and via constraints.

1. Minimize the total wire length.
   Minimize $\sum_{i=1}^{n} L(T_i)$, where $L(T_i)$ is the length of the steiner tree.
2. Minimize the maximum wire length.
   Minimize $max_{i=1}^{n}\ L(T_i)$, where $L(T_i)$ is the length of the steiner tree.
3. Minimize the number of vias used.
4. Optimize the number of routing layers.

TABLE 8.4. Comparison of IC and MCM Routings

Factors	IC	MCM
Grid size	Small	Large
Memory requirements	Small	Large
Number of layers	Fixed	Not fixed
Delay	Critical	Critical
Xtalk and tran line behavior	Not critical	Critical
Pitch	Small	Large

8.4.3 Comparison of IC and MCM Routings

In this section, we outline the difference in IC and MCM routings. Table 8.4 summarizes these differences.

1. **Grid Size:** The grid size used in MCM routing is very large. As a result, the memory requirements for MCM routing is very high. For example, in MCC2 [112], the grid size is well over 4,000. ICs use relatively smaller grids.
2. **Time Complexity and Space:** Due to the large input size and grid size, the MCM routing algorithms are relatively slow compared with the IC routing algorithms. In addition, the memory requirements for MCM routing is very large in comparison with the IC routing. One of the main reasons for large memory requirements in MCM routing is the large grid size.
3. **Number of Layers:** The number of layers is fixed in ICs and thin film MCMs. Therefore, routing algorithms for ICs and thin film MCMs may use the traditional *HVH* or *VHV* models for routing. For the thick film MCMs, the number of layers used is not a constraint. Additional layers can be added to complete the routing.
4. **Crosstalk and Transmission Line Effects:** Crosstalk and transmission line effects have to be considered while routing in MCMs. This is not very critical in IC routing.
5. **Three Dimensional:** As discussed above, routing in thick film MCMs is a three-dimensional problem. Otherwise, routing is generally carried out in one layer or a pair or adjacent layers at a time in ICs and thin film MCMs. This is because the number of layers is fixed in ICs and thin film MCMs. Several routing techniques are discussed elsewhere that can be used for routing in ICs and thin film MCMs [184].

8.4.4 Basic Routing Algorithms and Techniques

Routing algorithms can be classified into two types:

1. **Sequential Approach:** In the sequential approach, as the name suggests, nets are routed one by one. However, once a net has been routed it may block other nets that are yet to be routed. As a result, this approach is very sensitive to

the order in which the nets are considered for routing. Usually, the nets are sequenced according to their criticality, perimeter of the bounding rectangle, and number of terminals. The criticality of a net is determined by the importance of the net. For example, the clock net may determine the performance of the circuit and, therefore, it is considered to be a very important net. As a result, it is assigned a high criticality number.

The criticality number and other factors can be used to sequence nets. However, sequencing techniques do not solve the net ordering problem satisfactorily. In a practical router, in addition to a net ordering scheme, an important phase is used to remove blockages when further routing of nets is not possible. However, this also may not overcome the shortcoming of the sequential approach. One such improvement phase involves "rip-up and reroute" [27, 51] technique, while another involves a "shove-aside" technique. In "rip-up and reroute," interfering wires are ripped up and re-routed to allow routing of the affected nets. In "shove-aside," wires that will allow completion of failed connections are moved aside without breaking the existing connections. Another approach [50] is first to route simple nets consisting of only two or three terminals since there are few choices for routing such nets. Usually, such nets comprise a large portion of the nets (up to 75%) in a typical design. After the simple nets have been routed, a steiner tree algorithm is used to route intermediate nets. Finally, a maze routing algorithm is used to route the remaining long nets, which are not too numerous.

2. **Concurrent Approach:** This approach avoids the ordering problem by considering routing of all the nets simultaneously. The concurrent approach is computationally hard, and no efficient polynomial algorithms are known even for two-terminal nets. As a result, integer programming methods have been suggested. The corresponding integer program is usually too large to be employed efficiently. Hence, hierarchical methods that work top down are employed to partition the problem into smaller subproblems, which can then be solved by integer programming.

We now discuss the two basic maze routing algorithms.

8.4.4.1 Lee's Algorithm Lee's algorithm is based on the sequential approach. This algorithm, which was developed by Lee [128] in 1961, is the most widely used algorithm for finding a path between any two vertices on a planar rectangular grid. The key to the popularity of Lee's maze router is its simplicity and its guarantee of finding an optimal solution if one exists.

The exploration phase of Lee's algorithm is an improved version of the breadth-first search. The search can be visualized as a wave propagating from the source. The source is labeled "zero," and the wavefront propagates to all the unblocked vertices adjacent to the source. Every unblocked vertex adjacent to the source is marked with a label "1." Then, every unblocked vertex adjacent to vertices with a label "1" is marked with a label "2," and so on. This process continues until the target vertex is reached or no further expansion of the wave can be carried out. An example of the algorithm is shown in Figure 8.12. Due to

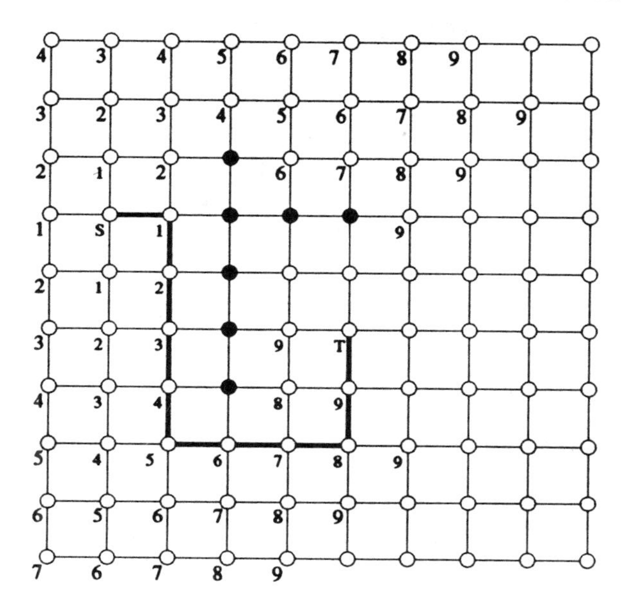

FIG. 8.12. A net routed by Lee's algorithm.

the breadth-first nature of the search, Lee's maze router is guaranteed to find a path between the source and target if one exists. In addition, it is guaranteed to be the shortest path between the vertices.

The input to the Lee's algorithm is an array B, the source (s), and target (t) vertex. $B[v]$ denotes if a vertex v is blocked or unblocked. The algorithm uses an array L, where $L[v]$ denotes the distance from the source to the vertex v. This array will be used in the procedure *RETRACE* that retraces the vertices to form a path P, which is the output of Lee's algorithm. Two linked lists *plist* (propagation list) and *nlist* (neighbor list) are used to keep track of the vertices on the wavefront and their neighbor vertices, respectively. These two lists are always retrieved from tail to head. We also assume that the neighbors of a vertex are visited in counterclockwise order, that is, top, left, bottom, and then right.

The formal description of the algorithm appears in Figure 8.23. The time and space complexity of Lee's algorithm is $O(h \times w)$ for a grid of dimension $h \times w$.

Lee's routing algorithm requires a large amount of storage space, and its performance degrades rapidly when the size of the grid increases. There have been numerous attempts to modify the algorithm to improve its performance and reduce its memory requirements.

Lee's algorithm requires up to $k + 1$ bits per vertex, where k bits are used to label the vertex during the exploration phase and an additional bit is needed to indicate whether the vertex is blocked. For an $h \times w$ grid, $k = \log_2(h \times w)$. Acker [1] noticed that, in the retrace phase of Lee's algorithm, only two types of neighbors of a vertex need to be distinguished; vertices toward the target and vertices toward the source. This information can be coded in a single bit for each vertex.

The vertices in wavefront L are always adjacent to the vertices in wavefront $L-1$ and $L+1$. Thus, during wave propagation, instead of using a sequence 1, 2, 3 . . . , the wavefronts are labeled by a sequence like 0, 0, 1, 1, 0, 0, ... The predecessor of any wavefront is labeled differently from its successor. Thus, each scanned vertex is either labeled "0" or "1." Besides these two states, additional states ("block" and "unblocked") are needed for each vertex. These four states of each vertex can be represented by using exactly two bits, regardless of the problem size. Compared with Acker's scheme, Lee's algorithm requires at least 12 bits per vertex for a grid size of $2{,}000 \times 2{,}000$.

It is important to note that Acker's coding scheme only reduces the memory requirement per vertex. It inherits the search space of Lee's original routing algorithm, which is $O(h \times w)$ in the worst case.

8.4.4.2 Soukup's Algorithm Lee's algorithm explores the grid symmetrically, searching equally in the directions away from target as well as in the directions toward it. Thus, Lee's algorithm requires a large search time. To overcome this limitation, Soukup [189] proposed an iterative algorithm in 1978. During each iteration, the algorithm explores in the direction toward the target without changing the direction until it reaches the target or an obstacle; otherwise, it goes away from the target. If the target is reached, the exploration phase ends. If the target is not reached, the search is conducted iteratively. If the search goes away from the target, the algorithm simply changes the direction so that it goes toward the target, and a new iteration begins. However, if an obstacle is reached, the breadth-first search is employed until a vertex is found that can be used to continue the search in the direction toward the target. Then, a new iteration begins.

Soukup's algorithm improves the speed of Lee's algorithm by a factor of 10–50. It guarantees finding a path if a path between source and target exists. However, this path may not be the shortest one. The search method for this algorithm is a combined breadth-first and depth-first search. The worst case time and space complexities for this algorithm are both $O(h \times w)$ for a grid of size $h \times w$.

8.4.5 Routing Algorithms for MCMs

The approach to routing MCMs depends heavily on the type of MCMs. MCMs with four to eight layers are called thin film MCMs. The routing problem remains the same for both advanced VLSI and thin film MCMs. In thin film MCMs, the number of routing layers used is fixed and, hence, is critical. Thick film MCMs have no restriction on the number of routing layers. Additional layers are added to complete the process of routing. For example, an MCM with more than 100 chips and 63 layers has recently been reported [201].

Few results have been reported for MCM routing problems. Several approaches [44, 96, 165] decompose the routing phase into a pin redistribution phase, where pins on the chip layer are first redistributed evenly. Next in the layer assignment phase each net is assigned a layer or a pair of layers. Finally, the detailed routing phase actually finds and assigns a signal path. Recently, a new approach was developed by Khoo and Cong [111], called *SLICE* which integrates pin redistribution and routing. The basic idea of SLICE is to perform planar routing on one layer at a time. The latest development in MCM routing is the V4R router, also developed by Khoo and Cong [112]. The V4R router has reduced the time complexity and the number of layers used for routing compared with the SLICE. The V4R router uses two adjacent layers and routes a column one at a time from left to right by using four vias for most nets.

A completely new routing methodology for MCM problems is presented in [210]. Rather than converting the three-dimensional routing problem into two-dimensional routing problems, the routing methodology decomposes three-dimensional routing into several smaller three-dimensional problems to achieve the best utilization of the three-dimensional routing resources.

Amont the MCM routers available, some of them are more suitable for thin film MCMs and the rest for thick film MCMs. In following section, we present the different routing approaches currently available for MCMs and discuss them in detail.

8.4.5.1 Maze Routing The most commonly used routing method is three-dimensional maze routing. Although this method is conceptually simple to implement, it suffers from several problems. First, the quality of the maze routing solution is very much sensitive to the ordering of the nets being routed, and there is no effective algorithm for determining a good net ordering in general. Moreover, since the nets are routed independently, global optimization is difficult, and the final routing solution often uses a large number of vias despite the fact that there is a large number of signal layers. This is due to the fact that the maze router routes the first few nets in planar fashion (using shorter distances); the next few nets use a few vias each as more layers are utilized. The nets routed toward the end tend to use a very large number of vias since the routing extends over many different layers. Finally, three-dimensional maze routing requires a long computational time and large memory space.

8.4.5.2 Topological Routing Dai et al. [44] developed a multilayer router based on rubber-band sketch routing. This router uses hierarchical top-down partitioning to perform global routing for all nets simultaneously. It combines this with successive refinement to help correct mistakes made before more detailed information is discovered. Layer assignment is performed during the partitioning process to generate routing that has fewer vias and is not restricted to one-layer one-direction. The detailed router uses a region connectivity graph to generate shortest-path rubber-band routing.

The router has been designed primarily for MCM substrates, which consist of multiple layers of free (channel-less) wiring space. Since MCM substrate designs have a potentially large number of terminals and nets, a router of this nature must be able to handle large designs efficiently in both time and space. In addition, the router should be flexible and permit incremental design process. That is, when small changes are made to the design, it should be able to update incrementally and not regenerate from scratch. This allows faster convergence to a final design. To produce designs with fewer vias, the router should be able to relax the one-layer one-direction restriction. This is an important consideration in high speed designs since the discontinuities in the wiring caused by bends and vias are a limiting factor for system clock speed.

To support the flexibility described above, the router must have an underlying data representation that models planar wiring in a way that can be updated locally and incrementally. For this reason, SURF models wire as rubber-bands [40, 130]. Rubber-bands provide a canonical representation for planar topological wiring. Because rubber-bands can be stretched or bent around objects, this representation permits incremental changes to be made that only affect a local portion of the design. This representation has been described elsewhere [45].

Once the topology of the wiring is known, the rubber-band sketch can be augmented with spokes to express spatial design constraints such as wire width, wire spacing, via size, and so on. [46]. Since a successful creation of the spoke sketch guarantees the existence of a geometrical wiring (Manhattan or octilinear), the final transformation to fixed geometry wiring can be delayed until later in the design process. This allows most of the manipulation to take place in the more flexible rubber-band format. Figure 8.13 shows different views of the same wiring topology. These represent various states of the rubber-band representation.

In this context, a topological router has been developed that produces multilayer rubber-band sketches. The input to this router is a set of terminals, a set of nets, a set of obstacles, and a set of wiring rules. These rules include geometrical design rules and constraints on the wiring topology. The topological constraints may include valid topologies (daisy chain, star, etc.) as well as absolute and relative restrictions on segment lengths. The output of the router is a multilayer rubber-band sketch in which all the points of a given net are connected by wiring. The routability of a sketch is not guaranteed until the successful creation of spokes. At each stage, the router uses the increasingly detailed information available to generate a sketch without overflow regions. This increases the chance that the sketch can be successfully transformed into a representation (the spoke sketch) that satisfies all of the spatial constraints. In addition, the router tries to

FIG. 8.13. Rubber-band representations.

reduce overall wire length and number of vias. A more detailed analysis of the routability of a rubber-band sketch is given elsewhere [46]. The routing of an MCM is a three-dimensional general area routing problem where routing can be carried out almost everywhere in the entire multilayer substrate. However, the pitch spacing in MCM is much smaller and the routing is much denser than in conventional PCB routing. Thus, traditional PCB routing algorithms are often inadequate in dealing with MCM designs.

Topological routing is done in two steps: global routing and local routing. The global router determines rough net topology and partitions the routing area into a set of bins. The interfaces between individual bins are specified by placing crossing points on the bin boundaries for each net that crosses the boundary. These cross points specify a crossing position and a layer. The local router then routes the individual bins independently.

The global routing approach employs two principles from the field of artifical intelligence: the least commitment principle and the notion of maximal use of information. The least commitment principle states that if the correct choice in a decision is not known, the decision should be delayed. This guards against making arbitrary decisions early that, if wrong, could adversely affect the outcome. Maximal use of information states that all available, relevant information should be applied to solve the problem.

Once the global router has partitioned the problem into bins, local routing is performed. In local routing one net is processed at a time within the limits of a bin. This limits the search to a single bin and improves the time and space efficiency of the router. The local router uses a region graph that represents the geometrical and topological adjacency of areas of the sketch. By using this graph, the router can efficiently find the shortest planar path through the partially routed bin. Both the global and local routers rely heavily on an underlying data structure built on constrained Delaunay triangulation [36]. The Delaunay triangulation of a set of points is the straight line dual of a veronoi diagram for the set. An important property of the Delaunay triangulation is that the circumcircle of each triangle contains no points in its interior. Based on the Delaunay triangulation, such problems as closest pair, all nearest neighbors, and euclidean minimum spanning tree (MST) in the plane can be solved in linear time. Constrained Delaunay triangulation includes edges that are forced to be part of the triangulation. SURF uses constrained edges to represent rubber-band segments, obstacles, and so on. The global router relies on fast MST generation for calculating the cost matrices, which are used to determine crossing point locations. The local router relies on the triangulation for performing shortest path calculations within its region graph.

8.4.5.3 Timing Driven Routing A new linear programming based timing-driven global routing method has been proposed by Huang et al. [101]. This timing driven routing problem is formulated as a multiterminal, multicommodity flow problem with integer flows and additional timing constraint. Two different steiner tree approaches based on the timing and one more for rerouting are used.

The authors have proposed this approach for gate array and standard cell routing models. As discussed earlier, we can map a gate array type of problem into a chip array based problem by replacing the gates by chips. Hence, it is desirable to understand this approach so as to extend this method for chip array based MCMs. In this section, we briefly discuss the gate array based approach.

This approach includes the interconnection delays into the routing and re-routing process while trying to minimize the area and at the same time meet the timing constraints. The delay model used in this approach is based on the model presented by Prasitjutrakul and Kubits [164] and by Sakurai [175]. The problem is formulated as a linear programming problem and its dual formulation.

A grid graph model $G = (V, E)$ is used with the following terminologies. Each vertex in V represents a block of region. E consists of two sets of edges E_H and E_V, horizontal and vertical edges. Each edge $e \in E$ is associated with a capacity $c(e)$. For gate arrays, the capacity is fixed. Let N be a set of nets to be routed while meeting the timing constraints. Each net $n_i \in N$ has a demand $b(n_i)$ associated with it. The capacity of each edge should be greater than or equal to the total demand of all the nets $n \in N$ using that particular edge. Each edge e is assigned a distance function $d(e)$. $d(t)$ represents the total distance of the tree. Each net $n_i \in N$ is also assigned a distance function $\lambda(n_i)$, which is the cost to route the net. The problem of routing is reduced to that of finding a tree t_i on the graph G covering all the vertices of a net n_i for all i such that the tree obtained satisfies the timing constraint.

Linear Programming Formulation In a gate array model the available resources are fixed. Hence, the objective of the global routing is to minimize the maximum density of the edges. The routing problem is formulated as a network flow problem. The following functions are used to facilitate the formulation and make it simpler;

1. $f(n_i, t_i)$ is used to denote the flow amount of the tree t for the net n_i.
2. $\delta(n_i, t_i, e) = 1$ if three t_i passes the edge e; otherwise the function is set to zero.

A factor s is used to scale the capacity $c(e)$. The aim of the network flow problem is to minimize s. If s results in being >1, then the solution is not feasible.

Minimize: s
Subject to:

$$\sum_{t_i \in T_n} f(n, t) \geq b(n), \quad \forall n_i \in N$$

$$\sum_n \sum_{t_i \in T_n} f(n, t)\, d(n, t, e) \leq s \times c(e)$$

$$f(n, t) \geq 0 \quad \forall \in T_n, \quad \forall n \in N$$

where T_n denotes the set of all possible trees of net n, which satisfies the timing constraint. The above formulated integer programming equations are simple, and they specify the flow of the sum of all its tree to be greater than or equal to the demand of the net and edge capacity to be maintained.

Dual Formulation $d(e)$ and $\lambda(n)$ are of nets that are used as dual variables. The dual problem formulation is as follows,

$$\textit{Minimize}: \sum_n \lambda(n)\, b(n)$$
$$\textit{Subject to}:$$

$$d(e)c(e) \leq 1 \quad \forall e \in E,$$
$$\sum_{e \in t} d(e) \geq \lambda(n) \quad \forall t \in T_n \forall n \in N$$
$$d(e) \geq 0, \quad \forall e \in E,$$
$$\lambda(n) \geq 0, \quad \forall n \in N$$

The first constraint is converted from the primal problem with respect to the variable s, and the second constraint is derived with respect to the variable $f(n, t)$.

For high speed designs, transmission line delays must be considered, and hence the constraint is the wire length. The approach assumes that a performance driven steiner tree is generated using the approach presented by Hong et al. [98]. An iterative improvement algorithm is then used to detour the path from the congested edges and maintain the timing requirement. Given the current tree, the improvement algorithm reroutes the congested portion to reduce the routing density. This is done by deleting the most expensive edge in a tree, and then Dijkstra's shortest path algorithm [55] is used to replace the deleted path with the shortest path.

8.4.5.4 Multiple Stage Routing In this approach, the MCM routing problem is divided into several subproblems. The close positioning of chips and high pin congestion around the chips require separation of pins before routing can be attempted. Pins on the chip layer are first redistributed evenly with sufficient spacing between them so that the connections between the pins of the nets can be made without violating the design rules. This redistribution of pins is done using few layers beneath the chip layer. This problem of redistributing pins to make the routing task possible is called *pin redistribution.* After the pins are distributed uniformly over the layout area using pin redistribution layers, the nets are assigned to layers on which the assigned nets will be routed. This problem of assigning nets to layers is known as the *layer assignment problem.* The layer assignment problem resembles the global routing of the IC design cycle. Similar to the global routing, nets are assigned to layers in a way such that the routability in a layer or in a group of layers is guaranteed and at the same time the total number of layers used is minimized. The layers on which the nets are distributed are called *signal distribution layers.* The detailed routing follows the layer assignment. The detailed routing may or may not be reserved layer model. The horizontal and vertical routing may be done in the same layer or different layers.

Typically, nets are distributed in such a way that each pair of layers is used for a set of nets. This pair is called the *x-y plane pair* since one layer is used for horizontal segments while the other is used for vertical segments. Another approach is to decompose the netlist such that each layer is assigned a planar set of nets. Thus, the MCM routing problems become a set of *single layer* problems. Yet another routing approach may combine the *x-y* plane pair and single layer approaches. In particular, the critical nets are routed in the top layers using single layer routing because *x-y* plane pair routing introduces vias and bends, which degrade performance.

We now discuss each of these problems in greater detail.

1. **PIN Redistribution Problem:** As mentioned earlier, in MCM routing, first the pins attached to each chip in the chip layer are redistributed over the uniform grid of pin redistribution layers. The pin redistribution problem can be stated as: Given the placement of chips on an MCM substrate, redistribute the pins using the pin redistribution layers such that one or more of the following objectives are satisfied (depending on the design requirements):

 a. Minimize the total number of pin redistribution layers.
 b. Minimize the total number of signal distribution layers.
 c. Minimize the crosstalks.
 d. Minimize the maximum signal delay.
 e. Maximize the number of nets that can be routed in planar fashion.

It is to be noted that the separation between the adjacent via-grid points may affect the number of layers required [37]. The pin redistribution problem can be illustrated as shown in Figure 8.14. The terminals of chips need to be connected to the vias as shown in Figure 8.14a. Usually, it is impossible to complete all the connections. In this case, we should route as many terminals as possible (as shown in Fig. 8.14b). The unrouted terminals are brought to the next layer and routed in that layer as shown in Figure 8.14c. This procedure is repeated until each terminal is connected to some via. Various approaches to pin redistribution problem have been proposed [37].

2. **Layer Assignment:** The main objective of layer assignment for MCMs is to assign each net into *x-y* pair of layers subject to the feasibility of routing the nets on a global routing grid on each plane pair. This step determines the number of plane pairs required for a feasible routing of nets and is, therefore, an important step in the design of the MCM. The costs of fabricating an MCM, as well as cooling the MCM, when it is in operation, are directly related to the number of plane pairs in the MCM, and, thus, it is important to minimize the number of plane- pairs. There are two approaches known to the problem of layer assignments [96, 191]. The problem of layer assignment has been shown to be NP-complete [96]. An approximation algorithm, for minimizing the number of layers, has been presented by Ho et al. [96].

3. **Detailed Routing:** After the nets have been assigned to layers, the next step is to route the nets using the signal distribution layers. Depending on the

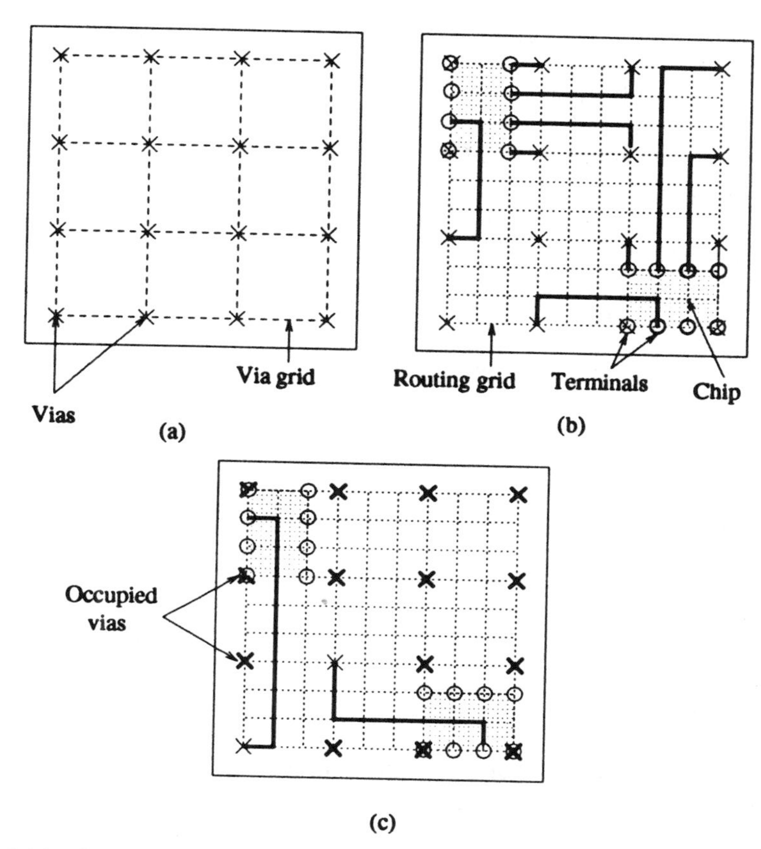

FIG. 8.14. Pin redistribution example: **(a)** via grid, **(b)** first layer, **(c)** second layer.

layer assignment approach, the detailed routing may differ. The routing process may be single-layer routing or *x-y* plane pair routing. Usually, a mixed approach is taken in which the single layer routing is first performed for more critical nets, followed by *x-y* plane pair routing for less critical nets. Two models can be employed for *x-y* plane pair routing, namely, the *xy* reserved and the *xy* free models. One advantage of the *xy* free model is that bends in nets do not necessarily introduce vias, whereas bends in nets introduce vias in the *xy* reserved model. The routing has been discussed in detail [132].

8.4.5.5 Integrated Pin Distribution and Routing Khoo and Cong [111] present an integrated algorithm SLICE for routing in MCMs. Instead of distributing the pins before routing, the algorithm redistributes pins along with routing in each layer. The basic idea of SLICE is to perform planar routing on a layer by layer basis. After routing on one layer, the terminals of the unrouted nets are propagated to the next layer. The routing process is continued until all the nets are routed.

The important feature of SLICE is how to compute the planar set of nets for each layer. The algorithm tries to connect as may nets as possible in each layer. For the nets that cannot be completely routed in a layer, the algorithm attempts to route those nets partially so that they can be completed in the next layer with shorter wires. The routing region is scanned from left to right. For each adjacent column pair, a topological planar set of nets is selected by computing a maximum weighted noncrossing matching, which consists of a set of noncrossing edges that extend from the left column to the right column. Then, the physical routing between the column pair is generated based on the selected edges in the matching. This process is carried out for each column from left to right.

After completing the planar routing in a layer, the terminals of the unrouted nets are distributed so that they can be propagated to the next layer without causing local congestion. Since the left to right scanning operation in the planar routing results in mainly horizontal wires in the solution, in order to complete the routing in the vertical direction a restricted two-layer maze routing technique is used. The unnecessary jogs and wires are removed after each layer is routed. The terminals of the unrouted nets are propagated to the next layer. Finally, the routing region is rotated by 90° so that the scanning direction is orthogonal to the one used in the previous layer. The process is iterated until all the nets are routed. Details of the planar routing, pin redistribution, and maze routing are discussed elsewhere [111].

8.4.5.6 MCM Router Based on Four-Via Routing Khoo and Cong [112] have presented an efficient multilayer general area router, named *V4R*, for MCM and PCB designs. It uses no more than four vias to route every net and yet produces high quality routing solutions. It combines global routing and detailed routing in one step and produces high quality detailed routing solutions directly from the given netlist and module placement. As a result, *V4R* is independent of net ordering, runs faster, uses less memory when compared to other general area routers.

In each routing layer, a horizontal grid line is called a *horizontal track* and a vertical grid line is called a *vertical track*. The terminals on the same horizontal track form a *row* and those on the same column form a *column*. For a terminal p, let $x(p)$ and $y(p)$ denote the x and y coordinates (in terms of grid point coordinates) of p, respectively, and let $row(p)$ and $col(p)$ denote the row and column number of p, respectively. Two adjacent rows form a horizontal channel, and two adjacent columns form a vertical channel. The channel formed between the ith and $i+1$th channel (rows) is named the ith vertical (horizontal) channel. It can be seen from this that there are no terminals present within the channel.

V4R has the capability of handling the multiterminal nets; however, in the following discussion it has been assumed that there are two terminal nets only. Each net is routed using up to five connected segments alternating between vertical and horizontal directions. A horizontal segment is called an h-segment

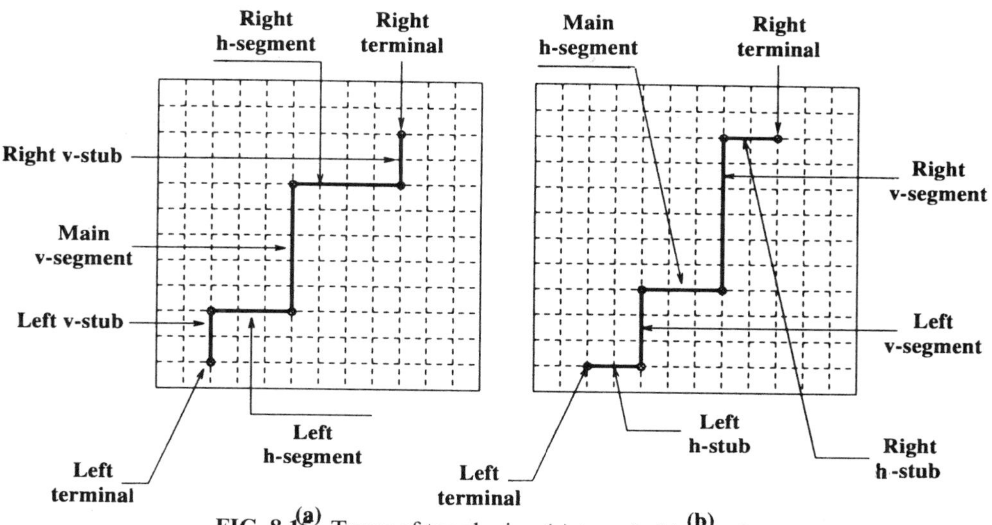

FIG. 8.15. Types of topologies: **(a)** type 1, **(b)** type 2.

while a vertical segment is called a *v*-segment. There are two possible routing topologies depending on the direction of the segment connected to the left terminal or to the right terminal, as shown in Figure 8.15. The topologies are called *type 1* and *type 2* and are as shown in Figure 8.15.

V4R routes two adjacent layers at a time, the odd-numbered layers are for v-segments and the even numbered layers are for h-segments. For each layer pair, columns are processed one by one starting from the left. At each column c, V4R executes the following four steps,

1. **Horizontal Track Assignment of the Right Terminals:** For each right terminal q_i whose left terminal p_i is in column c, the algorithm tries to connect q_i to an appropriate horizontal track that is free between $col(p_i)$ and $col(q_i)$, using right v-stub in column $col(q_i)$. The nets that are successfully assigned to feasible tracks are designated as *type 1* nets and, hence, will be routed using type 1 topology. The remaining nets are treated as *type 2* nets and will be routed using type 2 topology. Figure 8.16a shows the track assignment for the right terminals q_i and q_3, which implies that nets 1 and 3 will be routed using a type 1 topology. For a type 1 net i, its right h-segment will be routed in track t_i^r. For a type 2 net j, its right h-stub will be routed in row $row\,(q_j)$.
2. **Horizontal Track Assignment of the Left Terminals:** This step consists of two phases. In phase 1, it is attempted to connect each type 1 left terminal p_i in the current column c to appropriate horizontal track t_i^r using the left v-stub in column c. The left h-segment of a type 1 net i will be routed in track

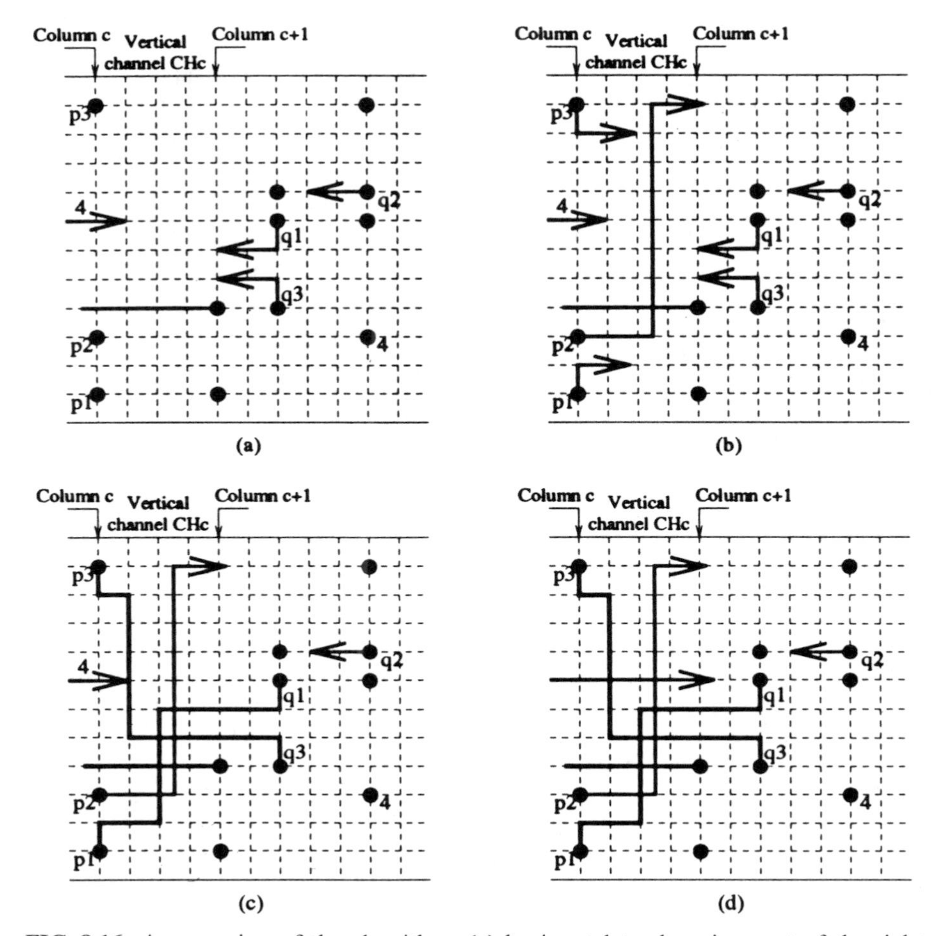

FIG. 8.16. An overview of the algorithm: **(a)** horizontal track assignment of the right terminals, **(b)** horizontal track assignment of the left terminals, **(c)** routing in the vertical channel, **(d)** extending to the next column.

t_i^r. Figure 8.16b shows the track assignment of the type 1 left terminals p_i. In phase 2, it is attempted to assign a horizontal track for the main h-segment for type 2 left terminals. Note that the main h-segment can be connected to the left terminal only after its left h-stub and left v-segment are routed. Figure 8.16b shows the track assignment for the type 2 terminal p_2.

A net is said to be *active* if its left and right terminals have been assigned the appropriate horizontal tracks and yet its routing has not been completed. A net is said to be *pending* if one of the following is satisfied:

a. It is the main v-segment of a type 1 active net.
b. It is an unrouted left v-segment of a type 2 active net.

c. It is an unrouted right v-segment of a type 2 active net, its left v-segment has been routed, and the row $row(q_i)$ is free between $col(q_i)$ and column c.

3. **Routing in the Vertical Channel:** Select a maximum subset of the pending v-segments and route them in the $c - th$ vertical channel CH_c. The density of the selected v-segments should not exceed the capacity of CH_c. In Figure 8.16c, the nets 1, 2, 3, and 4 are all active nets, but only the main v-segments of nets 1 and 3 and the left v-segment of net 2 are the pending v-segments, and all of them are routed in CH_c.
4. **Extending to the Next Column:** We extend the left h-segments of the remaining active nets to column $c + 1$. If the h-segment of an active net i is blocked, all the routed segments of net i are ripped up and i is added to the list $L_n ext$ (used to propagate nets to the next layer pair). In Figure 8.16d, the h-segments of nets 2 and 4 are extended to column $c + 1$.

At the completion of these four steps, the algorithm processes the $(c + 1)$th column. After all the columns in the current layer are processed, V4R processes the next layer pair using $L_n ext$.

It may be noted that V4R does not store a routing grid during the routing process. It stores only the assignment of the horizontal tracks and the vertical segments of the active nets, which requires a smaller amount of memory than the others. A detailed analysis of the algorithm is described elsewhere [112].

8.4.5.7 *Three-Dimensional Routing* Yu et al. [210] presented a new approach to MCM routing. Rather than converting the three-dimensional routing problem into a set of two-dimensional routing problems, the proposed routing methodology divides three-dimensional routing into several smaller three dimensional problems to achieve the best utilization of the three-dimensional routing space. Several different performance constraints are incorporated. The approach consists of different phases. First, the routing is distributed into a two-dimensional x-y plane as well as in the z-dimension. This distribution ensures that no specific region of the MCM routing space is overcongested or undercongested. This phase helps to satisfy the net-length constraints and the manufacturing constraints. This routing distribution also helps to partition the routing problem of the entire routing space into the routing problems of several smaller regions called *towers*. A good layer estimate is obtained after the completion of the z-dimension routing distribution. After the completion of z-dimension routing distribution, the exact positions of the terminals on the face of each tower are assigned.

During this phase, the net-separation constraints are considered to reduce crosstalk between nets. The terminal assignment phase also maximizes the number of planar nets in each tower. After the completion of this phase, the routing problem of a complex three-dimensional problem has been converted to that of routing several small towers. As the problem of routing in each tower is independent of the rest, all the towers can be processed in parallel. At this stage, a tower can be partitioned, and the above approach is applied to the tower again

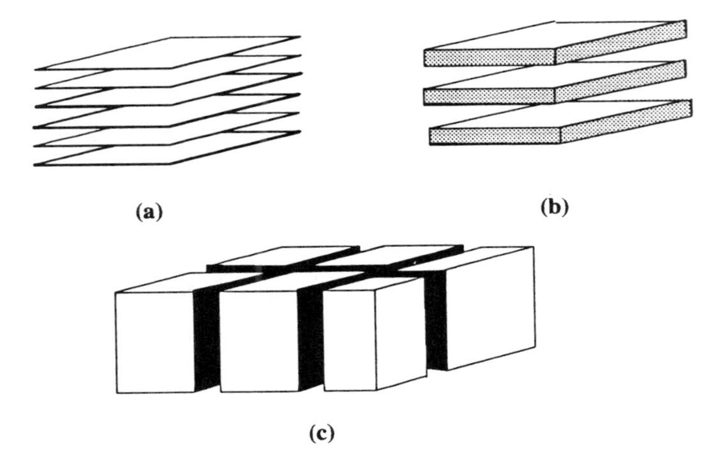

FIG. 8.17. Different routing methodologies for six layers: **(a)** single-layer routing, **(b)** two-layer routing, **(c)** three–dimensional routing.

if necessary. Thus, this approach is recursive and allocates more computing resources to the regions that require them. During the tower routing phase, a maximum set of planar nets for each layer is determined that can be directly routed on the layer using an approximation algorithm that guarantees 0.60 of the optimal solution. This reduces the running time of the tower routing. The tower routing is designed to obey net-separation and via constraints. During each phase, an increasingly accurate estimate of the net lengths and number of layers is obtained.

The three-dimensional routing approach differs from the other approaches primarily on problem decomposition. The three-dimensional approach maintains the characteristics of the three-dimensional routing problem while the other convert the three-dimensional problem into a two-dimensional or two-layer routing problem. Figure 8.17 compares the routing structure of this approach with the other existing approaches. The methodology is inherently parallel, which can be used to improve the time complexity and memory constraints.

1. **Performance Constraints:** The interconnections among terminals are made with the help of metal wires laid out on the routing layers beneath the chip layer. The minimization of the number of routing layers is the driving force for most VLSI and PCB routers. However, as more layers are readily available and a host of performance constraints are introduced by the MCM routing problems, the number of layers is not as critical as in VLSI routing problems. Instead, the objective of the MCM routing approach should satisfy the performance constraints while minimizing the number of layers required for routing. The performance constraints include the manufacturability constraint, net-length constraint, net-separation constraint, and via constraint.

The routing algorithm presented in the next section generates the routing for the netlist with a minimum number of layers while satisfying all the performance constraints.

2. **Overview of the Routing Algorithm:** MCM routing is carried out in a three-dimensional routing space using a recursive formulation. Due to the complexity of routing in the entire three-dimensional space, the routing is completed in several different phases:

a. Tiling
b. Off tile routing
c. Two-dimensional routing distribution
d. Z-Dimension routing distribution
e. Terminal assignment
f. Tower routing

An overview of the approach is shown in Figure 8.19. As can be seen, the routing problem of the entire routing space is converted into the routing problems of several smaller towers that are independent of each other. For complex tower routing problems, which are too large due to either the size of the tile or the large number of nets present in the tile, this approach is applied in a recursive fashion to break up the original problem into several smaller tower routing problems and solve them individually (see Fig. 8.18). As a result, the amount of computing resources spent on a routing region (tower, subtower, subsubtower) is directly proportional to its complexity. On the other hand, during the routing process, more information about the use of routing resources becomes available after each phase. This characteristic makes this approach very flexible, and it can be used not only for the routing of MCMs but also for providing routing information for the placement tools. Furthermore, as each of the tower routing problems is independent of the other, each tower can be routed by assigning one tower to a different processor and, hence, achieve a better time complexity if a

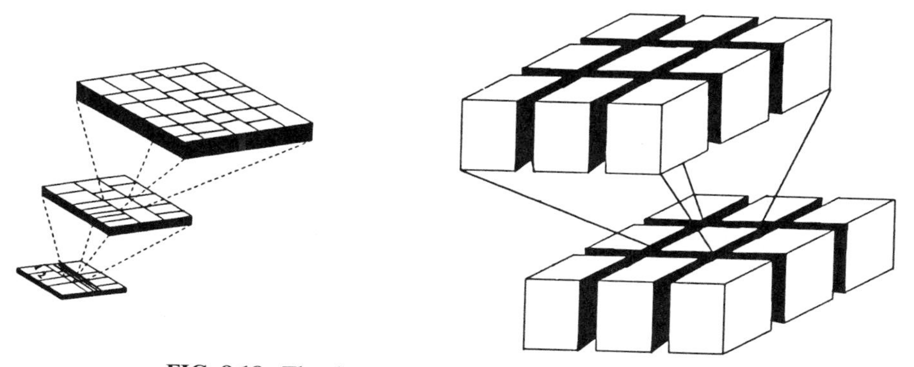

FIG. 8.18. The three-dimensional approach is recursive.

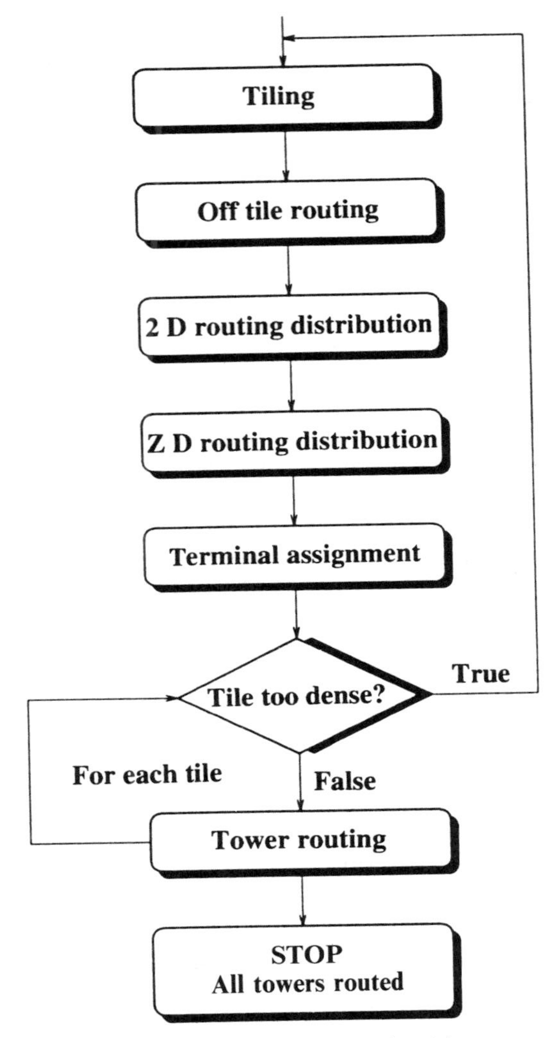

FIG. 8.19. Overview of the algorithm.

parallel computing environment is used. Furthermore, a host of performance constraints are met in this approach. Uniform distribution of wiring is possible due to the global and three-dimensional perspective of this approach. The net-length, net-separation, and via constraints are met due to dividing the routing into several phases and the constraints are handled in each of these phases.

3. **Z-Dimension Routing Distribution:** During this phase, the layer on which each net passes on each tower face is determined. The objective of this phase is to distribute the congestion uniformly along the Z-direction. At the completion of this phase, the total number of layers required for routing is known. In addition,

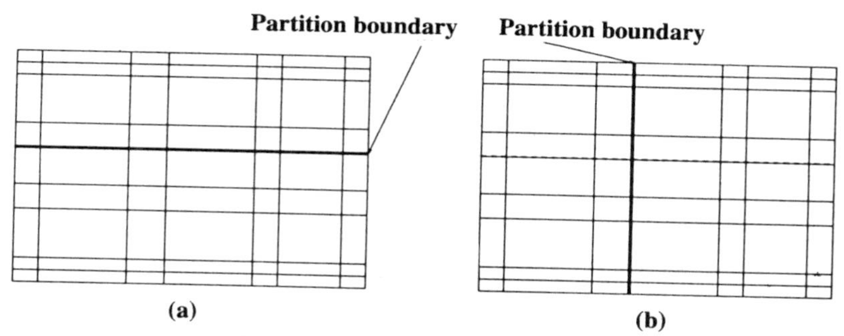

FIG. 8.20. Substrate partitioning.

since the length of each net in the Z-direction can be estimated, the total length of each net along the x-y dimension as well as the z-dimension can be estimated accurately.

4. **Terminal Assignment:** The exact location of the terminal of each net on the face of each tower is determined. The objective of this phase is to maximize the set of planar nets on each layer in a tile while assigning the terminals. After terminal assignment, the length of those nets in the planar set is known. Note that at the end of each phase we get a better estimate of the length of the nets in an incremental fashion, finally leading to the exact length. In addition, the net-separation constraints β are met by assigning each terminal separated from the other terminals by the required distance.

The terminal assignment is carried out by bipartitioning the substrate recursively. At each level of bipartitioning, a set of tiles is partitioned into two sets of tiles that have the same or similar size, and the location of the point where each net crosses the partition boundary is determined. Figure 8.20a shows the first level of bipartitioning, and Figure 8.20b shows the second level of bipartitioning. Since the tower faces that each net passes has been determined during the two-dimensional routing distribution phase and the layers where the terminals of the nets are located have been determined during the z-dimension routing distribution phase, we can only permute the terminals of the nets along a tile edge. In the following discussion, we assume that the nets under consideration pass a tower face on the same layer and the terminals of these nets on the tile edge can be permuted within the tile edge. In addition, the tile edge under consideration is along the partition boundary. The location of the terminals on a tile edge is determined in two steps. During the first step, the ordering of the terminals along a tile edge are found so as to achieve the maximum planar set of nets. Then the terminals are assigned to exact locations according to the ordering of the terminals.

5. **Terminal Ordering:** The ordering of the terminals is obtained by permuting the terminals on a tile edge. We define the intersection between the line segment connecting the terminals of one net and the line segment connecting the terminals of another net as a *crossing* between these two nets. The set of nets between which

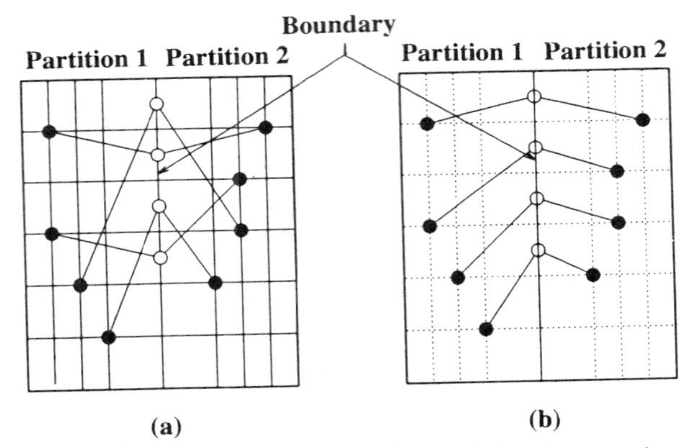

FIG. 8.21. Permutation of terminals minimizing crossing.

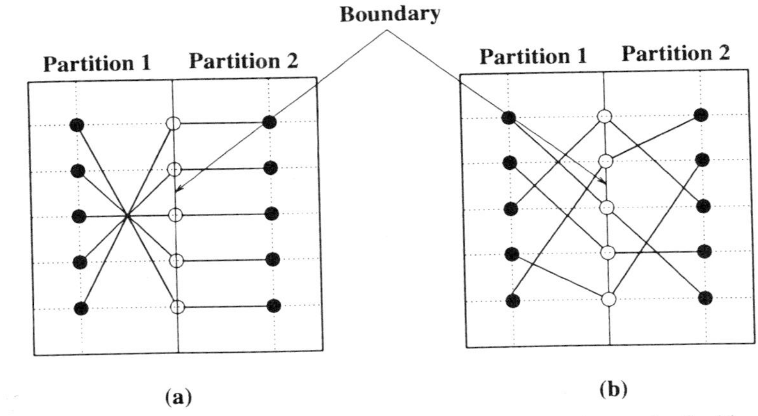

FIG. 8.22. Permutation of terminals minimizing crossing on both sides.

there is no crossing within a tile is called a *planar set of nets* in that tile. To maximize the number of planar nets, it is necessary to minimize the number of crossings between the nets.

One objective function is to achieve the minimum number of crossings between the nets. Another objective function is to balance the number of crossings on both sides of the tile edge. The objective function to be used in the terminal ordering is chosen according to the constraints. An example of minimizing the number of crossing is shown in Figure 8.21, and an example of balancing the number of crossing on both sides of the tile edge is illustrated in Figure 8.22. By achieving the minimum number of crossings, we can achieve the maximum planar set of nets, whereas the planar nets will be distributed evenly over the entire substrate if the number of crossings on both sides of the partition boundary is balanced at each level of bipartitioning.

6. **Finding Locations for Terminals:** After the ordering of the terminals has been obtained, the terminals are assigned to an exact location on the tile edge. First the terminals will be assigned next to each other according to the ordering of the terminals. If the net-separation constraint is not specified, the terminal assignment phase is completed. However, if the net-separation constraint is specified, the separation between the terminal of net i and terminal of net j should be no less than β_{ij}. In case the original terminal assignment cannot achieve the desired separation, either i or j will be permuted with some other terminal to satisfy the net-separation constraint. Note that this permutation of terminals may result in fewer planar nets as the original ordering of terminals has been changed. In addition, if β is specified for every pair of terminals, the net-separation constraint also affects the number of layers since an increase in β leads to a decrease in the number of nets that can be assigned to a tile edge, and, therefore, leads to an increase in the number of layers.

7. **Tower Routing:** The nets are routed within each tower. This approach uses an approximation technique to find a maximum set of planar nets that can be routed directly and quickly to reduce the running time of tower routing. The remaining nets are routed one at a time using the three-dimensional Soukup technique [189]. We specify the net-separation constraint β and via constraint γ for each net in the tower routing to ensure that net-separation and via constraints are satisfied. The net-separation constraints are satisfied by using "cable routing."

8.5 SUMMARY

The MCM approach to microelectronic packaging has improved the system performance significantly by bridging the gap between the existing PCB packaging approach and the advancing VLSI technology. The physical design of MCMs is considered an important ingredient of the overall MCM design cycle. The dense and complex VLSI/ULSI chips available today necessitate the automation of the physical design of MCMs. Research on developing the algorithms required for MCM physical design has been rather limited; as a result, further developments of MCMs are being obstructed. In an attempt to meet the CAD needs of MCMs, several CAD tools are now available. However, these tools do not satisfy all the requirements of the MCM physical design and are essentially extensions of PCB or IC design tools. This is mainly due to the fact that MCMs pose entirely new problems that cannot be solved by existing PCB tools or IC layout tools. Therefore, a considerable amount of research is needed to develop algorithms for MCM physical design automation (Fig 8.23).

```
Algorithm LEE-ROUTER (B, s, t, P)

Input:  B, s, t
Output: P
begin
        plist = s;
        nlist = φ;
        temp = 1;
        path_exists = FALSE;
        while plist ≠ φ do
            for each vertex v_i in plist do
                for each vertex v_j neighboring v_i do
                if B[v_j] = UNBLOCKED then
                    L[v_j] = temp;
                    INSERT(v_j,nlist);
                    if v_j = t then
                        path_exists = TRUE;
                        exit while;
            temp=temp + 1;
            plist = nlist;
            nlist = φ;
        if path_exists = TRUE then RETRACE (L, P);
        else path does not exist;
end.
```

FIG. 8.23. Algorithm LEE-ROUTER.

8.6 PROBLEMS

1. Partition the graph shown in Figure 8.25 using the Kernighan-Lin algorithm in Figure 8.24.
2. For the graph in Figure 8.26, let the delay for the edges going across the partition be 25 ns. Each vertex has a delay, which is given below. Consider vertex v_1 as the input node and vertex v_s as the output node. Partition the graph such that the delay between the input node and the output node is minimum and the partitions have the same size. The delays for the vertices are $d(v_1) = 2$ ns, $d(v_2) = 3$ ns, $d(v_3) = 4$ ns, $d(v_4) = 1$ ns, $d(v_5) = 2$ ns, $d(v_6) = 4$ ns, $d(v_7) = 11$ ns, $d(v_8) = 6$ ns, $d(v_9) = 4$ ns, $d(v_{10}) = 1$ ns, $d(v_{11}) = 2$ ns, $d(v_{12}) = 2$ ns, $d(v_{13}) = 6$ ns, $d(v_{14}) = 3$ ns, and $d(v_{15}) = 4$ ns.

```
Algorithm KL(G)

Input:  Graph, G = (V, E)
Output: Graph decomposed into two partitions with minimum number
        of crossings.
begin
      INITIALIZE();
      while( IMPROVE(table) = TRUE ) do
      (* if an improvement has been made during last iteration,
      the process is carried out again. *)
         while ( UNLOCK(A) = TRUE ) do
         (* if there exists any unlocked vertex in A,
         more tentative exchanges are carried out. *)
            for ( each a ∈ A ) do
               if (a = unlocked) then
                  for( each b ∈ B ) do
                     if (b = unlocked) then
                        if (D_max < D(a) + D(b)) then
                           D_max = D(a) + D(b);
                           a_max = a;
                           b_max = b;
            TENT-EXCHGE(a_max, b_max);
            LOCK(a_max, b_max);
            LOG(table);
            D_max = -∞;
         ACTUAL-EXCHGE(table);
end.
```

FIG. 8.24. Kernighan-Lin algorithm.

3. Heat generated in an MCM can be removed in several ways. From the physical design aspect, we can classify them in two ways, namely, *chip based heat removal* and *area based heat removal*. Chip based heat removal refers to dissipating heat individually from a chip, and area based heat removal is the process of removing the heat globally from the whole substrate. Modify the Kernighan-Lin algorithm described in Section 8.2 to consider the chip based heat removal method so that it can be used suitably in MCM partitioning.

4. Consider the *thermal-driven* placement problem for a *chip array based* approach. Develop an efficient algorithm considering an area based heat removal method.

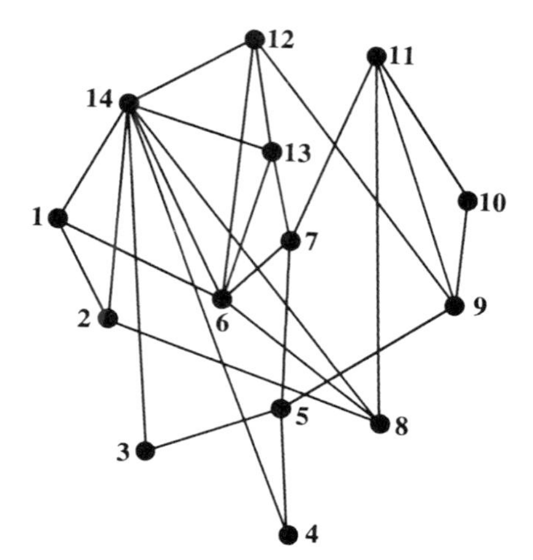

FIG. 8.25. A graph partitioning problem.

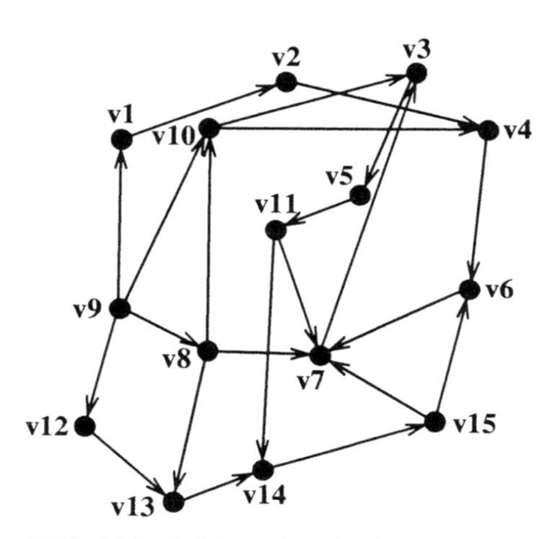

FIG. 8.26. A delay minimization problem.

5. Develop a simulated annealing based algorithm for the MCM placement problem using the *full custom* approach.
6. Is thermal management important in MCM routing? Why?
7. Develop an efficient global routing algorithm for MCMs so as to distribute the density of nets over the third dimension.
8. Delay is a major consideration in all the steps of the physical design. Physical design algorithms have to consider the factors affecting delay so as to optimize performance and yield. Develop an efficient placement algorithm

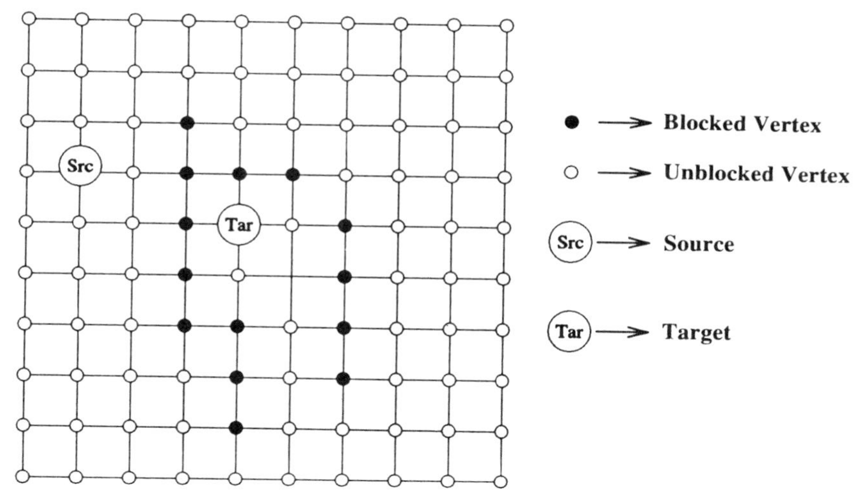

FIG. 8.27. Grid graph for Soukoup's algorithm.

for the MCMs such that the length of the critical nets connecting different chips is minimized for the full-custom approach. (Assume two terminal nets.)

9. Develop an efficient global router for the MCMs such that the router has the capability of minimizing the delay due to the transmission line effects.
10. What would be the effects of the different substrates discussed in Chapter 2 on the MCM routing phase.
11. Apply Soukup's algorithm on the grid shown in Figure 8.27. Perform a step by step analysis.
12. Develop an efficient *crosstalk-driven* routing algorithm for thick film MCMs.

BIBLIOGRAPHIC NOTES

Several CAD tools for partitioning at the chip level [123, 171] are now available. Wilson [209] discusses the topological relationship. Kernighan and Lin [110] proposed a graph bisectioning algorithm for a graph. Schweikert and Kernighan [181] proposed the use of a net model for partitioning hypergraphs. An algorithm using a vertex-replication technique to reduce the number of nets that cross the partitions was presented by Kring and Newton [119]. Krishnamurthy [120] proposed an extension to the Fiduccia and Matheyses algorithm [66] that accounts for high order gains to get better results and a lower dependence upon the initial partition.

Simulation based partitioning algorithms [35, 78, 113, 173] carry out the partitioning process by using a cost function, which classifies any feasible solution, and a set of moves, which allows movement from solution to solution. Early work on simulated annealing used the Metropolis algorithm [146]. Some results are based on the combination of MCM partitioning and placement. Shin, et al. [186] presented an integrated partitioning and placement algorithm, and Saucier et al. [179] presented a new approach for circuit partitioning with a constraint on the number of I/O pins and containment of the critical path within the partitions.

Placement algorithms can be classified in several ways. Researchers have produced some useful results in simulation based algorithms. These type of algorithms simulate some natural phenomenon. Two approaches to MCM placement problems have been discussed by LaPotin [126] as part of the early design analysis, packaging, and technology trade offs. The concept of a *2.5-D integration scheme* is derived from ideas postulated by McDonald [142] and Tewksbury [196] for wafer scale integration, which can be applied in full custom style placement techniques.

Acker [1] proposed a new strategy for the retrace phase of Lee's algorithm. Lee's algorithm requires a large search time. To overcome this limitation, Soukup [189] proposed an iterative algorithm in 1978. Few results have been reported for MCM routing problems. Several approaches [44, 96, 165] divide the routing phase into a pin redistribution phase, layer assignment phase, and detailed routing phase.

CHAPTER 9

CAD TOOLS FOR MULTICHIP MODULES

The MCM design cycle consists of several steps and is indeed long and complex. Each step is tedious, time consuming, and driven by a multitude of electrical, thermal, and testing considerations. As a result, it is not feasible to complete the MCM design process manually. For example, consider the placement phase of the physical design cycle. As discussed in the previous chapter, the placement algorithms must consider timing, routability, and thermal constraints while meeting the placement objectives. These constraints increase the complexity of the placement phase, necessitating the automation of placement to improve performance and efficiency. To improve performance, reduce cost, and increase reliability it is necessary to automate the MCM design cycle as much as possible. Without CAD tools, it is impossible to design and analyze MCMs efficiently.

The existing tools used in the physical design automation of ICs and PCBs cannot be used for MCMs. This is mainly because MCMs have a large number of dies in a relatively small area and a high operating frequency. This introduces difficulties in package design, thermal management, signal integrity, testability, and cost. CAD tools used for ICs cannot be used for MCMs because of their significantly different substrate sizes. For example, typically the substrate size of an IC ranges from 2 mm a side to 20 mm a side, while that of an MCM is of the order of 150 mm a side. In addition, consider the transmission line behavior as discussed in Chapter 4. CAD tools for MCMs must consider this effect, as the length of the signal wires affect the timing. Transmission line effects are not significant for ICs because of relatively shorter interconnect lengths. In comparison, CAD tools used in the design of PCBs cannot be used for MCMs due to the density considerations. For example, PCBs use a pitch of 250 λ and thicker signal lines, while MCMs use a pitch of 75 λ(MCM-Cs use 10 λ) and relatively thinner signal lines.

CAD tools are required in the design of MCMs to perform design computation, early estimation, design verification, physical design automation such as place and route, thermal analysis, manufacturability process, electrical analysis, reliability analysis, feasibility analysis, and report and statistics generation.

To meet these demands, several commercial CAD packages for the design of MCMs have been developed. These CAD packages make it possible to integrate a wide variety of tools from different vendors into one package. CAD packages are available with all the necessary tools for complete design of MCMs. In addition, CAD tools are also available from universities and research institutions that are capable of handling certain aspects of MCM design. In this chapter, we discuss the important features of an MCM CAD package and several commercially available CAD packages and summarize existing research CAD tools developed at universities and research institutions.

9.1 ESSENTIAL FEATURES OF MCM CAD TOOLS

The high complexity of MCM designing has resulted in division of the entire MCM design into several phases, as shown in Figure 9.1. Each phase of the design cycle is further split into several steps.

Conceptual design is the first step of the MCM design cycle. As discussed in Chapter 3, the *conceptual design* or system level design refers to the design activities that produce the system specifications. This phase is responsible for creating a high level system representation while considering performance, functionality, cost, and physical dimensions. At the end of this phase, a behavioral representation of the system is obtained.

The second phase of the MCM design cycle is referred to as the *logic design*. Logic design is the process of converting a behavior specification into circuit representation. In this phase, the functional design is represented by a set of boolean expressions. These expressions are optimized to achieve the smallest logic design that conforms to the functional design. The logic design of the system is simulated and tested for correctness. The output of this phase is the circuit representation of the system.

Logic design can be performed either by structural design or by behavioral level design. Design at the structural level begins to approximate the actual hardware. Structural design includes all the details for the complete design, including component level semantics and necessary information required for analysis. Behavioral level design is used to describe the system, including inputs and outputs. VHDL are used to describe the behavioral design. The objective of this level of design is to specify at a very high level while ignoring low level details. By using VHDL, the designer has a fast and accurate way of verifying designs at an intermediate level to take corrective actions.

The next step in the design cycle is the verification phase, which checks the validity of the logic, analog, thermal, and signal integrity. Verification is essential after each step in the design phase. However, the basic objective of this phase is

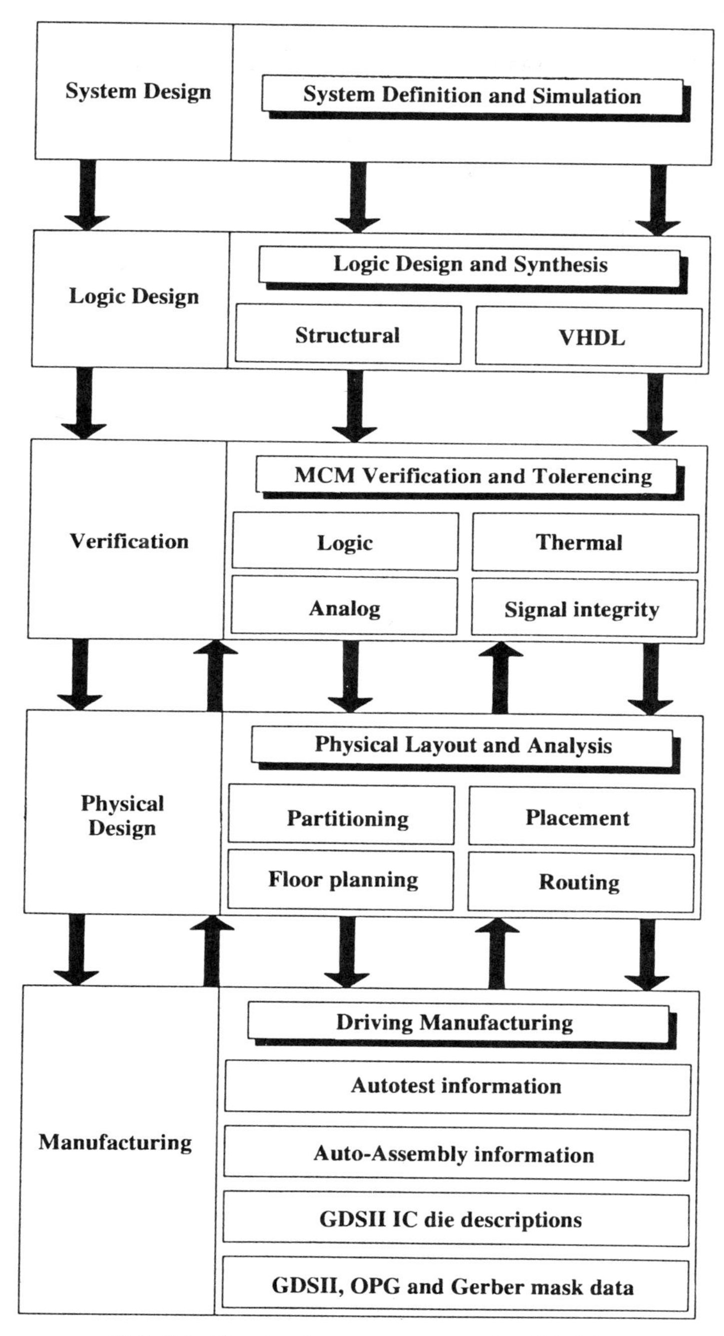

FIG. 9.1. The development process of an MCM.

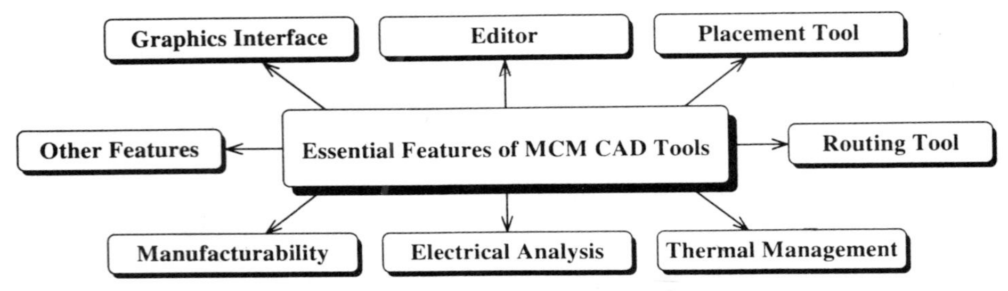

FIG. 9.2. Essential features of MCM CAD tools.

to guarantee the system specification. At the completion of this phase the system is functionally correct. As a result, it is not required to return to this phase during the design cycle to verify or correct the specifications. The output of this phase is a problem free circuit design meeting the design specifications.

In the next phase, the physical design of the circuit is performed. This phase has been discussed in detail in Chapter 7. The output of the physical design phase is a layout that is ready for fabrication.

Finally, the last phase of the design cycle is to convert the layout obtained from the physical design phase into a suitable format such that it can be directly used to produce masks. Once the masks are manufactured, the fabrication process can begin.

It follows from the above discussion that all the phases of the MCM design cycle are tedious, time consuming, and complex. This necessitates the use of CAD tools. Hence, CAD tools must be capable of performing all the design activities discussed above. In addition, all the tools must be integrated into a single CAD package as much as possible.

Based on the above discussion, we present six tool sets (see Fig. 9.2) that are necessary to design MCMs.

1. Editing
2. Placement
3. Routing
4. Thermal management
5. Electrical analysis
6. Manufacturability

We now discuss some of the important features that are required for these tools.

9.1.1 Editing Tools

An editor is basically used to define, move, align, rotate, pivot, and swap individual or groups of components on a user defined placement grid. In addition, the editor is used to assign predefined values and specify the constraints.

Once the constraints have been specified all the tools must adhere to these values until the end of the design process. Normally, editing tools are integrated into the framework or the graphical interface. The graphical interface allows invocation of all tool sets in an integrated fashion.

9.1.2 Placement Tools

CAD tools for placement must be capable of handling a large number of dies, meet performance criteria, estimate routability, and satisfy thermal constraints efficiently. The placement tools must be built with fast and efficient algorithms so as to speed up the placement phase. Efficient placement algorithms improve system performance, reduce cost, and increase reliability. CAD tools for placement must be capable of supporting automatic placement. The designer must be allowed to specify various parameters and constraints based on the design rules before beginning the placement phase. Once the parameters and the constraints have been specified, the automatic placement tool can be invoked. The automatic placement tool must adhere to the parameters and constraints that have already been specified throughout the placement phase while meeting its objectives. In case some of the constraints cannot be satisfied, the placement tool must be capable of informing the designer of the specific details so that the designer can make suitable changes. Designers typically have preferred locations for certain critical dies. An interactive feature allows a designer to place a few blocks manually at the desired location and then run the automatic placement routine to place the remaining blocks. Placement tools must be capable of estimating thermal profiles, routability, and delay considerations so as to predict the outcome of the design while satisfying the constraints. This helps the designer to take corrective measures and make changes at an early stage. As a result, the delay and cost associated with the design phase can be significantly reduced.

9.1.3 Routing Tools

CAD tools for routing in MCMs must be capable of handling large and complex designs and satisfy several constraints such as delay, crosstalk, noise, skin effects, and via constraints. The routing tools must be efficient so as to speed up the routing phase while meeting the performance criteria. As in the case of placement, efficient CAD tools for routing are useful in reducing the overall time to design an MCM, decrease the cost of fabrication, and increase the overall reliability of the system. In addition, as the memory requirements are significantly high, it is important that the CAD tools for routing in MCMs must use good data structures. Once the constraints and design rules have been specified, the routing tools must adhere to the specified constraints until the completion of routing process.

The CAD packages for MCMs must be capable of supporting auto route facility. By *auto route* we mean the capability to route all the nets automatically while satisfying the constraints and meeting the design rule specifications. The auto route feature simplifies the tedious and time consuming job for the designer. Interactive routing is also an essential feature. As in the case of the placement tool, interactive routing helps in manually routing critical nets such that these nets meet their design considerations, and then the automatic router is invoked to complete the routing. As stated before, this facility helps in optimizing the routing. Routing tools must also support rip-up and reroute features so as to achieve routing optimizations. In addition, the routing algorithms must be able to guarantee high performance, speed, and a 100% completion rate.

9.1.4 Thermal Management Tools

As discussed in Chapter 5, thermal management is a very important consideration in the design of MCMs. The reliability of a system greatly depends on the thermal management system. A large amount of heat is generated by high speed ICs. When a number of high speed ICs generating heat are placed close to each other on an MCM, heat generated from an MCM is very large. To overcome the problem of heat it is essential to consider the thermal effects during the design process. As a result, it is desirable to have a CAD tool for efficient thermal management and analysis.

Thermal management at the prototype stage is redundant and expensive and impairs the designer's ability to reach a timely solution. The early evaluation of thermal performance allows design teams to take corrective steps at critical stages during the design process to optimize cost, quality, and reliability. CAD tools for thermal management must support the modeling of all heat transfer mechanisms, handle all conventional cooling effects, and support transient simulations with the user specified timing functions and many more related thermal analysis features.

In Chapter 5, MCM thermal management and thermal design issues such as the basic concepts of thermal management in MCMs and SCMs, thermal resistance and thermal management technologies, thermal management methods such as thermal vias, heat sinks, cooling pistons, cooling bellows, and cooling channels are discussed in detail. In addition, these technologies are also compared and presented in Chapter 5 to show their advantages and disadvantages. In Chapter 8, the effects of thermal considerations on MCM physical design are presented.

9.1.5 Electrical Analysis Tools

As discussed in Chapter 6, electrical simulation and analysis play an important role in the design of MCMs. The layout that is realized at the end of the physical design phase has to be simulated and analyzed. This ensures that layout meets the design specification such as delay and takes into account transmission line effects, skin effects, noise, and various other electrical considerations. As the complexity involved in the simulation of electrical characteristics is very large, CAD tools are necessary to perform these tasks.

CAD packages for the MCMs must be able to test, analyze, and produce results and statistics of the sophisticated high speed electrical and physical requirements that drive the layout process. The system should be able to adhere to the given constraints and apply them wherever necessary so that it does not become tedious for the designer to remember and also specify the constraints repeatedly. The system should be highly capable of estimating different electrical characteristics such as delay, noise, and other related effects while adhering to the industrial standards.

9.1.6 Manufacturability Tools

MCM CAD packages must include tools capable of generating outputs in several formats accepted by the industries that are useful in the manufacture of masks. These manufacturing tools must also be capable of generating formats that can be redirected to pen, photo, raster and laser plotters, NC automatic insertion/onsertion, NC drilling machines, and optical pattern generators. In addition, the manufacturing tools must also be capable of loading data files of different formats and permit the designers to make changes on them. This is particularly useful when designing something similar to an already existing design and making modifications to it to obtain the new design. Additional outputs such as plots and documentation, location of bonding wires, and location of vias must also be supported as these outputs are useful in manufacturing.

9.2 COMMERICAL CAD TOOLS

In response to the growing needs of MCM technology, several vendors have developed CAD tools. In this section, we review some of these tools. The purpose of this review is to study desirable attributes in existing CAD packages and expose limitations for further research and development. We discuss the following commerical CAD packages.

1. MCM Design Suite
2. MCM Engineer
3. Allegro
4. MCM Station

Table 9.1 gives the vendor names and other attributes of some packages, including the ones listed above.

9.2.1 MCM Design Suite

MCM Design Suite is developed specifically to handle the design, construction, layout, and manufacture of next generation high speed and high density MCMs

TABLE 9.1. Commercial CAD Tools

Company	Package Name	Platforms	OS
Cadence Design Systems Inc.	Allergo-MCM	DEC, HP, IBM, Sun	Unix
Cooper & Chyan Technology Inc.	Specctra SP50	IBM RT RS/6000, Sun SPARC Stations	HP-UX, IBM AIX, Sun OS 4
Dazix	MCM Engineer	Intergraph, Sun	Unix, Intergraph OS, Sun OS
Harris EDA	Finesse MCM	Sun SPARC stations	Sun OS
HEM Data Corp.	Snap-Master for Windows	IBM PC/AT 386/486, PS/2	Windows 3.×
Mentor Graphics Corp.	MCM Station	HP, Sun, and UP	Unix
Racal-Redac Ltd.	MCM Design Suite	DEC, HP, IBM, Sun (UK)	Ultrix, Unix SPARC stations
Zuken Inc.	CR 3000	HP 300. 400, and 700	Unix, Sun

[168]. The MCM Design Suite is built to encompass completely the entire development process from conceptual design through analysis to manufacture. The MCM tool set provided by MCM Design Suite is capable of performing complete thermal and signal integrity analyses using CAD Expert's integrated thermal analysis tool set and the High Performance Engineering tool set (HyperScan and ScopeProbe). Unified within the VISION tool framework, the MCM Design Suite is able to communicate directly with all other Expert Series tools. The MCM Design Suite provides the user with complete control over the flexible rules and constraints-driven architecture that underpins the Expert Series philosophy.

In this subsection, we briefly discuss the features of MCM Design Suite. Figure 9.3 gives a global perspective of the features built into MCM Design Suite.

1. **Editing Tools:** MCM Editor is the key tool of MCM Design Suite that helps in communicating between different tool sets integrated into the system and is also primarily responsible for loading the circuit design and specification. MCM Editor is, in fact, a graphical interface with editing features and a host of functionalities grouped together for the MCM Design Suite.

Wire bonded components are well manipulated by the MCM Editor. The bonded components are called from the library and realized with bond pads in default or user-defined positions. Connectivity between the bond pads and the component is maintained within the design, and the wire bonds are automatically displayed on screen. The bond pads can be easily moved and changed in shape by the designer to facilitate placement and routing.

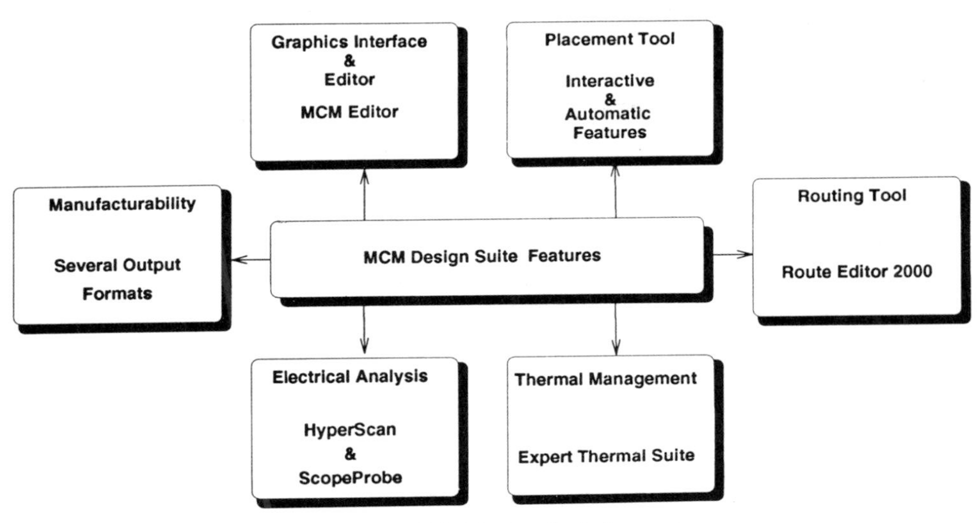

FIG. 9.3. MCM Design Suite features.

MCM Editor has the capability to transfer subcircuit floorplanning areas that are defined during the design specification automatically. These floorplan areas contain the subcircuit associated components. As a result, the auto-interactive floorplanning routines can manipulate the individual subcircuits to achieve an optimal result.

2. **Placement Tools:** MCM Editor has automatic, manual, and auto-interactive capabilities to enable the designer to realize optimum placements. During placement, the designer can specify components by reference designators from the component library and either respond to automatic prompts for placement or intervene manually to reposition components for optimum routing. The connective data structures within MCM Editor allow the designer to move components around the substrate in real-time without having to redraw the connections.

The tree-based automatic placement algorithms consider the entire net during placement, thus allowing the designer to optimize floorplanning. Subcircuit floorplanning is supported through automatic independent placement on each side of the substrate using one or more interacting placement areas simultaneously. A set of components can be defined by a variety of techniques, including wildcard attributes, picking individual components with a cursor, "boxing" components with the mouse, naming components, and a combination of any of these methods. Once defined, the set can be interactively manipulated to suit the layout and to define the placement order. These sets can be moved, stored, rotated as a group, or each block handled individually. MCM Editor's auto-interactive placement tools give the user the ability to pack and unpack components in rows, in columns, or by area. These capabilities are useful when the designer has to reprocess or re-design.

MCM Editor's interactive placement tool set allows the designer to realize optimum placement using subcircuit clusters and height restriction areas created during circuit capture. Interactive design features included are tools for powerful conductor area manipulation, easy to use placement routines, flexible component pad creation and modification, and group manipulations.

3. **Routing Tools:** MCM Editor is fully integrated with CAD Expert's Route Editor 2000. This routing tool delivers fast grid-free, error-free routing in a user controlled environment that is constraint driven but free of limitations. In addition, group manipulation facilities enable the designer to use previously tested routing patterns for speeding up the design process. MCM Editor's interactive design features consist of powerful conductor area generation and manipulation, including advanced power plane routines that contain the choice of creating complex solid, hatched, or cross hatched conductor areas with a wide range of user definable variables to tailor the result fully for optimum manufacturability. Full interactive control and online electrical and manufacturing checks ensure that MCM Editor minimizes the time taken to prepare designs for manufacture.

The maximum number of layers that can be handled by the Route Editor 2000 of the MCM Design Suite is 256. A maximum substrate size of 10 m^2 with the number of leads per component being unlimited gives an additional advantage to the designers. The routing facilities include eight different symbol alternatives, resolution of 0.01 μm, auto-routing on all the 256 layers simultaneously, auto-interactive routing, design rule checking, group manipulations, and unlimited number of nets to be routed. The routing tools are also flexible by using mixed units. Octagonal dynamic delay routing, which is also another feature, plays an important role as a visual guide while manually routing the MCM using minimum and maximum delay rules.

4. **Thermal Management Tools:** The thermal management is handled by the CAD Expert's Thermal Analysis Suite from Quantic Laboratories and Quad Design. The Expert Series thermal analysis tool set offers a multilevel approach at all design stages; for example, an immediate and advance warning of hotspots during component floorplanning up to a detailed boundary condition is made available to the designer by the thermal tool. The thermal manager also performs a full three-dimensional enclosure level analysis.

5. **Electrical Analysis Tools:** The MCM Design Suite is integrated with physical analysis tools such as HyperScan and ScopeProbe transmission line analysis tools and full analysis tools from Quantic Laboratories and Quad Design. Full integration provides complete data transfer and invocation of the tool from within the MCM Design Suite. This high level of integration allows early analysis of the design while minimizing design interruption. Full interactive control and manufacturing checks ensure that MCM Editor minimizes the time taken to prepare designs for manufacture.

An online easy to use transmission line analysis is available with the probe and simulates the ScopeProbe simulator, which gives fast graphical feedback on the analog effects of critical signals. This tool can be driven from either schematics or the MCM Design Suite giving the engineer/designer totalflexibility to work in the optimum environment.

Factors that can be specified to the MCM Design Suite, in order to monitor the electrical behavior of the layout, are hole proximity; bond wire length; spacing rules by layer and by signal class; pad to pad spacing; via type per layer pair; no via area/no conductor areas; blind, buried, stacked, or staggered vias; wire bond proximity; acid trap check; via type per net; routing direction; layer usage/ material specification; conductor width by layer; and via/bond pad spacing.

6. **Manufacturability Tools:** A total manufacturing integration for a wide range of manufacturing links from GDS2 stream or OPG data to ACM and ATE links for complete automated assembly and test is supported by MCM Design Suite. Use of the integral powerful two-dimensional drafting options and the built-in report generation capability reduces the risk that manu-facturing documentation needs will create. All design rules are available for the designer so that the designers can modify and suit their design to match their layout technology and manufacturing methods and processes. MCM Design Suite is flexible and is capable of generating outputs in several formats accepted by the industries.

GDSII input for component graphics and output for mask generation are well supported. MCM Design Suite is capable of generating formats that can be redirected to pen, photo, raster and laser plotters, numerically controlled automatic insertion/onsertion, numerically controlled drilling machines, and optical pattern generators.

The Interfaces associated with the MCM Design Suite are Schematic editor, Mechanical CAD, and several more.

7. **Other Features:** The MCM Design Suite can handle MCM designs up to a resolution of $0.01 \mu m$. Its integration with VISION framework enables it to share data with a wide variety of tool kits. MCM Editor also allows up to eight alternative component mounting representations to be defined, that is, TAB, flip-chip, solder bump, wire bond, PLCC.

9.2.2 MCM Engineer

MCM Engineer is a complete system for the design, verification, and manufacture of MCMs. MCM Engineer links automatic and interactive layout functions, providing a flexible tool set for addressing complex MCM technologies, specifically MCM-L, MCM-C, MCM-D, MCM-D/C, and MCM-Si. These design technologies are related to various manufacturing methods that incorporate different substrate, conductor, and dielectric materials. In addition, it also addresses the complex system-level design environment of MCM designs. Pre-layout and post-layout simulation tools are also integrated to provide support for various analysis capabilities such as analog and digital simulation, transmission line, thermal, parasitic, and reliability analysis [49].

In this subsection, we briefly discuss the features of MCM Engineer. Figure 9.4 gives a global perspective of the features built into MCM Engineer.

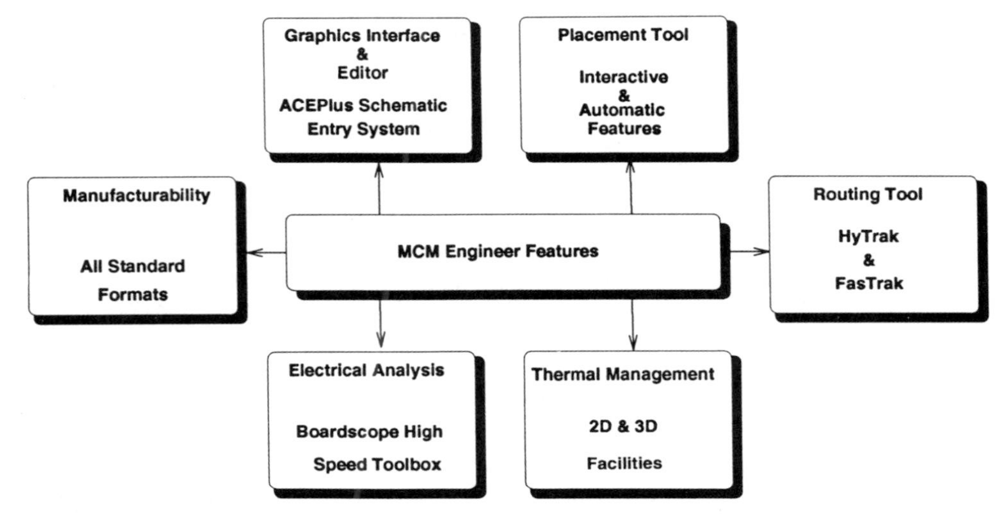

FIG. 9.4. MCM Engineer features.

1. **Editing Tools:** The ACEPlus schematic entry system is the chief control unit of MCM Engineer as it controls a majority of the package features. ACEPlus provides top-down as well as bottom-up design techniques, offering greater flexibility in creating schematics to the design terms.

The ACEPlus schematic entry system is useful in specifying the design and is driven by data provided by extraction from a schematic or entered from a keyboard. The tool also has the flexibility of checking the design rules and the electrical design and also facilitates in accepting different types of input through system level, functional level, and gate level diagrams.

Using the top-down structure the designer can partition the system level description into a set of modules. A system generated design map is used to track the design architecture as it progresses. At the completion of different phases reports, graphs and statistics are generated. With the help of ACEPlus, designers are allowed to partition the whole system into blocks and represent each of the blocks by a schematic symbol. After partitioning the designer can enter a behavioral VHDL symbol. These are verified by digital simulations. This system also facilitates testing either each individual block in an incremental fashion or the complete system.

The schematic entry system and the MCM design tools operate in a high performance graphics environment that is flexible and can display simultaneously views of the schematic and layout editors. When the designer selects a component symbol or a signal in the schematic window, then the system is built up such that the appropriate trace or block or the related item is highlighted in the MCM Editor window. This feature is bidirectional that is, the same is valid while performing any operation in the MCM layout editor. ACEPlus also provides extensive symbol libraries for the MCM design.

An online library on the workstation or on the network stores several details, and this database is evidently helpful for a designer. A user may also create a new library using command macros and store them in the database for later use. Once the circuit is laid down, online standard simulators that meet the industry standards can be used to test the circuit.

2. **Placement Tools:** MCM Engineer provides the designer with interactive, semiautomatic, and automatic placement features. As discussed in Section 9.1.2, all these features help in optimizing the design.

Interactive placement is facilitated in different ways. All components to be considered are made available to the designer in an easy way by means of a scrolling list. The designer may choose the device to be placed and drag the same onto the screen or tag the reference number from the list. Also several other features are present to perform a select operation.

MCM Engineer provides several ways to place an object. A place-by-net option in the interactive placement allows the specification of a particular net or nets to be considered for placement. This option is useful for the critical device placement, as by using this option the designer is presented with only those devices associated with the specific net selected. Place-by-reference allows devices to be placed for viewing in alphabetic order. This approach is mainly useful while placing parts that have predefined locations such as connectors and I/O pads. Constructive initial placement is another interactive approach in which the tool will calculate placement order and initial component location based on the minimum length of interconnects. By doing so each component is displayed one after the other in geometrically ideal locations with respect to components to which they are connected. The devices are then placed based on the designer's choice. Schematic pick and place facility is also a part of the interactive package. As each component is selected for placement all its related and connected objects are also displayed.

Automatic placement is another important feature that is provided by the MCM Engineer. Critical devices are placed first, using the interactive routines, and then the auto-place routine is invoked. MCM Engineer allows the designer to specify the design specifications before invoking the auto-placement routine. The auto-placement routine adheres to the specified design considerations until the end of the design phase.

3. **Routing Tools:** The routing facilities of MCM Engineer are provided by an automatic router called HyTrack and FasTrak, a high speed circuits router. The HyTrack automatic router is specifically designed to address the requirements of MCM interconnect technologies. HyTrack takes care of multiple layers, placing vias in staggered, stacked, or spiral way. HyTrack also automatically generates fanout vias in user definable patterns. The designer is allowed to specify certain options, such as number of routing layers, design clearances, trace restrictions (45° or 90° bends), via stagger distances, multiple trace widths, and multiple embedded planes, and then activate the automatic router. The routing phase is completed in two steps as follows. First, the specified design is evaluated and routed based on achieving maximum completion. After routing all the nets, the

next step is to facilitate rip-up and re-route so as to optimize the routes. The ability to improve design is cost based.

In FasTrak, the designer is allowed to preset certain conditions required for high speed MCM circuits. Transmission lines are routed by using a net structure (star, daisy chain). The structures are recognized by the router as a design rule. Crosstalk in parallel lines or among adjacent layer lines is also taken care of or minimized by iterative routing passes. Differential pairs are identified on input and routed correctly. Matched lengths of nets are achieved by specifying the minimum or maximum length of the nets or by simply setting them to be equal. In addition, several more routing options are available. They are rip-up and re-route, notch, stretch, and neckdown. When high speed signal lines are encountered, the designer has the option of using curved tracking to make changes of direction. These features are built into HyTrack coupled with the specific high speed options of FasTrak.

4. **Thermal Management Tools:** Thermal analysis of the module allows a designer to detect potential problems before the act of fabrication. The MCM Engineer products can be combined with industry standard software, enabling complete control to be maintained during the design cycle. Optional packages can be used to provide design and analysis in two and three-dimensional thermal management.

5. **Electrical Analysis Tools:** The electrical characteristics and analysis of the MCM Engineer is provided by the Boardscope High speed ToolBox. Signal degradation caused by factors such as large etch delays, signal reflections, crosstalk, attenuation, and many others is much more evident during high speed circuit operation and is handled efficiently by the Boardscope High Speed ToolBox. Boardscope High Speed ToolBox is provided to analyze the critical factors and give feedback. Nets are analyzed against user specified criteria, and violation reports are generated aiding the designer to take corrective action. The designer can specify appropriate control parameters to analyze circuit characteristics such as track length, parallel tracks, etch delay, characteristic impedance, impedance, line capacitance, and load capacitance.

6. **Manufacturability Tools:** The output generated is adaptable to the industry standards. GDSII stream data is supported. The designer specifies the different components or objects to appear on a given mask, and the tool generates the data in GDSII format.

GDSII data files can also be loaded onto the MCM Engineer and changes can be made on it. This is particularly useful while designing something similar to an already existing design. Additional outputs that are generated and useful in manufacturing are

a. Check plots and documentation.

b. *x-y* locations of bonding wires.

c. *x-y* locations of vias on a layer pair basis.

d. Gerber output.

7. **Other Tools:** When the interconnections are completed, the designer is given the flexibility of running a batch DRC on the entire system or a specified subsystem. The DRC outputs all the errors and goes into an interactive mode with which the designers can correct their errors. Mechanical checks are also possible using the powerful three-dimensional capabilities of the graphical system on which MCM Engineer is based. MCM Engineer supports the integration of third party design tool sets into the CAD package.

9.2.3 Allegro

Allegro is a constraint-driven design-layout system providing an advanced environment for the physical design and analysis of PCBs, MCMs, hybrids, and multiwire board designs. Allegro addresses performance criteria while meeting reliability, testability, and manufacturability concerns. Many of the features in Allegro have resulted from the MCM contract work with Microelectronics and Computer Technology Center Corp. (MCC), Austin, Texas [136]. Details are provided elsewhere [193].

Allegro's fully integrated set of capabilities include an editor for defining and checking design rules, automatic and interactive place and route capabilities, capacity to handle simultaneously up to 48 signal layers, and a range of in-process analysis tools for electrical, thermal, reliability, testability, and manufacturability analysis. Allegro extends the concepts of rules-driven design and in-process analysis to create a system in which important design or technology considerations are defined in advance and are automatically checked and enforced throughout the design process to shorten design cycles and optimize design, performance, quality, and cost. Figure 9.5 gives a pictorial view of the features integrated into Allegro.

1. **Editing Tools:** One of the most important and powerful features of Allegro is the Constraints Editor, which acts as the command center for the correct-by-design process. This editor is used to establish rules or constraints regarding electrical, physical, or thermal considerations required to ensure signal integrity, design reliability, testability, and overall manufacturability. Allegro uses the information contained within these constraint definitions to drive the layout process.

The Constraints Editor is forms-driven and allows users to create a matrix of technologies and constraints matched to specific fabrication environments. Through the forms-driven matrix, the Constraints Editor is useful to the design team for presenting constraints options that must be considered for a particular design type, depending on the interconnect technology involved. Designers can assign constraints and design rule specifications at the start of the design cycle so that the interactive and automatic tools in Allegro can adhere to them throughout the design cycle and can assist in evaluating trade off during different phases of the design cycle.

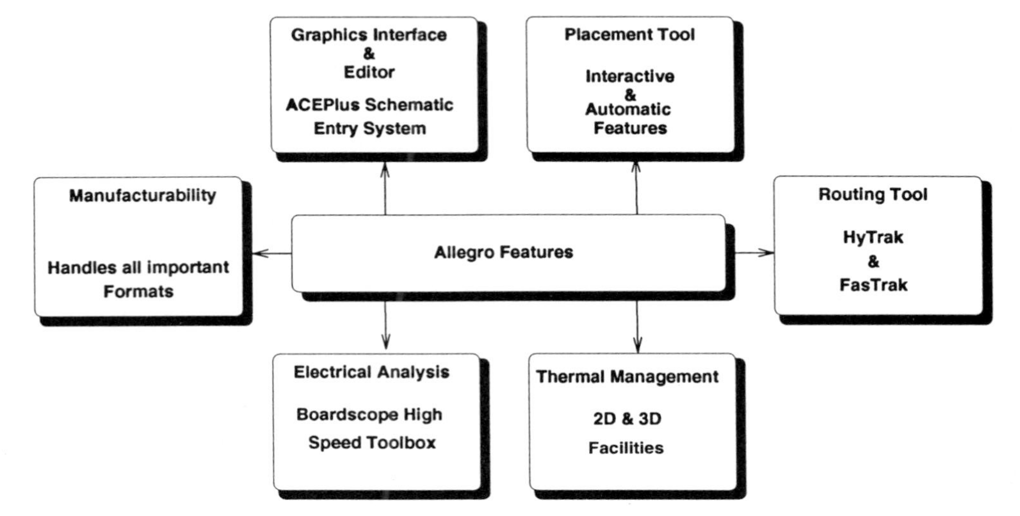

FIG. 9.5. Allegro features.

The correct-by-design system provides process control but does not limit the creativity of design team members. Critical constraint sets may be locked, helping users to avoid common errors and prohibiting them from changing or overriding constraint values. However, users can turn off or delay most constraint checks based on their expertise and technology requirements.

2. **Placement Tools:** To address the issue of signal integrity concerns, such as noise, timing, clock skews, and so on, Allegro provides a timing-driven placement and crosstalk-driven routing. Allegro supports both interactive and automatic placement capabilities.

Allegro's design methodology drives a powerful and flexible set of placement tools, including floorplanning, auto-interactive placement, and a unique placement evaluator, that automatically analyzes the routability of a design at any stage of the placement. Allegro's placement software automatically considers "rules-driven" engineering criteria such as signal delays, critical paths, and terminator assignments as factors for determining the positioning of critically linked components.

The auto-interactive placement mode computes and displays ideal or optimum locations for each component, while enabling experienced designers to make modifications based on their own judgment. In addition, Allegro features a unique placement evaluator that automatically analyzes the routability of a design at any stage of the placement. The placement evaluator uses actual routing parameters from the system's auto-router to analyze available routing channels and create a density report in the form of a color map. The graphic display, which is more accurate than histograms based solely on rats-nest information, highlights overflow areas and location where there are too few available routing channels.

3. **Routing Tools:** Allegro's advanced placement tools are coupled with the *Prance-XL* and *INSIGHT* expert routing systems to ensure correct quality routing with high completion rates and fast routing times for novice and experienced users alike.

Prance-XL is a high performance, high completion auto-routing system that incorporates WARP-4 technology to provide the completion and flexibility of gridless routing systems and higher speed grid-based routing algorithms. WARP-4 maps the available routing space into contours that flow around pads and other obstacles to achieve maximum trace density. Since Prance-XL views these channels as straight, it is free to proceed at full speed. Prance-XL provides a solution that supports constraints such as route-to-length, matched length, string order, stub length, and minimizing parallelism.

Allegro's INSIGHT autorouter also supports critical electrical design considerations in addition to the physical design constraints. The system is monitored throughout the routing stage to ensure correctness. Between Prance-XL and INSIGHT the guesswork of parameter setting is removed, and it runs multiple routing algorithms effectively. Users specify a few directives such as spacing rules and line widths, and the system automatically sets over 40 routing parameters to provide the best results. Advanced users can interrupt, make adjustments, and reenter the routers at any time, providing complete control over the routing process. This flexibility and control allows users to use the INSIGHT router for critical multiconstraint nets and Prance-XL to route all remaining nets quickly. Both routers support a variety of through-hole, SMT, and MCM design requirements. INSIGHT's scheduling algorithm automatically sequences sources and loads, while a terminator assignment routine matches the appropriate terminator to minimize reflections.

4. **Electrical Analysis Tools:** Allegro is integrated with flexible features for shape handling and analog design. Allegro's flexible interactive and automatic tools ease the development of power and ground planes. For example, when user-created shapes are superimposed on traces, the system automatically generates thermal relief for connected pads and clearances around pads and nets not attached to the shape. Nets can be attached to shapes manually or by the autorouter, facilitating the generation of split planes. Voids can be interactively or automatically merged.

For analog boards, a variety of special layout features is included to support the analog designer. For example, components may be rotated at predefined or user-defined angles of 0.001 degrees to optimize board area. Signal reflections are minimized by the systems ability to route with smooth curved traces. Copper areas with curved boundaries can also be created to minimize resistance.

Allegro's online electrical parameter calculator provides access to electrical information for any etch segment. Values for propagation delay, capacitance, resistance, inductance, and impedance are displayed interactively. Modifications to etch width, via types, or terminator values result in immediate feedback of the new values, helping designers to create circuits that meet the desired electrical performance.

5. **Thermal Management Tools:** *Thermax* is an optional suite of three-dimensional thermal analysis tools integrated with Allegro. A thermal expert or layout designer can use the tool to gain feedback quickly on thermal aspects of the layout decisions. The system uses an isothermal map to display heat distribution on the substrate graphically and highlights hot components whose junction temperatures exceed user defined values.

Thermax is integrated with the Allegro correct-by-design system. Viable reliability analysis and Analog Workbench deliver a thermal analysis solution that helps the designer to make appropriate decisions. The Thermax environment

a. Provides three-dimensional modeling with the flexibility to model most arbitrary structures using automatic and adaptive mesh generation
b. Models all heat transfer mechanisms
c. Handles all conventional cooling methods and analyzes the effects of heat sinks mounted on components or boards or embedded within the boards
d. Supports transient simulation to improve thermal management of designs by simulating thermal characteristics with user-specified timing functions
e. Integrates thermal analysis into the concurrent design engineering environment by facilitating quick "what-if" analysis to satisfy thermal constraints during the design process
f. Addresses both the thermal Expert's modeling flexibility and layout designer's ease-of-use requirements
g. Combines robust models and hierarchical libraries at the material, package, and component levels with an easy to use, intuitive graphical user interface
h. Uses the flexible extension language for customization into user design environments

6. **Manufacturability Tools:** Allegro's design for manufacture capabilities allows system designers to design easily and to audit against factory-specific or unique manufacturing process requirements. The optional design for assembly (DFA) and design for manufacture (DFM) tool sets feature user-definable design rule support, a complete set of analysis and audits, and rules-driven DFM generators for the fabrication process. DFA provides design verification for the assembly process, including component placement to accommodate automatic assembly, soldering, inspection, and repair. The DFM software automatically generates factory fabrication data such as solder mask, solder paste, and registration coupons. A comprehensive report generator automatically produces documentation such as the final netlist, placed and unplaced components by location, components by pin, function summaries, bill of materials, padstack summaries, DRC errors, ECL loading reports, and more.

A full range of Gerber compatible artwork including etch, drill, silkscreen, solder, resistor paste, and test materials drawings can be produced. The CALS-

OUT documentation package enables the designer to produce documentation to meet standards including ANSI, ISO, DIN, BIS, JIS, and AFNOR.

7. **Other Tools:** Allegro features a wide range of standard and optional analysis tools to monitor design constraints. The viable reliability analyzer provides estimated mean-time-between-failures (MTBF) for assemblies and absolute failure rates for components as soon as packaging is done.

Allegro's optional integrity analysis tools *SigNoise* and *SigDelay* enable users to screen their designs for potential noise or timing problems. Based on user-specified design rules, transmission line simulation and crosstalk flag networks with excessive reflections, crosstalk, thermal shift, and ohmic loss can be detected. The tools provide a waveform display to portray graphically the simulation results of either a rising or falling edge, reporting both overshoot and undershoot, as well as backward and forward crosstalk.

Powerful interactive features provide "controlled automation" to maintain user control while maximizing productivity. All interactive tools are supported by Allegro's correct-by-design methodology, automatically alerting users when predefined constraints are exceeded. These tools also ease the development of power and ground planes. Designers are allowed to pre-route critical connections and analyze their electrical performance before routing the entire design. When adjustments are needed, special editing capabilities simplify the tasks. For example, Allegro's push and shove capability makes room for the new connections automatically. Or, when vias are needed, Allegro's bubble feature will automatically bend traces around the new via in accordance with the design rules. In addition to reliability, electrical, and thermal analyses, a unique testability analyzer is available to monitor physical testability requirements, such as test method and probe size, and to make sure that considerations for bare board and in-circuit tests are met.

9.2.4 MCM Station

MCM Station is a design system for MCMs and supports all major substrate technologies, chip-mounting techniques, complex via structure, and various other features helpful in the design of MCMs. MCM Station covers the design process from design entry and analysis through constraint-driven placement and routing, thermal and signal-integrity analysis, to manufacturing outputs and drawings.

MCM Station is based on Mentor Graphics Falcon Framework for concurrent design, which supports openness and integration of tools from third party vendors and in-house tool developers. The graphical environment is based on OSF/Motif standard. With the use of INFORM, Mentor Graphic's interactive CD-ROM based online information system, users have fast access to integrated reference help and full online manuals and tutorials.

All the logical and physical aspects related to MCM design are integrated and available in MCM Station. During the front-end design process, the designer can

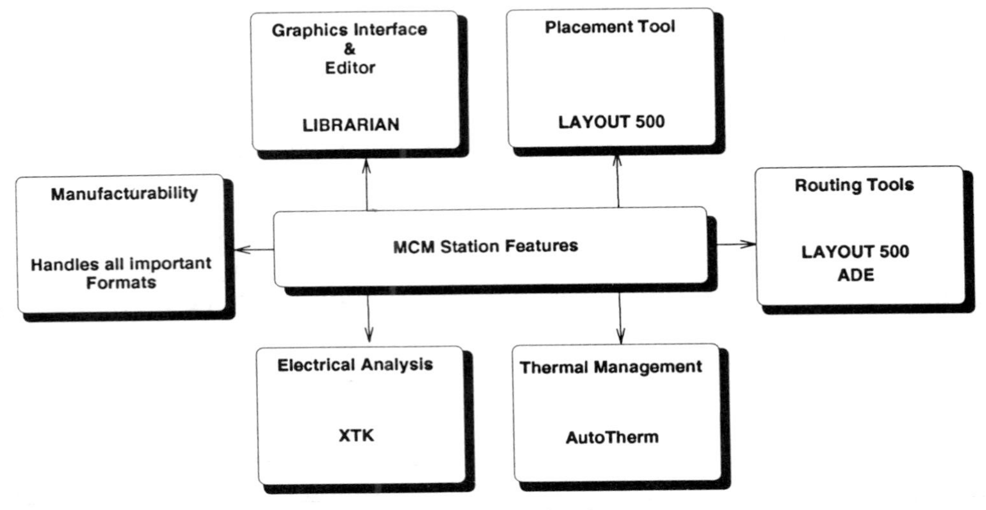

FIG. 9.6. MCM Station features.

specify timing, topology, and physical layout rules. Using the material property information, such as layer thickness, dielectric constants, and conductivity, MCM Station calculates the signal propagation effects of these properties and maps them into physical constraints used by placement and routing algorithms. User feedback of the system helps the designer to meet users' electrical requirements during interactive design editing. MCM Station supports both metric and English unit system handling with up to 0.01 μm resolution. In this subsection we briefly discuss the features of MCM Station. Figure 9.6 gives a global perspective of the features built into MCM Station. Please refer to Mentor Graphics [77] for more details.

1. **Editing Tools:** MCM Station is provided with a graphics editor called LIBRARIAN for defining and modifying geometric parts such as die, device, layer specific material attributes, and so on. LIBRARIAN allows the designer to define design rules such as spacing and width, placement, routing, via keep out areas, and so on. Multiple windows are integrated in LIBRARIAN to support logical-to-physical mapping definitions. MCM Station handles TAB, flip-chip, SMT, and chip-and-wire footprint definitions with the help of LIBRARIAN. Component highlighting from schematic for easy part identification, dynamic display of guidewires during placement, and Histogram display for Manhattan wiring congestion are some of the other features of LIBRARIAN.

LIBRARIAN aids the user to define number of layers and attributes, specify minimum separation between the pads, give via descriptions, and list the parts data. In addition, the designer can also define the type of via structures to be used, such as blind, buried, or staggered, and their rule capabilities. LIBRARIAN also stores information regarding via and routing keep out areas or substrate definition, substrate outline, layer names, and devices of thermal-power dissipation.

LIBRARIAN is capable of importing die-geometry footprints from the MCM intermediate Die Table format, which can come from other CAD tools. In addition, LIBRARIAN also supports in-corporation of substrate design rules and automatic updates of parts files and is capable of reading GDSII input for device geometries.

2. **Placement Tools:** To overcome the challenges of the placement phase of the physical design cycle, MCM station supports both interactive and automatic placement, helping the designer to achieve design optimizations.

Interactive and automatic placement in MCM Station is provided by *LAYOUT 500* in MCM Station. LAYOUT 500 is an automatic and interactive placement and routing environment for MCM design. This tool is capable of generating reports, graphs, statistics, histograms, force vectors, and guidewires most of which can be used to predict the output of the placement and routing phase and hence is useful to the designer when taking corrective measures. In addition, the information generated serves to give graphical information about connectivity and routability, once again helping the design team to take corrective measures.

LAYOUT 500 supports interactive and connectivity-driven placement while providing capabilities to move, align, rotate, pivot, and swap individual or groups of components and user-optional placement grid. In addition, several more features are integrated with LAYOUT 500. They are component placement preferred regions and keep out regions based on user-defined properties, any-angle device placement in one degree increments, net priority support for highspeed design, ability to protect device locations, dynamic display of guidewires during placement, and automatic and interactive pin and device swapping.

With the help of automatic constructive placement from LAYOUT 500, the MCM Station can be requested to exercise its knowledge of connectivity and net priority in placing components. The designer can also specify rough floor plans by defining placement regions and preferred circuit groups for these regions. To further refine and improve placement, the tool allows the designer to look at the thermal, timing, and signal integrity effects.

3. **Routing Tools:** As in the case of placement, MCM Station supports both interactive and automatic routing features. The interactive routing facility from LAYOUT 500 supports point-to-point addition of vertices to define wire paths on any layer and toggling between layers. An online design checker ensures the correctness of the MCM layout.

The primary routing tool in MCM Station is a maze router with rip-up, retry, and shove-aside algorithms. User-defined or default costs guide the router. The aims of the costs are based on number of bends, vias, grids, and wrong way wiring to minimize the total cost of the wiring activities per connection.

The router provided by the MCM Station is capable of routing as many layers as required. Automatic net topology routing, timing-constrained routing, rip-up and retry, blind, buried, and staggered via support, and layer allocation to horizontal, vertical, and diagonal preffered routing to prevent crosstalk are also well supported. In addition, several more features integrated into LAYOUT 500 router are memory routing, squeeze-through/shove-aside algorithms, any-angle interactive routing, real-time drag of actual trace width plus clearance during interactive editing, orthogonal/diagonal auto-routing at pin entry/exit, user-assigned design rules by net or net class and/or by layer, user-specified net length for complete net or by layer, user-directed control of routing costs, simultaneous auto-routing of multiple trace widths and via types, interconnect protection by net name or area, automatic critical net priority with subsequent rerouting protection, automatic template creation for correct via widths and clearances during interactive routing, and user-definable net-length reports for length violations.

MCM Station is also supported by an Advanced Dynamic Editor (ADE), which allows the designer to freehand sketch a path through immovable objects such as vias, pads, and pins that lie between the path's beginning and end points. ADE then creates the routed path with full design rule compliance by pushing, shoving, and re-routing other traces as necessary. ADE allows maximum utilization of space with its grid-less capabilities. However, ADE is also capable of adhering to a user-specified grid.

MCM Station highlights all violations on screen within LAYOUT 500. Reports can be generated at any time to describe electrical and design rules. The reports include information such as Manhattan distance versus actual lengths, electrical rules versus actual physical violations, and all other relevant information about high speed design requirements.

4. **Thermal Analysis Tools:** MCM Station is supported with a tool called AutoTherm for thermal and signal analysis. AutoTherm lets the designer optimize MCM reliability by testing the thermal behavior of the design before fabrication. The thermal analysis is based on finite element analysis (FEA) methods. This tool provides fully automatic meshing and support for conduction, convection, radiation, and forced-convection cooling mechanism. AutoTherm lets a user quickly iterate part placement, device mounting parameters, airflow, and heat removal methods, such as different materials or the addition of thermal vias and/or heat sinks, in a matter of seconds. It also reads substrate and device material properties and power parameters from LAYOUT 500 and the parts library. AutoTherm then allows a user to evaluate quickly and easily the effects of convection, forced-air convection, conduction, and radiation cooling mechanisms. To evaluate manufacturing processes, the effects of power cycling, loss of cooling, or other time-dependent effects, the designer can use the transient analysis capabilities of AutoTherm. To summarize, AutoTherm supports the following features:

a. Conduction, convection, radiation, and heat transfer mechanisms under steady state and transient conditions
b. Component models that account for three-dimensional thermal effects

c. Automatic generation of FEA mesh from model geometries, eliminating the need to learn this tedious step
d. Accurate meshing of irregular shapes
e. Potential flow modeling for computing velocities, heat transfer coefficients, and forced air convection cooling
f. Support for standard and user definable convection models.
g. Automatic calculation of heat transfer coefficients
h. Reporting capabilities

5. **Electrical Analysis Tools:** For high speed timing and signal integrity analysis, MCM Station contains *XTK*, an integrated version of Quad Design Technology's crosstalk tool kit. XTK includes modules for both crosstalk and transmission line analysis. Using field solver algorithms, XTK predicts the effects of placement, routing, and circuit termination on system level timing and signal integrity for an entire MCM quickly. With Design Architect's physical layout module, the user is allowed to specify net topology control, maximum interconnect delay, impedance control, matched length rules, shielding rules, and crosstalk rules from the schematic. Once these constraints are specified, they are automatically forwarded to the physical layout. Topology constraints and controls include the following:

a. Minimum/maximum net or segment lengths
b. Stub length by net or segment
c. Layer restrictions by net or segment
d. Via limits and usage
e. Parallelism limits by segment or cumulative, same or different layers
f. Balanced pair/bus routing constraints

The impact of interconnect delay introduced in the physical design of an MCM is significant and must be considered as early as possible. MCM Station supports the minimum and/or maximum interconnect delays allowable for given nets. The designer can also match delays between nets or within branches of the same net to limit layout induced signal skew. MCM Station automatically converts delay constraints (in nanoseconds) specified in Design Architect into net and segment length constraints (in microns or mils) using the material properties of the substrate.

6. **Manufacturability Tools:** MCM Station automatically generates manufacturing data and documentation. The data are formatted automatically for drilling and milling NC machinery and common artwork standards such as GDSII and Gerber. From the artwork, the user can verify GDSII mask or film accuracy by previewing the output onto the workstation screen. The application also includes panelization of substrates.

7. **Other Tools:** MCM Station supports both thin and thick film designs with silicon, ceramic, and other substrates.

TABLE 9.2. Noncommercial CAD Tools

Tool Name	Developed At	Purpose
AUDiT	Cornell University	Automated design and trade off simulator for integration of computer structures
MICON	Carnegie Mellon University	Chip set synthesis system
PEPPER	IBM	Functional partitioning, chip placement, package congestion, pin assignment, and coupled noise analysis
SUSPENS	Stanford University	System-level circuit model for central processing units
The MSDA Tool	MCC	Software tool to reduce design risks associated with the selection of packaging technologies for integrated circuits

9.3 NONCOMMERCIAL CAD TOOLS

In this section, we discuss the following noncommercial CAD tools:

1. AUDiT
2. MICON
3. PEPPER
4. SUSPENS
5. The MSDA tool

Each of these noncommercial tools attempts to treat concurrently a subset of the interdisciplinary system synthesis/specification space using a mixture of estimation, simulation, and synthesis techniques [178]. Table 9.2 gives a global perspective of the noncommercial CAD tools.

9.3.1 The Multichip Systems Design Advisor Tool (MSDA)

The MSDA tool has been developed at MCC [176, 177]. MSDA is a software tool for enhancing the manufacturability and decreasing the design risk associated with the selection of packaging technologies for integrated circuits. MSDA is primarily designed to compute the physical, electrical, thermal, reliability, testability, and cost performance metrics for multichip systems. The technologies supported by the MSDA tool are

1. Traditional and fine-line PCBs
2. Low temperature co-fired ceramic
3. Thin film MCMs

The tool is built from multiple set design rules that are capable of handling all the technologies. In addition, the user is allowed to customize the properties or add their own proprietary technologies. The component assembly approaches that are well supported by MSDA are

1. Wire bonding
2. TAB
3. Flip-chip
4. Single chip packages

In addition, materials are also available for bare die attach, encapsulation, heat exchanger attach, and for defining the bonding and interconnect technologies.

First, the tool begins by capturing design, performance budgets, and constraints. Component descriptions include the entry of physical dimensions, material, I/O constraints such as number of signals, power, ground, unused, electrical constraints such as bias level, logic swing, simple driver and receiver descriptions or driver and receiver SPICE models, thermal constraints, and manufacturing constraints. All these details are associated with the chip class. In addition, the chip class contains details of the type of bonding, die attach material, encapsulation, and single chip package associated with a particular component. Furthermore, the characteristics of heat spreaders and thermal vias are captured as properties of each component rather than the module or board so that they can be easily manipulated on a per component basis.

After the components are described to the tool, they are placed into partitions. Partitions consist of any subset of defined components grouped into a few-chip package, MCM, or PCB. It is through the creation of partitions that complex systems can be built using the tool. The characteristics of partitions include interconnect technologies, heat exchangers, and connectors. A netlist corresponding to the partition under design, a group of partitions, or a design larger than that under consideration is an optional input to the trade off tool. However, conceptual design often takes place before the synthesis of a netlist. The partitions are designed to be hierarchical and organized in a tree structure so that they can be placed within each other to build systems and study trade offs associated with alternative distributions of components among system interconnects.

Besides providing a concurrent performance estimation through closed-form algorithms, the MSDA tool supplements its internal analysis by linking to detailed simulation through model building and simulator management activities. Detailed simulation links are presently used as part of the electrical performance analysis and reliability evaluation links are presently used as part of the electrical performance analysis and reliability evaluation in the tool. The analysis of critical net delays is one example of the fusing of estimation level analysis and simulation and the effective use of simulators in conceptual design activities. The evaluation of critical net delays at the conceptual design level consists of the formulation of a packaging delay that includes time-of-flight, RC charging, and contributions from reflections. Packaging delays along with noise contributions and the delays associated with gates, receivers, and memory access times determine the maximum operating frequency of multichip systems. Dewey and Director [54], provide a brief description of the process by which critical net delays are evaluated and of the MSDA tool.

9.3.2 Stanford University System Performance Simulator (SUSPENS)

In this section, we discuss a system level circuit model for CPUs referred to as SUSPENS [7]. This model emphasizes the interactions among devices, circuits, logic, packaging, and architecture.

The SUSPENS model is primarily an analytical model that takes some material, device, circuit logic, package, and architecture parameters as inputs and uses them to calculate the clock frequency of the system. The SUSPENS model has to be complemented by a system level architectural model to calculate the organizational efficiency of the system (CPI ratio). The cycle time and cycles-per-instruction ratio together yield the system performance. In addition, SUSPENS provides chip sizes, power dissipation, and package dimension.

The derivation of the SUSPENS model, results obtained by SUSPENS on the clock frequency, power dissipation, and chip and module sizes of existing microprocessors, gate arrays, and miniframe and mainframe computers are discussed in detail by Bakoglu [6]. In addition, details of the model being used to predict the performance of future microprocessors and to compare CMOS, bipolar, and GaAs technologies in a system environment are also discussed in detail. SUSPENS supports the following performance-related factors:

1. **Device Properties:** Device properties include current drive capabilities and input capacitance of transistors.
2. **On-Chip Interconnections:** Electrical properties such as resistance and capacitance, packaging density, and number of levels on the on-chip wires are the factors that are considered as a part of the on-chip interconnections.
3. **Chip-to-Chip Interconnections:** The factors are the same as that in the on-chip interconnections except that the number of levels of module level wiring is considered.
4. **Cooling Capability:** Power dissipation of the module is considered.
5. **Machine Architecture, Organization, and Implementation:** The factors determined by the machine architecture, organization, and implementation considerations are interconnection requirements of the system, average interconnection lengths, and the logic-depth and fanout of the circuits. In addition, they also determine the machine cycles required per instruction (CPI ratio).

The SUSPENS model also supports various chip and packaging technologies. The chip technology can be silicon *n*MOS, CMOS, BJT, GaAs, MESFET,

HEMT, or HBT. Differences due to basic transistor properties and in-chip wiring densities of these technologies can be compared using SUSPENS. The module, as defined in the SUSPENS model, can be a PCB, a ceramic multilayer hybrid, a silicon-on-silicon package, or global connections in a wafer scale integrated system. The major differences among these packages are the electrical properties and packing densities of their interconnections, electrical properties of the connections between the module and chips (capacitive and inductive discontinuities), and the power-dissipation capability of the module. The basic principles are general enough so as to be applied to different packaging schemes.

SUSPENS can be used to calculate the optimal chip integration level for a multichip CPU for various packaging techniques. Experimental results show that as the minimum feature size is scaled down, interconnections and packaging become major performance bottlenecks. Another important application of SUSPENS is to model the hierarchical nature of interconnecting a digital system and to determine the best way of partitioning the interconnections between package and silicon. This can be used to establish the optimal level of integration (number of logic gates per chip). The effects of technology parameters, such as minimum feature size, number of interconnection levels, and interconnection pitch, are also examined by varying them independently. The overall performance of systems built from different technologies can be compared for different minimum feature sizes. The relative importance of chip and package level limitations and their interrelation are also studied.

The SUSPENS model is developed in a very general way and is also well structured, thereby allowing easy modifications. This model can be easily tailored and tuned to examine a particular design style or product line accurately. The important aspect of SUSPENS is the approach adopted, which ties together the material, device, circuit, logic design, packaging, and system architecture parameters.

9.3.3 AUDiT

AUDiT is an automated design and trade off simulator for integration of electronic systems developed at Cornell University [121, 161]. AUDiT provides a methodology for package design and trade off studies. The packaging hierarchy from a circuit on an IC to a full system can be described with the physical and architecture models of AUDiT.

AUDiT represents a more detailed version of the SUSPENS work with extensions into module and multiple module estimations. Like SUSPENS, AUDiT starts with a circuit count and/or bit count along with a technology database that includes information about each logic type and feature size dictated design rules. AUDiT estimates the number of I/Os per component, component dimensions, on-chip electrical properties including crosstalk and delta-I noise, on-chip delay quantities, and power dissipation. AUDiT's module level abilities include delay estimation (and associated metrics: clock rate, cycle time, MIPS), module size, and power dissipation. The delay and timing estimations in

AUDiT are considerably more sophisticated than those in the earlier SUSPENS tool. In addition, AUDiT includes a simulated annealing numerical optimizer.

AUDiT can be used effectively in the testing of MCMs. MCMs are at the center of advanced packaging research due to their added advantages of high density, high input/output, and high speed chip packaging. Despite their progress, MCM design and trade off issues have largely remained uncharted. These issues have to be answered for computer types of applications in order to learn more about the conceptual, structural, material, and performance issues involved. AUDiT aids in achieving these objectives. The system characteristics are computed by solving a large set of equations. Optimization techniques are explored using simulated annealing algorithms. AUDiT can run under UNIX, C Fortran, and X Windows environments [122, 162].

The use of AUDiT has been demonstrated over an MCM model, and the overall MCM has been optimized using the simulated analysis results [121]. The MCM model is assumed with a descriptive analysis of all its logical and physical parameters.

9.3.4 MIcroprocessor CONfigurer (MICON)

MICON was developed at Carnegie Mellon University [19–21]. MICON differs from SUSPENS, AUDiT, and the MSDA tool in the sense that it is netlist based versus physically based, that is, the goal of the MICON is to select an appropriate set of components and generate a netlist for the module using power, area, and cost constraints.

MICON is a library-based chip set synthesis system. MICON does not support the physical definition or technology knowledge in the packaging area that the SUSPENS, AUDiT, and MSDA tools support. However, the netlist orientation of MICON and it's behavorial level interface, bringing it closer to being a "system compiler" than any of the other tools.

The MICON tool is being commercialized. The commercial version of the tool is called *Fidelity*. Fidelity provides a VHDL link for synopsis for ASIC synthesis. Several advisor modules are also being developed including design-for-test and design-for-reliability for Fidelity.

9.3.5 PEPPER

PEPPER was developed at IBM [127]. PEPPER focuses on functional partitioning, chip placement package congestion, pin assignment, and coupled noise analysis. PEPPER requires a technology definition, wiring strategy choice, and a design description as inputs, and it performs a net topology analysis (routing prescreen) and a congestion analysis. A more accurate description of the PEPPER tool is as a constraint-driven pre-router (the constraints being delay and noise). PEPPER can provide significantly more accurate routing, chip placement, and layer requirement predictions than any of the other trade off tools discussed here; however, PEPPER requires significantly more inputs than the other tools, including net-topology constraints and wiring rules. PEPPER has a built in optimization capability like AUDiT.

9.3.6 Yoda

Yoda was developed at Carnegie Mellon University [54]. Yoda is a VLSI system planning tool targeted at the domain of digital signal processing and in particular at the task of designing digital filters. Yoda starts with the functional specifications of the desired frequency response and allows the designer to examine a wide variety of possible filter design plans. Yoda works almost exclusively at the chip level and may not seem to fit with the other five trade off tools discussed above; however, the philosophy developed within the Yoda is closer to the approach discussed in this book.

9.4 SUMMARY

Without CAD tools, it is impossible to design and analyze MCMs efficiently. The existing tools used in the physical design automation of ICs and PCBs cannot be used for MCMs because MCMs have a large number of dies in a relatively small area and a high operating frequency. This introduces difficulties in package design, thermal management, signal integrity, testability, and cost. CAD tools are required in the design of MCMs to perform design computation, early estimation, design verification, physical design automation, thermal analysis, manufacturability process, electrical analysis, reliability analysis, and feasibility analysis and to generate reports and statistics. As a result, several commercial CAD packages for MCMs have been developed. These CAD packages make it possible to integrate a wide variety of tools from different vendors into one package. Research CAD tools are also available from universities and research institutions.

BIBLIOGRAPHIC NOTES

Several commercial as well as noncommercial CAD tools for the physical design of MCMs are now available [43]. Technical specifications of MCM Design tools such as *MCM Design Suite, MCM Engineer, Allegro,* and *MCM Station* have been published [49, 77, 168, 193]. Various CAD packages for the design of MCMs are discussed by Donlin [59], as are different problems faced in the MCM designs. Maliniak [136] has discussed the evolution of EDA tools to suit MCM designers, and Dai [43] has discussed multichip routing and placement. Several noncommercial tools have been outlined [6, 178].

A system level circuit model for CPUs referred to as SUSPENS [7] emphasizes the interactions among devices, circuits, logic, packaging, and architecture. AUDiT, an automated design and trade off simulator for integration of electronic systems, was developed by the research group of J.P. Krusius at Cornell University [121, 161]. MICON, a library-based chip set synthesis system, was developed at Carnegie Mellon University [19–21]. PEPPER, a tool focusing on the functional partitioning, chip placement package congestion, pin assignment, and coupled noise analysis, was developed at IBM [127]. Yoda, a system planning tool was developed at Carnegie Mellon University [54]. The MSDA tool was developed at MCC [176, 177]. MSDA is a software tool for enhancing the manufacturability and decreasing the design risk associated with the selection of packaging technologies for ICs.

CHAPTER 10

PROGRAMMABLE MULTICHIP MODULES

The MCM approach to electronic packaging has improved the system performance by bridging the gap between the current packaging technology and the advanced IC technology. However, the time required to complete the design of a full-custom MCM can be long, and the process can be expensive. A complete custom design of an MCM is a task that usually requires many man months of engineering effort. As a result, programmable multichip modules have been introduced in order to minimize the time-to-market and reduce cost. In addition, programmable MCMs are also used as system prototypes, which is useful in evaluating the merits and demerits of the MCM design.

Programmable multichip modules (PMCMs) are in a way similar to field programmable gate arrays (FPGAs). FPGAs are devices that are programmed by a user to implement a logic function, while PMCMs are programmed to implement only the interconnection. The first step in the development of a PMCM is to prefabricate the substrate in bulk. Each substrate is then customized to provide interconnection based on the specific circuit of the user. The customization process is carried out by using electrically programmable switches or by mask programmable technique. Recently, the use of laser programmable technology has also been demonstrated [15].

The main difference between a PMCM and an MCM lies in the way the interconnect wiring is fabricated. The programmable version of the MCMs comes with prefabricated interconnect layers with provisions to customize the design based on the user's needs. Depending on the customization process, the PMCMs are classified into two types: fully programmable MCMs and semi-programmable MCMs. The customization process in fully programmable MCMs can use either electrically programmable or laser programmable technologies.

The semi-programmable version requires customization in only last few layers (typically one to three layers).

PCMCs are increasingly becoming attractive alternative to MCMs where fast turn-around and less cost are critical factors. The main trade off is in performance due to the programmable nature of the substrate. Hence, PMCMs are not suitable for very high performance applications.

In this chapter, we first present the concepts of programmable interconnects, the properties of which have led to the availability of PMCMs. Next, the general design concepts of PMCMs are discussed. In the subsequent sections, the different customization processes such as the fully programmable, laser programmable, and semi-programmable MCMs are discussed in detail.

10.1 PROGRAMMABLE INTERCONNECTS

PMCMs are basically MCMs with a "programmable routing substrate." A programmable routing substrate is a substrate that can be user programmed to establish connectivity between different terminals of each net in a given netlist. A programmable substrate can be abstractly viewed as a matrix of segments (a permanent connection between any two given points) with programmable junctions. Hence, by programming a junction, a path is established between two previously unconnected points. Such a matrix can be called a *matrix with programmable interconnects.*

The concept of a programmable interconnect architecture for the MCM substrate closely follows the routing resource technology available in FPGAs. In the FPGA technology, several methods are used to simulate a programmable interconnect. The main influencing factor is based on the application of FPGAs. For example, a reprogrammable interconnect approach is desirable in applications requiring reconfigurable systems, while a "one time" programmable technology is used in nonreconfigurable applications. The most widely used technologies for routing programmable interconnects are

1. SRAM technology
2. Floating gate programming technology (FGPT)
3. Antifuse technology

In the following sections, we briefly discuss these technologies for programmable interconnects, with special emphasis on the types of antifuses used.

10.1.1 SRAM Technology

SRAM technology uses electrically active devices in the routing resources such as Static RAM cells to control pass gates or multiplexers. When a "one" is stored in the cell, the pass gate acts as a closed switch and can be used to make a connection between two wire segments. When a "zero" is stored, the switch is open and the transistor acts as a high resistance between two wire segments. In a multiplexer implementation, the SRAM acts as a select logic. The state of the SRAM selects the specific input that must be connected to the output.

The main advantage of SRAM technology is its re-programmability. A major disadvantage, however, is its large size. At least five transistors are required to implement a SRAM cell, and an additional transistor is needed to act as a programmable switch. Moreover, the SRAM cells are volatile and hence must be loaded and configured each time at "power up."

10.1.2 Floating Gate Programming Technology (FGPT)

FGPT closely follows the technology used in ultraviolet erasable PROMs (EPROMs) and electrically erasable PROMs (EEPROMs). In PROMs, the programmable switch is a transistor that can be permanently "disabled." This is accomplished by injecting a charge on the floating gate using a high voltage between the control gate and the drain of the transistor. This charge increases the threshold voltage of the transistor, switching it off. The charge can be removed by exposing the floating gate to ultraviolet light. This lowers the threshold voltage of the transistor, making it function normally. In the EEPROM technology, the charge is removed electrically in the circuit, without UV light. Hence, it is easier to reprogram EEPROMs than EPROMs. However, the EEPROM cell is about twice the size of a normal EPROM cell.

A major advantage of FGPT over SRAM technology is that no external memory is needed to program the chip at "power up." However, this technology requires three additional processing steps over an ordinary CMOS process. Moreover, the EPROM transistor has a very high "on-resistance" and a high static power consumption due to the pull-up resistor used.

10.1.3 Antifuse Architecture

An "antifuse" is a two terminal device with an "unprogrammed" or "off" state presenting a very high resistance between its terminals. When a high voltage (typically from 11 to 20 V, depending on the type used) is applied across it's terminals, the antifuse will "blow," creating a low resistance ("on-resistance") link. This link is permanent. Antifuses are used in FPGAs as well as in PMCMs.

One type of antifuse, called the PLICE, is discussed by Handy [88]. It can be described as a square structure that consists of three layers. A positively doped silicon layer (n + diff), a dielectric layer (oxy-nitro-oxy insulator), and a layer of polysilicon. This antifuse is programmed by placing a high voltage (18 V) across the terminals of the antifuse and driving a current of about 5 mA through the device. This generates enough heat to melt the dielectric insulator and form a low resistance (about 300Ω) connection between the two metal wires. The fabrication process for this type of antifuse requires three specialized masks in addition to the normal CMOS process.

Another type of antifuse called ViaLink is discussed in [18]. ViaLink also uses three layers, similar to PLICE. It uses a metal layer (bottom), an alloy of amorphous silicon (middle), and another layer of metal (top). The ViaLink has an "on-state resistance" of about 80 Ω, while in the "off state" it presents over a GigaOhm of resistance. The ViaLink requires a potential difference of about 10 V across it's terminals and a 5 mA current drive to switch to the "on state." In the fabrication process, a normal via is created, but is filled with the amorphous silicon alloy instead of metal. This process again requires three extra masks.

10.2 GENERAL DESIGN CONCEPTS OF PMCMs

The PMCM approach is rapidly gaining ground because of its fast turn-around time, high yield, and low cost. The fast turn-around time is achieved since a substantial fraction of the interconnects are prefabricated, and most of the design, tooling, and fabrication time is avoided and, hence, minimized for each application. The MCM approach is similar to IC technique of prefabricating all or most of the circuits and later customizing the device to meet the application specific needs of the user. Essentially, a PMCM is composed of a prefabricated blank substrate with programmable interconnects. The wiring segments are fabricated with electrically programmable switches. This type of technique results in the manufacturing of a generic substrate that can be mass produced with high yield. Next, the substrate is customized by making connections with the help of suitable voltages applied across the programmable switches. The programmable switches can use either electrical voltages or laser programmable techniques. In this section, we present some of the general design concepts of PMCM technology.

PMCM technologies provide the user with a pre-simulated or a precharacterized interconnect that aids significantly in the design and simulation phase. As a result, PMCM technology provides a predictable and well-characterized interconnect. On the other hand, substantial effort is required in a full-custom design environment to characterize and simulate the expected behavior of the interconnect, which ultimately translates into higher cost, additional design time, and probably lower yield. As a result, the semicustom MCMs provide a manufacturing leverage (lower cost) and rapid product realization. However, the following factors have to be considered in the design of PMCMs.

1. Electrical design performance
2. Thermal design
3. Routability
4. Manufacturability

These factors are discussed in detail in the subsequent sections.

10.2.1 Electrical Considerations

A key ingredient to any PMCM design is electrical performance. If suitable electrical performance cannot be realized with this approach, it will not be able to meet its application objectives. However, it is difficult to define suitable electrical performance broadly. The wide array of circuit designs that users may want to map onto a PMCM provides the manufacturer of the PMCM with a dilemma. The manufacturers have to consider a multitude of factors in order to achieve the desired performance. They must ascertain the level of performance that can be designed for a given technology. Since the substrate is prefabricated, it must satisfy a broad range of market requirements. Finally, the manufacturer must consider the effect of scaling on electrical performance.

In many circumstances, electrical performance of the signal interconnect will be relatively good even without rigorous design for characteristic impedance, low loss, stub minimization, and so on. This is because the electrical length of the signal wiring in most MCM environments is short compared with the wavelength/rise times of the IC signals. The main issue, in many cases, is capacitive loading reduction for CMOS systems to minimize delay caused by RC time constraints. In other words, a large fraction of system designs will need to address signal delay more than high bandwidth signal fidelity. This may not be the case in a more conventional signal chip packaging/PCB implementation, where physical/electrical lengths of interconnect are longer and more significant. Perhaps the more compelling issue related to signal fidelity is power distribution. Many signal noise problems are related to the absence of clean power and ground supplies from which noise is fed forward through output drivers and noise margins are diminished at the receivers. This places an additional demand on the design of a PMCM. Not only must the power distribution scheme be flexible, it must also support high performance. Due to prefabrication of substrate, the power distribution network of the PMCM design is predefined and must accommodate a variety of supply voltages, varying numbers of supply potentials, and a variety of both ac and dc current requirements.

In existing PCMCs, the power distribution is predefined along with the signal wiring network. As an option to the designer, one could define a family of part types to address differing electrical performance requirements for both signal and power distributions while still building on the same basic design concept. The obvious danger of such a strategy lies in proliferation of part types that begin to diminish the advantage of the PMCM approach.

10.2.2 Thermal Design

The MCM technology used for manufacturing PMCMs is called *wafer scale technology* to distinguish it from much smaller MCMs containing significantly fewer components. For example, the PSCB substrate is 100 mm wide and 75 mm high, and it exhibits superior density and ability to withstand stress [52]. Achieving circuit densities of this magnitude results in the problem of removing heat from the tightly packed ICs. Criteria for successful thermal performance include thermal conductivity at all steps along the path from IC junction to system boundary and the match of thermal expansion characteristics between the IC and the substrate. As discussed in Chapter 5, the key parameter for a thermal design is the temperature rise of the IC junction above the ambient system temperature per unit operating power dissipated by the system.

Usually, large, thin silicon substrates are directly attached to thermally conductive mechanical bases, thereby surpassing current norms for volumetric density in both circuitry and power consumption. Also, conductive heat transfer to an external air or liquid cooled heat exchanger could be adopted so that direct contact between the coolant and the circuitry is avoided. When multiple modules are arrayed parallel to each other in a higher level packaging system, the thermal path is lateral through each module to common perpendicular heat exchangers. For advanced systems seeking to package even larger circuits on tighter module pitches, a long lateral heat path through a thin module base could be used. This, then, would represent the largest single component of temperature rise in the system. Module bases with high thermal conductivities are important in containing this temperature rise.

The second problem in thermal design of PMCMs is thermal expansion of substrate and ICs. The substrate is usually made of silicon, and, hence, its thermal coefficient of expansion (TCE) matches perfectly the TCE of the unpackaged silicon circuits that are attached to the substrate. This thermal expansion match lessens the mechanical stress on the die-to-substrate joints and is an important factor when considering large sized dies. Figure 10.1 charts the weight-normal-

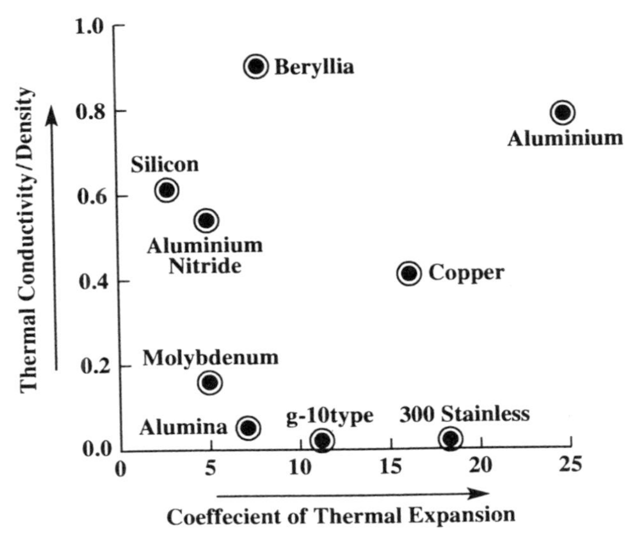

FIG. 10.1. Weight-normalized thermal conductivity and thermal coefficient of expansion for various substrate materials and representative metals.

ized thermal conductivity and thermal coefficient of expansion for various substrate materials. Silicon has the best combination of thermal, mechanical, and processing qualities for advanced MCM substrates. Moreover, the physical, chemical, and electrical properties of an elemental monocrystalline slice are inherently more consistent than those of the very best ceramic blending, pressing, and sintering operations. Recently, galium arsenide was considered as a substitute for silicon in high performance substrates, but this technology is still in its infancy, and a ready supply of irrefutably pure GaAs wafer from suppliers cannot be guaranteed.

Thus, the low resistance thermal path achieved with PMCM based packaging enables the development of systems with extremely high power densities. Most systems utilizing this technology dissipate 36 W per cubic inch of system volume without expensive thermal management components.

10.2.3 Routability

Like gate arrays, routability is considered an important factor in the design of semi-custom or programmable MCMs. In PMCMs, most if not all of the masking or phototooling steps are defined at the start, before the system is designed. A substrate is produced in a generic fashion and then customized to meet the specific needs of the user. The ability to route complex and dense multichip designs imposes the requirements for a prior design of a highly routable wiring topology. There are basically two ways through which good routability can be achieved: first, by designing highly efficient layout of the base substrate and high wire accessibility, allowing great flexibility in configuring wire resources; and second, by modest routability and wire access combined with an abundance of wiring made possible by the lithographic techniques used in most MCM manufacturing process. These two techniques imply that the method chosen to achieve good routability involves a series of trade offs. For example, choosing to fabricate more wiring to overcome routing limitations will push design rules and manufacturability. Choosing an efficient layout and high degrees of wire access may require more vias to achieve near arbitrary wire turns and/or connection modes and predefined wiring.

An important component in achieving efficient semi-custom designs is the design tool that can understand the semi-custom wiring structure and perform the actual customization (routing) needed to realize an application specific MCM. It should be noted that the routing efficiency is a factor of both the base wiring density (typically measured in inches of wire per square inch of the substrate area) and the resource utilization (i.e., what fraction of the available wiring can be utilized in routing a design). The total wire length used, relative to the minimum theoretical routing length, must be accounted for. This difference describes any deviation produced by the routing and avoids including any wire length associated with "stubs," which are essentially wasted resources even though they may be used in the routing.

10.2.4 Manufacturability

The structure of a PMCM is similar to that of an IC in thickness of layers, pad pitch, and thermal coefficients. Hence, usually mature, standard thin-film fabrication techniques and equipment are used to manufacture the wafers. Metal deposition, chemical vapor deposition, wet and dry etch, photolithography, and test equipment are all standard to that of IC manufacturing. The following discussion summarizes the standard fabrication process.

In the first phase, using both visual observation and thin film test and inspection equipment, each deposition and etch is monitored and checked for proper thickness, planarity, and overall quality. After wafer fabrication, electrical tests are conducted to verify metal deposition, dielectric depositions, etch, step coverage via connection, antifuse threshold and on-resistance, power dissipation, metal resistance, capacitance, metal-to-metal opens, and substrate resistance values. Then each line is checked to guarantee that all criteria for performance, line isolation, and antifuse integrity are satisfied. Test data are recorded in a substrate database for use by a design layout system, and all test data are back annotated to the wafer fabrication process quality and reliability database. The following four phases can be identified:

1. PMCM wafer fabrication
2. Module design
3. PMCM programming
4. MCM assembly

In the wafer fabrication phase, metal and silicon deposition is done first. Then, wafers are inspected using film tests and parameters tests. Finally, substrate resources are tested and validated. Output of the PMCM wafer fabrication phase is the PMCM substrate inventory.

Input to the *module design* is the *net and parts list* along with the packaging requirements. The design is partitioned, and die placement is done. Then, dies are connected to substrate. The critical path nets are routed first using simulation techniques and CAD tools. Then, the remaining nets are routed and simulated to study their effects on the substrate and to verify the overall integrity of the design. The output of this phase is a list of the *programming connections* to be effected. Module design is usually a semi-automatic process. A CAD system guides the wire bonding of the individual substrates and incorporates late engineering changes. Final partitioning is a designer decision, with an automatic component placement operation to aid in that design. The CAD tools usually check for design rules and provide limited automation in auto-bonding and auto-routing, wherein recommended bonding patterns are first set. This could then be changed manually. Automatic routing of the nets is done as the die pads are graphically wire bonded onto the simulated diagram.

Once the substrate is tested, a list of valid interconnections is generated that becomes the input to the programming phase and is referred to as the *program-*

ming connections list. The generation of this programming connections list is helpful in improving the yield because the substrate is not scrapped as in the case of IC design but used with legal connections.

In the programming phase, the substrates are first filtered or selected as necessary. Then, the substrate is programmed by burning the appropriate antifuses. Finally, the substrate is verified and inspected for validity and design integrity. The output of this phase is a *programmed substrate.*

This is input into the last phase, MCM assembly. Here, various processes such as die attachment to substrate, incremental testing, wire bonding die to substrate, full functional testing, and, finally, sealing the package occur. The output of this phase is a working PMCM module. Incremental testing and inspection can be done at every phase of production flow earlier, resulting in higher yields.

10.3 CLASSIFICATION OF PMCMs

PMCMs can be classified as (see Fig. 10.2)

1. Fully programmable MCMs (FP-MCMs)
2. Semi-programmable MCMs (SP-MCMs)

The architecture described by Banker [8] falls under the category of fully programmable MCMs. This architecture is composed of four inch silicon wafers that serve as *programmable silicon circuit boards* (PSCBs). The key difference between this MCM and others is that it supports four daughter pieces of silicon, called *segments*, that provide bonding sites for the various ICs used in implementing a module's functionality. The technique of combining four daughter segments with a base wafer allows the increase in overall yield because of incremental testing in comparison with the yield of all structures were to be placed on a single monolithic wafer. PSCBs are discussed in detail in Section 10.4.

Recently, a fully programmable MCM that uses a laser programmable technique to connect two interconnects was reported [15]. A laser programmable

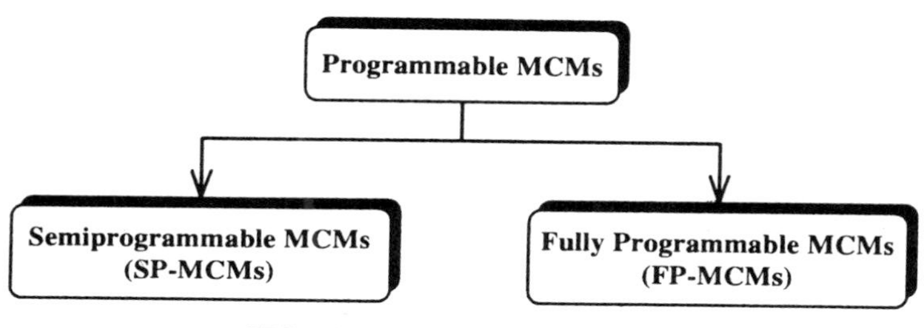

FIG. 10.2. PMCM classification.

MCM consists of a silicon substrate with a dense, predefined array of pads, tracks, and links. Integrated circuit chips are mounted, circuit side up, on the substrate, and the chip pads are wire bonded to the substrate pads.

Programming is accomplished by forming links (usually by burning antifuses) at the interconnections, which are formed by the intersections of vertical tracks with horizontal tracks. In section 10.4.2, laser programmable MCMs are discussed in detail.

The MCM architecture described in by Carey [30] is categorized as semi-programmable. The design of semi-programmable MCMs consists of customizing the last few layers (typically one to three layers) by programming the interconnects. The initial layers are mask fabricated. In Section 10.5, semi-programmable MCMs are discussed in detail.

10.4 FULLY PROGRAMMABLE MCMs

A fully programmable MCM (FP-MCM) process is one in which the substrate is mass fabricated and all the interconnects on the substrate are completely programmable. Fully programmable MCMs provide a single base substrate that supports any design that can be implemented using the routing resources available on the base substrate. The interconnects are laid out in a two metal layer of rows and columns across the wafer and are electrically connected at intersections via programmable switches called *antifuses*. The design specific information is added after wafer fabrication by an electrical programming process that converts the high resistance antifuses into low resistance conductive paths. Hence, the bare die wafers and segments on the substrate can be used together for quick turn-around design implementation at lower fabrication and development cost. In addition, fully programmable MCMs can be used as system prototypes for high performance systems.

In the following section, we discuss the details of *programmable silicon circuit boards* (PSCBs), which fall into the category of fully programmable MCMs. Also, a relatively new technology, called the laser *programmable multichip module*, is briefly reviewed in the subsequent section.

10.4.1 Programmable Silicon Circuit Board (PSCB)

The FP-MCM architecture [52] contains 4 inch silicon wafers called *programmable silicon circuit boards* (PSCBs). They mainly provide I/O buffer sites for chips to interface externally with the rest of the system, electrically programmable connectivity between these buffer sites and a generalized die attachment area on the wafer, and a base to which all components can be attached.

The FP-MCM supports four "daughter" pieces of silicon called *segments*, which provide bonding sites for the various ICs. These segments, one per quadrant, are attached to the base wafer and are surrounded by the I/O buffer sites and the interconnection area, called a *trench*. The interconnection area

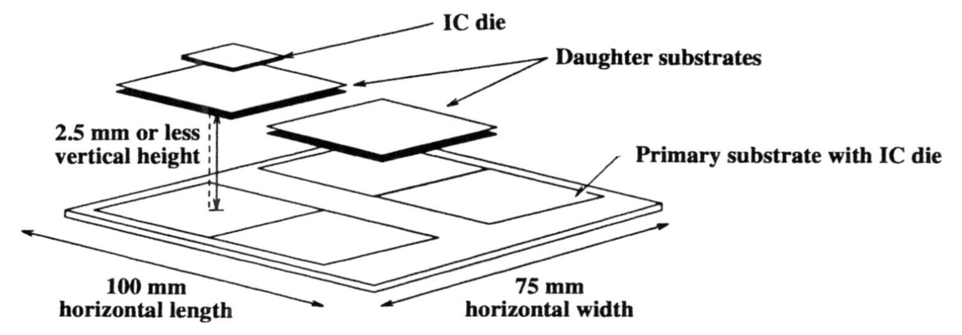

FIG. 10.3. Triple-decker architecture of PSCBs.

supports intersegment connections and limited chip bonding sites. Combining four segments with a base wafer leads to increased yield of the assembled wafer compared with the yield possible when all structures are placed on a single monolithic wafer. Figure 10.3 shows a three-dimensional view of the PSCB, which is also termed *triple-decker architecture.*

The key feature of the PSCB is that the interconnections on the substrates and the segments are programmable. The PMCM process employs wire bonding for connecting chips to the programmable substrate. The interconnections in the substrate are laid out in (two metal) layers of rows and columns across the wafer and are electrically connected at the intersections via a programmable antifuse.

10.4.1.1 Functional Description The PSCB is an off-the-shelf, electrically programmable interconnect system deposited on a silicon substrate. The universal nature of the PSCB permits circuit designers to mount and interconnect ICs of diverse technology, pin count, and size without the expense of an application specific MCM substrate.

The PSCB is made of four layers. The first two layers correspond to the ground and voltage planes. The voltage and ground planes are solid conductive "plates" of aluminum that provide excellent noise suppression because their capacitance is virtually free of series inductance. The voltage and ground planes are interchangeable. However, in most applications, the plane closer to the chip layer is used for ground.

The next two layers are used for signal wire routing. These two wiring layers support an xy layer pair plane type of routing. The wiring layers consist of a network of 11–21 μm wide aluminum traces on a silicon substrate. The traces are connected to one or more pads on the surface of the PSCB, forming a "sea of pads" for electrically attaching circuit dies to individual conductors. A single PSCB has space to mount a maximum of 64 dies and is referred to as a *segment* (see Fig. 10.7 below). Several segments may be interconnected when larger circuit modules are required.

Antifuses are available at all intersections of the traces in the signal wire routing layers. An electrical programming process that converts high resistance

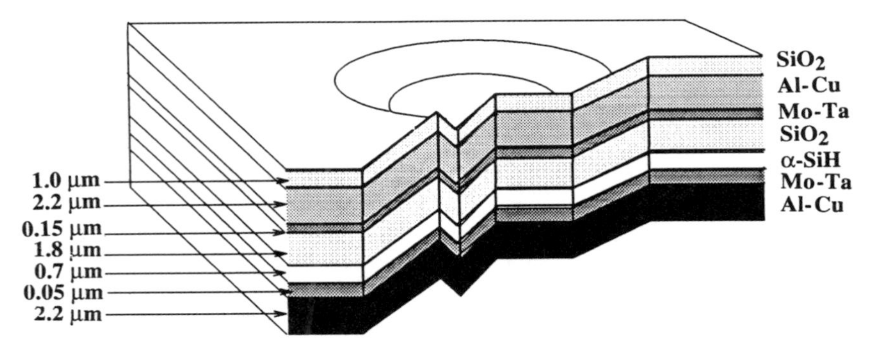

FIG. 10.4. Cross-sectional view of a PSCB antifuse.

(1 Gohm) antifuses to low resistance (1 Gohm) conductive paths adds the design specific information after wafer fabrication. ERIM uses these PMCMs to implement ASIC designs. Recently, designers demonstrated this technology on a 100,000 gate image memory controller and a 120,000 gate pipeline processing element. Various electrical, test, speed, and thermal design issues have been considered for implementing the ASIC designs [8].

10.4.1.2 Antifuse Operation The heart of the PSCB is the thin film amorphous-silicon antifuse. Traces on one metal layer can be connected to traces on the other metal layer at points of intersection. This interconnection is accomplished by changing the physical state of the electrically programmable "antifuses" that exist at specific intersecting points in the matrix.

Antifuses normally act as insulators, but can be electrically switched to behave permanently as low-resistance conductors by applying a threshold voltage of approximately 20 V dc. This feasibility of programming the circuit signal nets via the antifuse material allows designers to form chip-to-chip or chip-to-I/O circuit connections. The antifuse is constructed as shown in Figure 10.4. Its normal programming cycle can be understood by examining the current voltage characteristics as shown in Figure 10.5. An antifuse would have to be considered unreliable for use if it were subject to either spontaneous programming from operational segments or spontaneous de-programming under any circumstances. The nominal "off" resistance of an antifuse is approximately one billion ohms (1 Gohm), while the typical "on" resistance is 4 ohm.

10.4.1.3 Electrical Performance The electrical parameters considered in achieving the PSCB technology include speed, reflection, attenuation, and noise immunity. Several circuit models have been developed that allow detailed circuit simulation, ensuring that miniaturized circuits operate as designed. The interconnects with dimensions typical for silicon circuit boards can be considered as transmission lines for quasi-transversal electromagnetic waves (TEM). Thus, these lines can be modeled as distributed circuits with longitudinal

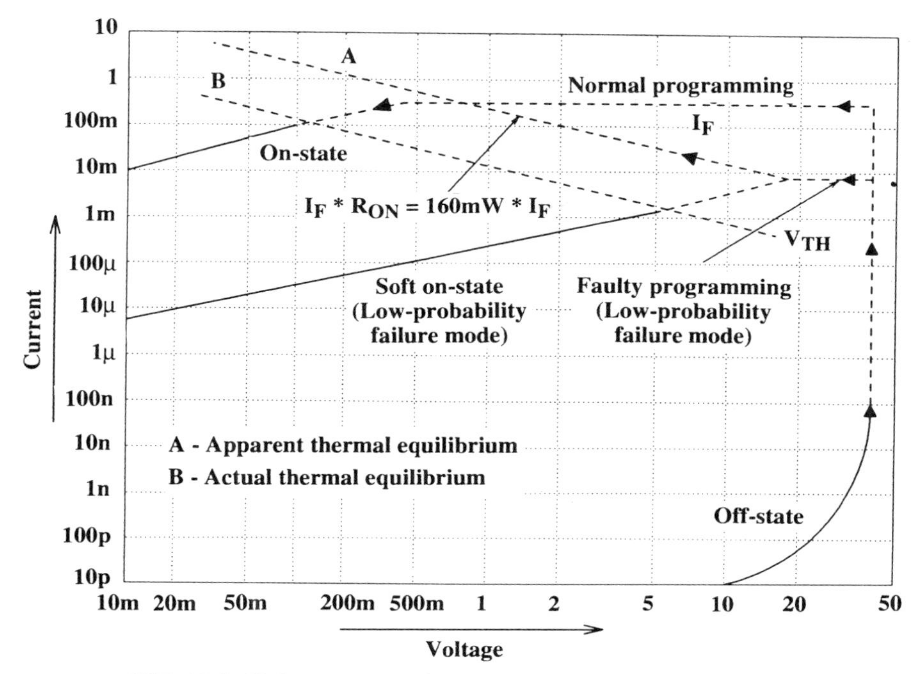

FIG. 10.5. Voltage-current characteristics of the PSCB antifuse.

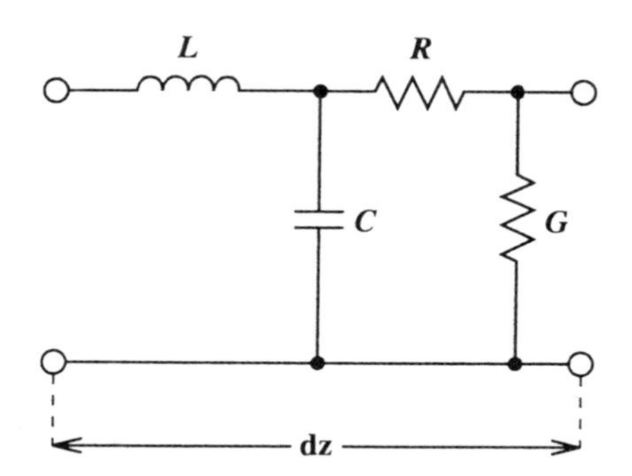

FIG. 10.6. Equivalent circuit for an infinitely short section of a TEM transmission line.

inductance, longitudinal resistance, transversal capacitance, and transversal conductance.

The transmission line modeling is shown in Figure 10.6. The distributed elements have been replaced by the lumped elements of an infinitely short section of the line. The infinitely short sections can be approximated by short sections of 1 mm in length. The PSCB behaves like coaxial cables, which also operate in the

TABLE 10.1. Electrical Characteristics of Two Typical PSCB Lines

Parameter	M3	M4	Unit
Capacitance (C) to ground, per length	0.29	0.18	$\frac{pF}{mm}$
Capacitance to a parallel line, per length	0.006	0.004	$\frac{pF}{mm}$
Inductance (L) with ground return, per length	210	330	$\frac{pF}{mm}$
Resistance (R) with ground return, per length	1.0	1.6	$\frac{\Omega}{mm}$
$\sqrt{L/C}$	27	43	Ω
$2 \ln 2 \sqrt{L/C}/R = \ln 2/\alpha$ (critical distance)	42	43	mm
Crossover capacitance	0.008	0.004	pF
Antifuse off-state capacitance	0.008	0.008	pF
Bonding pad capacitance	0.12	0.12	pF
Antifuse on-state resistance	1	1	Ω

TEM regime. However, there are some characteristic differences in magnitude when considering the losses caused by the transversal conductance and longitudinal resistance. The transversal conductance is negligible when compared with the losses caused by the longitudinal resistance and, therefore, can be removed from the circuit. The resistance of the "shield" versus the resistance of the "center conductor," which contributes about 15% to the total resistance, cannot be ignored. Also, 30 of the total inductance for a PSCB wire is from the "internal" inductance arising from internal magnetic fields.

Table 10.1 shows the composite values found for the two most prominent lines, the predominantly north–south trace layer (M3) and the east–west trace layer (M4). They include base parameters R, L, C, which dominate ground resistance and internal inductance for some derived parameters. Also, the values for the discrete coupling capacitance or resistance between orthogonal transmission lines and the distributed coupling capacitance between parallel transmission lines are given.

The validity of the PSCB interconnect has been demonstrated experimentally using fabricated test lines. Signal delays were measured and compared with the simulated interconnects by SPICE. The results from the tests show the following:

1. Experimental data on clock delays have verified PSCB SPICE simulation models.
2. Total transmission line delays across the entire wafer are less than 2 ns.
3. PSCB substrate delays for wafer scale MCMs are well within design margins for a 40 MHz system.

For high speed circuits, noise immunity is a very important factor. Typical sources of noise are parallel line crosstalk, orthogonal line crosstalk, and power supply crosstalk. In all practical cases the first two are considered negligible. The values for parallel coupling capacitance over total line capacitance and line crossing capacitance over the net capacitance are shown in Table 10.1. Composite noise levels from all sources have been confirmed by experiment to be less than 3. At signal frequencies, the power supply crosstalk is shown to be negligible.

PSCBs employ "distributed termination" instead of conventional termination schemes. The distributed termination employs series resistors distributed along the length of the line. Attenuation and end-of-line reflection compensate for each other so that neither is present for interconnects whose distributed resistance is 1.4 times as large as the characteristic impedance. The signal rise time is preserved, and signal propagation time is equal to that of a lossless line. For shorter lines with lower resistance, there will be some reflections, while a resistance larger than the critical resistance means that the delay is increased over the ideal amount. The resistance in a PSCB is usually larger than critical resistance. Therefore, there are no measurable reflections even when branch lines are added to the trunk line, a situation that causes significant bouncing in conventionally terminated transmission lines. Ringing effects (reflections) have never been observed in PSCBs.

In long PSCB lines (greater than 50 mm in length), the distributed resistance significantly exceeds the critical resistance. Consequently, the propagation delay time of the longest line (120 mm) rises to nearly twice the ideal value, and the signal rise time increases accordingly. Also, as there is no steady state current in the line, signal attenuation does not exist. From the simulation results, it has been observed that the delay falls into the range of practical applications. If unacceptable delays occur due to the required fanouts, then additional trunk lines can be programmed to bring the fanout per trunk line down to acceptable levels.

10.4.1.4 Line and Pad Types The PSCB segment has a repeating pattern of pads on its surface. The pads are logically organized to form cells. There are 64 such cells on the surface of the segment, arranged in an 8 × 8 pattern (see Fig. 10.7). All cells have identical pad layouts. There are several different types of

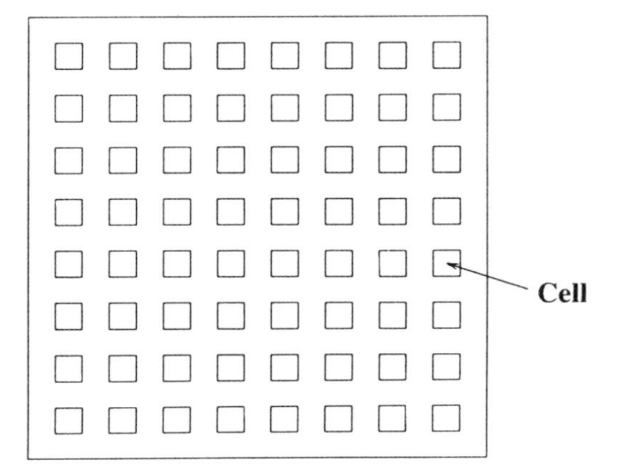

FIG. 10.7. PSCB daughter segment layout.

pads, and, with the exception of the J pad, each is connected to a unique type of line.

The conductors in the trace layers do not form a simple x-y grid, but instead consist of different types of lines of varying lengths and characteristics. Signal net resources that run north and south consist of three types of lines: full N (NF) lines, half N (NH) lines, and B lines. NF lines run between I/O pads on the north and south edges of the segment. NH lines also run north–south but terminate halfway through the segment. Two NH lines can be programmed together to provide the interconnect potential of a full N (NF) line. N lines, in general, are used to provide the north–south link in a signal net that requires either north and/or south edge I/O and/or connection between two ICs that do not lie in the same horizontal cell row. N line connections to B lines are established by programming the N line to an H or S line, which are discussed later. B lines also run north–south but have a maximum travel of one cell height. B lines provide the connection between a signal bonding pad and the signal net's east–west link. A B line can be programmed to any of 20 H/S lines that run through the associated cell.

Signal net resources that run east and west consist of two types of lines: segment (S) lines and half-segment (H) lines. S lines run from either side of the segment across to the opposite side, with an I/O pad on one end and one end only. H lines run from either side of the segment and terminate in the center. Similar to an NH line, two H lines can be programmed together to provide the interconnect potential of an S line, with the additional aspect of having an I/O pad on each end of the line. S and H lines provide the east–west signal net link to the B lines. An S or H line is programmable to 10 B lines in every cell that the particular S or H line travels through. S and H lines are also programmable to every N line crossed. Furthermore, S and H lines provide segment I/O through their associated east and west edge pads.

There are eight types of pads on a segment – B, J, H, NF, NH, S, V, and G. B (bonding) pads are used as the segment point of interconnect with the ICs. There is one B pad per B line. A B pad may or may not have an auxiliary pad associated with it. Twenty-four of the 40 B pads within a single cell have an auxiliary J pad. J pads add flexibility in wire bonding a particular IC to the PSCB segment. Figure 10.8 gives a layout plan showing different types of pads, ground, and power connections. H, NF, NH, and S pads are connected to the H, NF, NH, and S lines, respectively. V (voltage) pads are connected to the power plane customarily used to distribute voltage. G (ground) pads are connected to the power plane customarily used to distribute ground.

10.4.1.5 Cell Layout in PSCBs Each cell on the segment is large enough to hold a small die such as an SSI/MSI IC (i.e., a quad NAND gate or dual flip-flop). Because the cells are small, larger dies may span several cells. It may be possible in some instances for two different ICs to overlap and share the same cell as long as there are enough required types of pads to satisfy the needs of both ICs. In most cases, epoxy die-attach and ultrasonic wire bonding techniques are used to mount dies to the SCB.

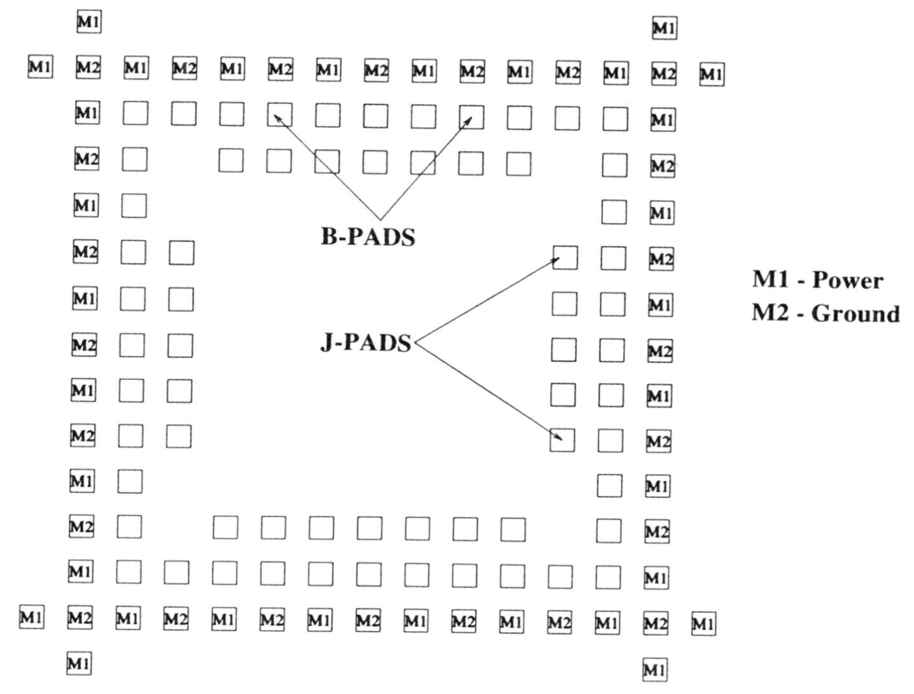

FIG. 10.8. PSCB cell layout.

10.4.2 Laser Programmable MCMs

A laser programmable MCM is a silicon substrate with a dense, predefined array of pads, tracks, and links [15]. ICs are mounted, circuit side up, on the substrate, and the chip pads are wire bonded to the substrate pads. Horizontal tracks are on one level of metal and vertical tracks on the other. Links can be formed at any of the intersections of vertical tracks with horizontal tracks. Links consist of a silicon nitride layer sandwiched between the two metal layers. When a laser pulse of the correct power and duration is directed to the link region, the metal and the nitrate fuse to form a conductive vertical path, with a resistance of typically 2 Ω. Tracks can also be cut by laser energy to isolate distinct nets and trim unwanted stubs from a net. Cut points are the narrow places in the tracks. The substrates are manufactured first and then quickly programmed by laser technique to build a user-specified system.

The module itself is built from a 10 cm wafer and contains pads and tracks spread over an area of 5×5 cm^2. It has two layers of metal and no active circuits. Power, ground, and signals share the tracks on the two metal layers. The module consists of an array of 258×258 pads with a pitch of 200 μm, and five tracks between any two adjacent pads. The first, third, and fifth tracks between two pads are used for signals, while the second and fourth are used for power and ground.

The latter are preconnected into power grids with direct, metal-to-metal vias (small squares with diamonds). Although there is no solid ground plane, every signal path is adjacent throughout its length to a power track. Berger 15 has presented the experimental means to determine the thickness of the metal and insulating wires. Thicker metal gives lower resistance, and thicker insulators give lower capacitance, both of which improve the speed of propagation of signals across the module. The target thickness for intermetal insulation is 1.5 μm of nitride at the link areas, augmented by 2.0 μm of oxide in the remaining places. The target thickness for metal is 2.0 μm. These values would give a resistance of about 140 Ω and a capacitance of about 15 pF for a 5 cm track 16 μm wide. Such a track should allow propagation across the whole wafer in about 1.1 ns, which should support signals switching at 100 MHz.

10.4.3 Test Vehicles of Laser-Programmable MCMs

The newly developed laser programmable MCM is tested with an evaluation module. Its only chip types will be CMOS inverting bus transceivers and decoupling capacitors. The signal tracks of the MCM are configured into compact meshes (capacitive loads), long steady state paths (resistive heaters), and long fast paths (RC transmission lines). At frequencies below 1 GHz, the inductance of the tracks are expected to be negligible compared with their resistance. Measurements will be made of

1. Speed of propagation of a signal across the MCM
2. Degradation of a signal along an extended path
3. Transients in the power and ground grids when heavy currents are drawn
4. Thermal resistances between chip-to-wafer and wafer-to-packages

Ring oscillators, whose frequencies vary with temperature, will be used to measure temperature at their own chips. A more substantial application module will probably contain one or two digital signal processing (DSP) chips, four static memory chips, electrically programmable read-only memory, and some control logic.

10.5 SEMI-PROGRAMMABLE MCMs

Semi-programmable MCMs are made of prefabricated blank interconnect arrays in copper polyimide (CuPi) substrate. The substrate is then automatically routed by using a large number of interconnects based on a description of the die pad placement and a netlist. The major design criteria of semi-programmable MCMs are to prefabricate the initial layers using mask fabrication and to customize the last few layers of interconnection by user programming. The advantages of using the semi-programmable MCMs are to achieve high yield and quick turn-around time. In this section, we discuss the Quick Turn-Around Interconnect (QTAI) technique, which is categorized as a semi-programmable MCM [30].

10.5.1 Quick Turn-Around Interconnect (QTAI)

In an effort to reduce cost and turn-around time problems, a high density interconnect technique called the *quick turn-around interconnect* (QTAI) has been developed [30]. The QTAI technique begins with a prefabricated, blank-interconnect array in a copper polyimide (CuPi) substrate. This substrate is then automatically routed by using a large number of interconnects. The routing is software driven based on the chip pad placement and netlist description. This technique is used for mass production of generic substrates for high volume cost reduction. The above stated technique is classified as a semi-programmable MCM technique as it uses only the last few layers (typically 1–3) for customization purpose.

The use of an interconnect array (customization mask) provides opportunities for high volume manufacturing of the generic components, thus reducing cost, improving yield, accelerating learning curves, and allowing process optimization. Since the customization represents a fraction of the overall interconnect fabrication, rapid turn-around time is achieved in comparison with the full-custom design. Figure 10.9 shows the sequence of steps in full-custom and semi-custom

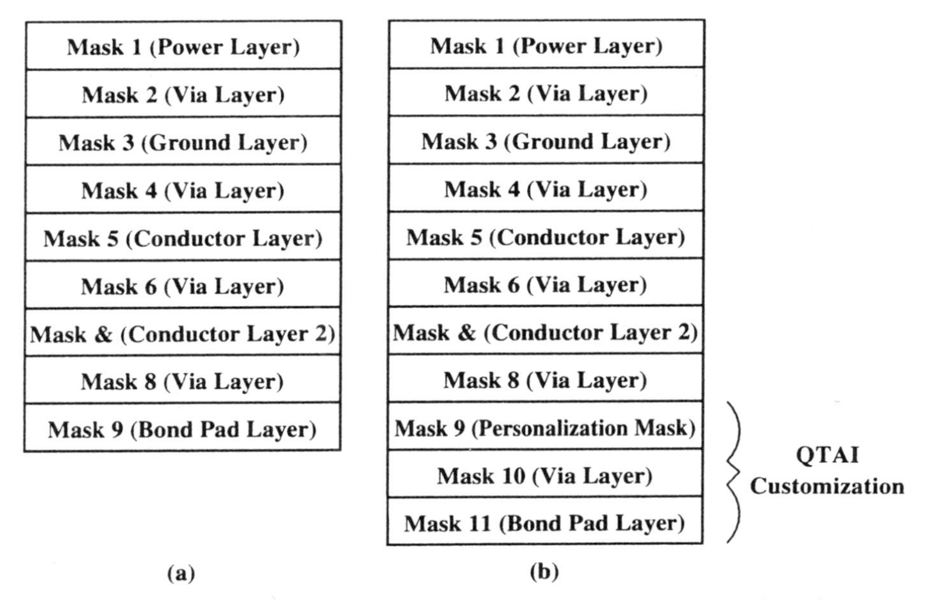

FIG. 10.9. Comparison of manufacturing sequences for full custom and semi-custom: **(a)** full custom approach, **(b)** QTAI (semi-custom) approach.

approaches. Designers of the QTAI approach claim that the technique can also reduce turn-around time by approximately 60%–70%. The QTAI design consists of a prefabricated substrate that contains power, ground, and an x-direction wiring layer and a y-direction wiring layer (see Fig. 10.9). Two voltages and a ground/reference layer are combined into the first two metal layers. This combination provides a flexible power distribution system, which is compatible with CMOS, BiCMOS, ECL, and GaAs. The wiring layers are composed of short buried wire segments whose terminal ends are brought to a third separate plane, used to customize designs. With wire ends available on this plane, the process can use short conductor links to configure the underlying wire segments into user-specific design routes.

Customizing the chips can involve from one to three additional patterned layers, which depend on the desired bonding technology, electrical characteristics, and device packaging density. Metal must be added to configure the underlying wire segments. Moreover, if bond pads can be selectively placed to avoid those regions where programming occurs, only a single mask step may be needed to define personalization and chip I/O pads on a single metal level. Adding a separate bond pad layer can make device placement and chip-attached footprints totally arbitrary (the equivalent of full-custom design). As a result, with the addition of a layer, the QTAI process involves three additional steps necessary for customization: personalization, vias, and bond pads. As the user defines the bonding layer, the QTAI can employ wire bond, TAB, flip-chip, or any combination of these technologies to connect the chips. There are three main issues related to QTAI:

1. Functional description
2. Electrical performance
3. Routing efficiency

We discuss these factors in the subsequent sections.

10.5.1.1 Functional Description QTAI substrate fabrication currently uses a copper polyimide process, which includes a plated copper interconnect, stacked vias, and a planar dielectric. This process is used to support both custom designed as well as QTAI functional CuP substrates, in addition to a variety of parametric test substrates. The process readily supports the QTAI design rules of 15 μm lines, 20 μm minimum spaces, and 30 μm via pads. Currently, all existing substrates have lines on a 75 μm grid (3 mil) that provide a substrate wiring density of approximately 670 in/in^2. While the process currently used for fabrication is well suited to the QTAI design requirements, the programmable substrate design can be fabricated using other process technologies for high density interconnects, including the staggered/staircase via processes in use at captive and commercial locations. This is an important feature of the QTAI approach, since the most (cost/performance) effective fabrication technology may be chosen to implement the programmable interconnect design.

10.5.1.2 Electrical Performance The electrical performance of the QTAI also has been studied using two test vehicles that were customized from the generic substrate. Measurements on propagation delay, impedance, crosstalk, and bandwidth were taken from these test vehicles. The only obvious electrical penalty found for the QTAI substrate versus a custom device, using similar design rules and construction, is an 8% time-of-flight delay increase per unit length. This is because the buried wire segments contain minor perturbations from a straight line x-y path. Thus, Manhattan distances are traversed through somewhat longer path lengths than in a custom design that would yield pure x-y distances to router from point to point. A nominal propagation delay of 73 ps/cm has been measured in the QTAI design using polyimide as the dielectric. Depending on the customization process, impedance and crosstalk can be varied. With no reference plane metal added to the personalization layer, impedance of 75–90 Ω and saturated backward crosstalk of 4% and 7% for the x and y layers, respectively, was realized. This is due to the dual-microstrip construction and the resulting differences in wiring plane height from lower reference (ground) plane. When reference plane metal is added to the personalization layer, a balanced dual-stripline construction results, and an impedance of 60 Ω and crosstalk of 2.5 is observed for both x and y layers. Rise times of 200 were supported for 2 inch long lines, and rise times of 500 ps were possible for lines 4 inches long.

With these performance characteristics, clock rates in excess of 100 MHz are readily supportable with the QTAI technology. These electrical characteristics apply to the substrates fabricated with the MCM process, using the design rules described. Substantial variations in electrical properties are possible, depending on design rules and/or processes chosen for QTAI implementation.

10.5.1.3 Routing Efficiency An early effort was undertaken to develop an auto-router that could address fundamental routability of QTAI design and serve as a tool for substrate customization. Routing efficiency of the QTAI design is characterized in two ways.

One FOM is the difference between the theoretical minimum amount of wire necessary to route a design assuming infinite Manhattan resources and the amount of wire needed to route the design under the constraints of wire resources in the actual substrate. This difference describes the deviation of routing. Using a net half perimeter estimate of the minimum theoretical wire needed, all routing algorithms have exhibited less than 10% meander, although actual meandering is probably less since the net half perimeter method underestimates the required net length by more than three nodes.

A second FOM used in routing is percentage of resources used, or utilization. This figure describes what percentage of the available wire (wiring channels) has actually been utilized by the router. The higher the number, the more dense the design and the less the substrate area required to implement the design. Custom auto-routers have achieved utilizations of 40–50% (i.e., less than half of the theoretically available wire channels can be used/filled before routing becomes intractable). The QTAI router has demonstrated utilization in excess of 65%.

Therefore, the QTAI design provides achievable densities that are competitive with a full custom approach. A series of case studies have been performed to examine the trade offs between the QTAI and the full-custom approach. No significant routability penalty has been observed using the QTAI design router, relative to custom design. The auto-router output is used to provide customization manufacturing data (mask types) for the MCM design.

10.5.2 QTAI Test Vehicles

Various test vehicles were used to perform hardware verification and integration of the QTAI design, the QTAI router, and in-house face-up lead from TAB technologies. Two designs were manufactured to test design approaches related to repair, rework, yield, bonding technologies, and multidesign substrates.

The first vehicle fabricated was a 96 × 96 expendable crossbar switch, based on a configurable Harris VHSIC CMOS switch chip design, and having a 24 × 24 architecture with expansion inputs. Die size is 0.3 × 0.36 inch, with 103 lead and 0.007 inch minimum lead pitch. The overall 96 × 96 crossbar system was achieved by a 4 × 4 matrix arrangement of the individual switch chips. An additional six SSI devices completed the system by providing all necessary multiplexing, chip selects, and scan test functions. Discrete capacitors were also mounted on the substrate to improve power regulation. The switch chips were bumped using a straight-wall gold bump process suitable for thermocompression gang inner lead bonding (ILB). Following bumping, wafers were returned for parametric and low speed testing to allow screening of devices for final module assembly. All dicing, ILB, excise/leadform, and single point outer lead bonding (SPOLB) were performed at the design site. The SSI devices were received from a commercial source in die form and bonded to the substrate using standard gold wire bonding practices.

Prior to device attachment, capacitance probe test was performed on the customized but unpopulated substrate. The testing was used to identify defective nets in the substrate. The customization of the substrate included spare surface traces between devices that were used in combination with short wire bonds to stitch together replacement nets, based on the known defects after all device bonding was complete. A separate bond pad layer was used in the design. The assembled and repaired module was returned to Harris for functional verification using a 200 MHz HP 82000 test system. The module ran to the limit of the CMOS chip set with crossbar data rates in excess of 75 MHz.

A second design that functioned as a high density solid state storage card was implemented in collaboration with Eastman Kodak Corporation. This design did not stress either performance or routing density, but did test two unique design elements. Redundant routing was used to increase module level yield by routing all nets twice. The redundancy provided significant yield improvements on finished substrates and demonstrated the ability to use redundancy in QTAI with the same relative ease as a full-custom design. The test vehicle also incorporated two separate designs in the base 2.25 × 2.25 QTAI blank area.

These two designs were later singulated using standard wafer sawing practices, illustrating the potential for a multiple-design-per-wafer capability from the base QTAI substrate layers. As with the first design, a separate bond pad layer was produced, along with the customization level, and turnaround time from receipt of masks to finished interconnect was 1 week.

In addition to QTAI hardward design, software has been developed to route the interconnections in the substrate automatically based on a description of the chip pad placement and netlist [30]. This software helps the user to complete most of the interconnects available in a high density substrate. The software is an integral part of the QTAI design approach and a necessary component to take advantage of the QTAI technology.

The test vehicles described above were fabricated using a standard fabrication process and on the identical set of underlying layers. However, the QTAI approach is general enough to allow for other designs and processes. For example, the incorporation of thermal vias into the design are being evaluated. These additional pillars are expected to provide a thermal impedance through the multilayer interconnect well below 1°C/W/cm^2. With the additional pillars, the thermal impedance is dominated by the die attach material and base substrate rather than the copper and polyimide layers. Also, research is underway to develop a low-cost laser-based tool that could be used for maskless personalization of QTAI substrates by the end-user. Also there is pursuit of alternate fabrication processes. The objectives are to lower the manufacturing costs of the interconnect dramatically and to expand the market for this technology.

10.6 SUBSTRATE TESTING AND SOFTWARE SUPPORT

Before PMCM substrates can be used, they must be tested for continuity of the tracks and freedom from shorts between crossing tracks. Also, a number of software tools will be needed to design and develop the modules efficiently. Here, we shall take a brief look at the facilities available for substrate testing and tools for software development for PMCM modules.

10.6.1 Testing of the PMCM Substrate

Testing is carried out to ensure continuity of the tracks and to check for shorts between crossing tracks. To allow probing for these tests, a pad is connected to each end of the 2,580 tracks by replacing some of the links along the edges and in the corners of the MCM by vias. Testing is done on a wafer prober by measuring the capacitance between the probed track and the silicon substrate. A capacitance lower than normal indicates a break in the track; one higher than normal indicates a short to some other structure. If a signal track is found to be open or shorted to an adjacent (power) track, it can be used for routing signals. Some other track can probably be used. However, if the two power grids are shorted together, the whole MCM is unusable.

10.6.2 Software Support

A number of software tools are used in the design and development of the modules. Most of the CAD tools are the ones that have been used previously on projects at Lincoln Laboratory for building wafer scale integrated circuits [3]. They facilitate in

1. Placement of chips on the substrate
2. Routing of the interconnect between the chips
3. Controlling the laser system in the linking and cutting operations
4. Generating test patterns

10.7 SUMMARY

Semi-custom or programmable MCMs may play a key role in speeding up the technology development, acceptance, and application. In recent years, the programmable approach to MCMs has been used. The main advantage of programmable MCMs is that one generic layout can be used for implementing several different designs based on the user's requirements. The programmable feature of this type of MCM obviates the need to fabricate designs at the mask level, offering fast turn-around time and high yield. The programmability of the interconnect is realized using electrical antifuse switches; however, the latest manufacturers are also investigating the use of laser-programmable switches. The programmable MCM technology reduces the cost of MCM development by a significant margin; however, it also degrades system performance compared with full-custom MCMs due to the electrical effects of the programmable elements.

10.8 PROBLEMS

1. How are MCMs different from PMCMs? Why are PMCMs necessary? Discuss briefly.

2. List the reasons why the performance of PMCMs is not comparable with that of MCMs.

3. List all the reasons that make the yield of PMCMs better than that of MCMs.

4. In a PSCB architecture, a "stub" is defined as the nonactive portion of a routed net associated with the predefined segments (see Fig. 10.10). Develop an efficient algorithm for routing on PSCBs such that the overall stub length of all the nets in a given netlist is minimized.

5. Develop an algorithm for routing nets on a PSCB while minimizing the length of the stub (see Fig. 10.10) for the *longest stub net.* A longest stub net

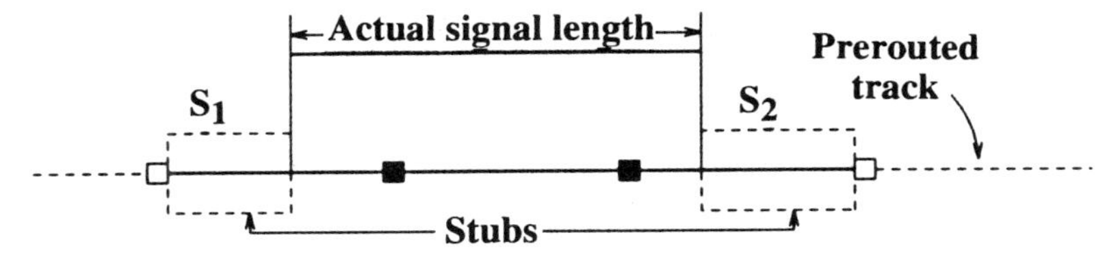

FIG. 10.10. Problems 5 and 6.

can be defined as a net that has the longest stub length of all the nets in a given netlist when routed.

BIBLIOGRAPHIC NOTES

Banker [8] has discussed ERIMs PSCB architecture and its application, and its electrical, thermal, performance, and manufacturability have been covered [52]. Also the details of the construction of the antifuses used and their voltage-current characteristics have been discussed. Carey [30] has presented the details of the semi-programmable MCM referred to as Quick Turn-around Interconnect (QTAI). In that paper, Carey discussed the routing efficiency, substrate fabrication, electrical performance, and a few test vehicles that have been run on QTAI. A new development in the field of programmable MCMs is the use of the laser technique to blow the antifuses in order to achieve interconnection. Berger [15] has demonstrated the use of the laser programming technique, and a few CAD tools useful in the software support have been discussed [3].

REFERENCES

1. S. B. Aker, "A Modification of Lee's Path Connection Algorithm," *IEEE Transactions on Computers*, pp. 97–98, Feburary 1967.
2. D. Akihiro, W. Tohihiko, and N. Hideki, "Packaging Technology for the NEC SX-3/SX-X Supercomputer," *Proc. 40th ECTC (Las Vegas)*, p. 525, 1990.
3. M. V. Alstyne, G. Young, R. Frankel, and J. Hunt, "An RVLSI CAD System," *Proceedings of International Conference on Wafer Scale Integration*, January 1989.
4. D. I. Amey, "Overview of MCM Technologies; MCM-C," *ISHM 92 Proc.*, pp. 225–234, 1992.
5. K. J. Antreich, F.M. Johannes, and F.H. Kirsch, "A New Approach for Solving the Placement Problem Using Force Models," *Proceedings of the IEEE International Symposium on Circuits and Systems*, pp. 481–486, 1982.
6. H. B. Bakoglu, *Circuits, Interconnections, and Packaging for VLSI*, Addison-Wesley Publishing Company, 1990.
7. H. B. Bakoglu and J. D. Meindl, "A System Level Circuit Model for Multi and Single Chip CPUs," *IEEE International Solid State Circuits Conference*, pp. 308–309, February 1987.
8. J. Banker, "Designing ASICs for Use with Multichip Modules," *Proceedings of International Conference on Computer Design*, 1992.
9. P. Bannerjee and M. Jones, "A Parallel Simulated Annealing Algorithm for Standard Cell Placement on A Hypercube Computer," *Proceedings of the IEEE International Conference on Computer Design*, p. 34, 1986.
10. A. Bar-Cohen, "Thermal Management of Air and Liquid-Cooled Multichip Modules," *IEEE Trans. Component, Hybrids, Manuf. Technol.*, CHMT-10(2), pp. 159–175, June 1987.
11. C. J. Bartlett, "Advanced Packaging for VLSI," *Solid-State Technol.*, pp. 119–123, June 1986.

12. C. J. Bartlett, "Multichip Packaging Design for VLSI-Based Systems," *IEEE Trans. Components, Hybrid, Manuf. Technol.*, 10, pp. 647–653, December 1987.
13. M. Beardslee and A. L. Sangiovanni-Vincentelli, "An Algorithm for Improving Par-titions of Pin-Limited Multi-Chip Systems," *ICCAD*, pp. 378–385, November 1993.
14. J. Ben-Meir and J. Weiss, "High-Density, High-Speed Design Focuses on Physical Performance Impacts," *Proceedings of the Technical Program. NEPCON West*, 2, pp. 525–532, 1992.
15. R. Berger, "A Laser-Programmable Multi-Chip Module on Silicon," *Submitted to the IEEE International Conference on Wafer-Scale Integration*, 1992.
16. K. Billings, *Switchmode Power Supply Handbook*, McGraw-Hill, 1989.
17. A. Bindra, "Prober Tests Chips at Mil Temperatures," *Electronic Engineering Times*, p. 11, February 1987.
18. J. Birkner, "A Very High-Speed Field Programmable Gate Array Using Metal-to-Metal Anti-Fuse Programmable Elements," *New Hardware Product Introduction at CICC '91 Custom Integrated Circuits Conference 91*, May 1991.
19. W. P. Birmingham, *Automated Knowledge Acquisition for a Hierarchial Synthesis System, Ph.D. Dissertation, Carnegie Mellon University*, Pittsburgh 1988.
20. W. P. Birmingham, A. P. Gupta, and D. P. Siewiorek, "The MICON System for Computer Design," *Proceedings of the 26th Design Automation Conference*, pp. 135–140, 1989.
21. W. P. Birmingham, A. P. Gupta, and D. P. Siewiorek, *The MICON Project*, Jones and Bartlett Publishers, Boston, 1992.
22. R. G. Biskeborn, J. L. Horvath, and E. B. Hultmark, "Integral Cap Heat Sink Assembly for the IBM 4381 Processor," *Proc. 1984 Int. Electronic Packaging Society Conf.*, pp. 468–474, 1984.
23. L. A. Bixby and D. T. Dimatteo, "Multichip Modules: An Alternative Packaging Technology," *Hybrid Circuit Technology*, pp. 9–13, December 1990.
24. A. J. Blodgett, "A Multilayer Ceramic Multichip Module," *IEEE Trans. Components, Hybrids, Manuf. Technol.*, 3 (4), pp. 634–637, 1980.
25. W. Blood, "ASIC Design Methodology for Multichip Modules," *Hybrid Circuit Technology*, pp. 20–27, December 1991.
26. W. Blood and A. Dixon, "MCM-L: Cost Effective Multichip Module Substrates," *Inside ISHM*, 20 (2), pp. 7–12, March/April 1993.
27. H. Bollinger, "A Mature DA System for PC Layout," *Proceedings of First International Printed Circuit Conference*, 1979.
28. M. Bregman, "A Thin Film MCM for Workstation Applications," *Proc. 42nd ECTC (San Diego)*, p. 968, 1992.
29. D. Burdeaux, "Benzocyclobutene (BCB) Dielectrics for the Fabrication of High Density, Thin Film Multichip Modules," *J. Electronic Materials*, 19(12), pp. 1357–1366, 1990.
30. D. Carey, "Programmable Multichip Module Technology," *Hybrid Circuit Technology*, p. 25, August 1989.
31. D. H. Carey, "Trends in Low-Cost, High-Performance Substrate Technology," *IEEE Micro*, 13(2), pp. 19–27, April 1993.

32. R. O. Carlson and C. A. Neugebauer, "Future Trends in Wafer Scale Integration," *Proc. IEEE*, 74, pp. 1741–1752, December 1986.

33. P. C. Chan, "Design Automation for Multichip Module-Issues and Status," *International Journal of High Speed Electronics*, 2(4), pp. 55–77, 1991.

34. C. S. Chang, "An Overview of Computer Packaging Architecture and Electrical Design," *IEEE Southern Tier Technical Conf*, pp. 239–257, April 25, 1990.

35. A. Chatterjee and R. Hartley, "A New Simultaneous Circuit Partitioning and Chip Placement Approach Based on Simulated Annealing," *Proceedings of Design Automation Conference*, pp. 36–39, 1990.

36. L. P. Chew, "Constrained Delaunay Triangulation," *Algorithmica*, pp. 97–108, 1989.

37. J. D. Cho and M. Sarrafzadeh, "The Pin Redistribution Problem in Multichip Modules," *Proceedings of Fourth Annual IEEE International ASIC Conference and Exhibit*, pp. 9-2.1–9-2.4, September 1991.

38. R. C. Chu, U. P. Hwang, and R. E. Simons, "Conduction Cooling for an LSI Package: A One-Dimension Approach," *IBM Journal of Research and Development*, 26(1), pp. 45–54, January 1982.

39. B. T. Clark and Y. M. Hill, "IBM Multichip Multilayer Ceramic Modules for LSI Chips: Design for Performance and Density," *IEEE Trans. Components, Hybrids, Manuf. Technol.*, 3(1), pp. 89–93, 1980.

40. R. Cole and A. Siegel, "River Routing Every Which Way, But Loose," *Proceedings of 25th Annual Symposium on Foundation of Computer Science*, pp. 65–73, 1984.

41. P. A. Collier, "Chip Attach for Silicon Hybrid Multi-Chip Modules," *8th IEMT 1990. International Electronic Manufacturing Technology Symposium*, pp. 53–62, May 1990.

42. Terry Costlow, "IBM Rolls Known-Good Dice to Market," *Electronic Engineering Times*, p. 18, October 1994.

43. W. M. Dai, "Multichip Routing and Placement," *IEEE Spectrum*, pp. 61–64, 1992.

44. W. W. Dai, T. Dayan, and D. Staepelaere, "Topological Routing in SURF: Generating a Rubber-Band Sketch," *Proceedings for the 28th Design Automation Conference*, pp. 39–44, 1991.

45. W. W. Dai, R. Kong, and J. Jue, "Rubber Band Routing and Dynamic Data Representation," *Proceedings for 1990 International Conference on Computer Aided Design*, pp. 52–55, 1990.

46. W. W. Dai, R. Kong, and M. Sato, "Routability of a Rubber-band Sketch," *Proceedings of the 28th Design Automation Conference*, pp. 45–48, 1991.

47. E. Davidson, "The Coming of Age of MCM Packaging Technology," *Proc 1st Int MCM Conf*, p. 103, 1992.

48. E. E. Davidson, "Physical and Electrical Design Features of the IBM Enterprise system/9000 Circuit Module," *IBM Journal of Research and Development*, 36(5), pp. 867–875, September 1992.

49. DAZIX, "MCM Engineer Product Description," *DAZIX*.

50. V. K. De, *A Heuristic Global Router for Polycell Layout*, PhD thesis, Duke University, 1986.

51. W. A. Dees, and P. G. Karger, "Automated Rip-Up and Reroute Techniques," *Proceedings of Design Automation Conference*, 1982.

52. Electronics Packaging Technology Department, *The Multichip Module Advantage*, Environmental Research Institute of Michigan (ERIM), 1992.

53. U. Deshpande, S. Swamolian, and G. Howell, "High Density Interconnect Technology for VAX 9000 System," *Proc. 10th IEPS*, p. 46, 1990.

54. A. M. Dewey and S. W. Director, *Principles of VLSI System Planning: A Framework for Conceptual Design*, Kluwer Academic Publishers, 1990

55. E. W. Dijkstra, "A Note on Two Problems in Connection with Graphs," *Numerische Mathematik*, 1, pp. 269–271, 1959.

56. D. DiMatteo and K. Willis, "High-Speed Digital System Design," *Printed Circuit Design*, pp. 17–24, 1994.

57. G. Dishon, "AOI for MCMs, Enhancing Quality and Yield," *Surface Mount Technology*, pp. 33–37, March 1991.

58. D. A. Doane and P. D. Franzon, *Multichip Module Technologies and Alternatives: The Basics*, Van Nostrand Reinhold, 1993.

59. M. Donlin, "MCMs Push Design and Test Tools to the Limit," *Computer Design*, pp. 59–65, 1992.

60. T. Donovan, M. Nealon, and K. Puttlitz, "MCM Technology: Design for Optimized Applications," *Proceedings of the Third Annual PCB Design Conference*, pp. 113– 138, 1994.

61. M. D. Edwards, *Automatic Logic Synthesis Techniques for Digital Systems*, McGraw-Hill, 1992.

62. R. C. Enck, "High Thermal Conductivity Substrate Materials," *Proc. NEPCON (Anaheim, CA)*, pp. 1153–1161, February 1990.

63. J. Enloe, "Design Concepts in AIN Packaging," *Proc. 1991 ISHM*, p. 455, 1991.

64. A. A. Evans and J. K. Hagge, "Advanced Packaging Concepts – Microelectronics Multiple Chip Modules Utilizing Silicon Substrates," *Proc. SAMPE Electronic Material and Process Conf.*, pp. 37–45, June 1987.

65. D. F. Fayette, "Reliability Technology To Achieve Insertion of Advanced Packaging (RELTECH) Program," *ICEMM Proceedings*, pp. 514–519, 1993.

66. C. M. Fiduccia and R. M. Mattheyses, "A Linear-Time Heuristics for Improving Network Partitions," *Proceedings of the 19th Design Automation Conference*, pp. 175–181, 1982.

67. P. D. Franzon and R. J. Evans, "A Multichip Module Design Process for Notebook Computers," *Computer*, 26(4), pp. 41–49, April 1993.

68. P. Garrou, "Thin Film Interconnection Technology Boosts MCM Performance," *Electronic Packaging and Production*, pp. 31–36, October 1992.

69. J. D. Giacomo, *VLSI Handbook*, McGraw-Hill, 1989.

70. J. D. Giacomo, *VLSI Handbook: Silicon, Gallium Arsenide, and Superconductor circuits*, McGraw Hill, 1989.

71. J. D. Giacomo, *Designing With High Performance ASICs*, Prentice Hall, 1992.

72. B. K. Gilbert and W. L. Walters, "Design Guidelines for Digital Multichip Modules Operating at High System Clock Rates," *The International Journal of Microcircuits and Electronic Package*, 15(4), Fourth Quarter 1992.

73. M. K. Goldberg and M. Burstein, "Heuristic Improvement Technique for Bisection of VLSI Networks," *IEEE International Conference on Computer Design*, pp. 122–125, 1983.
74. S. D. Golladay, "Electron-Beam Technology for Open/Short Testing of Multichip Substrate," *IBM Journal of Research and Development*, pp. 250–259, March/May 1990.
75. S. Goto, "An Efficient Algorithm for the Two-Dimensional Placement Problem in Electrical Circuit Layout," *IEEE Trans. Circuits Syst.*, CAS-28, pp. 12–18, January 1981.
76. D. Grabbe, "Modupak: A New, Low Cost, High Speed MCM Socketing System," *Proc. NEPCON (Anaheim, CA)*, pp. 1537–1538, February 1991.
77. Mentor Graphics, "MCM Station Product Description," *Mentor Graphics*, 1992.
78. J. Greene and K. Supowit, "Simulated Annealing Without Rejected Moves," *Proceedings of International Conference on Computer Design*, pp. 658–663, October 1984.
79. L. K. Grover, "Standard Cell Placement Using Simulated Sintering," *Proceedings of the 24th Design Automation Conference*, pp. 56–59, 1987.
80. T. Gucciardi and H. Green, "Bare Die Testing: Carrier Technology Aids MCM Quality," *Advanced Packaging*, pp. 26–28, March/April 1994.
81. J. Hagge, "State of the Art Multichip Modules for Avionics," *IEEE Trans. CHMT*, p. 25, 1992.
82. J. K. Hagge, "Mechanical Considerations for Reliable Interfaces in Next Generation Electronics," *Proc. NEPCON (Dayton, OH)*, 4, pp. 2021–2026, May 1989.
83. J. K. Hagge, "Ultra-reliable HWSI With Aluminum Nitride Packaging," *Proc. NEPCON (Anaheim, CA)*, pp. 1271–1283, March 1989.
84. J. K. Hagge, "Ultra-Reliable Packaging for Silicon-on-Silicon WSI," *IEEE Trans. Components, Hybrid, Manuf. Technol.*, 12, pp. 170–179, June 1989.
85. J. K. Hagge and R. J. Wagner, "High-Yield Assembly of Multichip Modules through Known-Good IC's and Effective Test Strategies," *Proceedings of the IEEE*, 80(12), pp. 1965–1994, December 1992.
86. J. K. Hagge and R. J. Wagner, "Known-Good ICs for MCM Assembly," *Thin Film Multichip Modules*, 1992.
87. B. Hajek, "Cooling Schedules for Optimal Annealing," *Oper. Res.*, pp. 311–329, May 1988.
88. E. Hamdy, "Dielectric Based Antifuse for Logic and Memory ICs," *International Electron Devices Meeting Technical Digest*, pp. 786–789, 1988.
89. M. Hanan and J. M. Kurtzberg, "A Review of Placement and Quadratic Assignment Problems," *SIAM Rev.*, 14(2), pp. 324–342, April 1972.
90. M. Hanan, P. K. Wolff, and B. J. Agule, "Some Experimental Results on Placement Techniques," *J. Design Automation and Fault-Tolerant Computing*, 2, pp. 145–168, May 1978.
91. B. Hargis, "Ceramic Multichip Modules (MCM-C)," *Inside ISHM*, 20(2), pp. 15–18, March/April 1993.
92. M. Hatamian, "Fundamental Interconnection Issues," *AT&T Tech. J.*, 66(4), pp. 13–30, July/Aug. 1987.

93. L. M. Higgins, "Material, Manufacturing and Assembly Considerations for Laminate Substrate-Based Multichip Modules (MCM-L)," *ISHM '92 Proceedings*, pp. 216–224, 1992.

94. C. Hilbert and C. Rathmell, "Design and Testing of High Density Interconnection Substrates," *Proc. NEPCON*, pp. 1391–1403, February 1990.

95. I. Hirabayashi, "Microelectronic Technology Takes on Hybrid ICs," *JEE*, pp. 58–61, October 1991.

96. J. M. Ho, M. Sarrafzadeh, G. Vijayan, and C. K. Wong, "Layer Assignment for Multichip Modules," *IEEE Transactions on Computer-Aided Design*, 9(12), pp. 1272–1277, December 1990.

97. T. L. Hodson, "Bonding Alternatives for Multichip Modules," *Electronic Packaging and Production*, 32(4), pp. 38–40, April 1992.

98. X. Hong, T. Xiong, C. K. Cheng, E. S. Kuh, and J. Huang, "Performance-Driven Steiner Tree Algorithms for Global Routing," *Research Report, EECS/ERL, UC Berkeley*, p. 1, 1992.

99. T. Horton, "Multichip Modules, Applications, Driving Forces and Future Directions," *Proc. NEPCON (Anaheim, CA)*, 487–494, February 1991.

100. C. Huang, "Silicon Packaging: A New Technique," *Proc. IEEE Custom Int. Circuit Conf.*, pp. 142–146, May 1983.

101. J. Huang, X. Hong, C. Cheng, and E. S. Kuh, "An Efficient Timing-Driven Global Routing Algorithm," *30th ACM/IEEE Design Automation Conference*, pp. 596–600, June 1993.

102. M. D. Huang, F. Romeo, and A. Sangiovanni-Vincentelli, "An Efficient General Cooling Schedule for Simulated Annealing," *Proceedings of the IEEE International Conference on Computer-Aided Design*, pp. 381–384, 1986.

103. T. Inoue, "Microcarrier for LSI Chip Used in the HITAC M-880 Processor Group," *Proc. 41th ECTC (Atlanta)*, p. 704, 1991.

104. A. Iqbal, M. Swaminathhan, M. Nealon, and A. Omer, "Design Tradeoffs Among MCM-C, MCM-D and MCM-D/C," *IEEE Proceedings*, pp. 12–17, 1993.

105. T. Isaac, "Staggered, Staircased, and Stepped (Vias for Multichip Modules)," *Printed Circuit Design*, pp. 42–48, February 1991.

106. R. J. Jensen, "Copper/Polymide Materials System for High Performance Packaging," *IEEE Trans. Components, Hybrid, Manuf. Technol.*, 7, pp. 384–393, December 1984.

107. R. J. Jensen, "Polyimides as Interlayer Dielectrics for High Performance Interconnections of Integrated Circuits," *American Chemists Society Symposium Series*, pp. 466–483, 1987.

108. R. W. Johnson, R. K. Teng, and J. W. Balde, *Multichip Modules: Systems Advantages, Major Constructions, and Materials Technologies*, IEEE Press, 1990.

109. H. Kent, "Multichip Module Connectors and Sockets," *Proc. IEPS (San Diego, CA)*, pp. 726–740, September 1991.

110. W. Kernighan and S. Lin, "An Efficient Heuristic Procedure for Partitioning Graphs," *Bell System Technical Journal*, 49, pp. 291–307, 1970.

111. K.-Y. Khoo and J. Cong, "A Fast Multilayer General Area Router for MCM Designs," *IEEE Transactions on Circuits and Systems*, 1992 (in press).

112. K.-Y. Khoo and J. Cong, "An Efficient Multilayer MCM Router Based on Four-Via Rputing," *Design Automation Conference*, 36, pp. 590–595, June 1993.

113. S. Kirkpatrick, C. D. Gellat, and M. P. Vechhi, "Optimization by Simulated Annealing," *Science*, 220, pp. 671–680, May 1983.

114. T. Kishimoto, and T. Ohsmaki, "VLSI Packaging Technique Using Liquid Cooled Channels," *Proc. 1986 IEEE Electronic Component Conf.*, pp. 595–601, May 1986.

115. M. Klein, "MCM Design: A Review," *Printed Circuit Design*, pp. 10–14, June 1994.

116. N. G. Koopman, *Microelectronics Packaging Handbook*, Van Nostrand Reinhold, 1989.

117. A. Kozak, "MCMs in Telecommunication Designs," *Surface Mount Technology*, pp. 46–49, Mar. 1991.

118. P. Kraynak and P. Fletcher, "Wafer-Chip Assembly for Large-Scale Integration," *IEEE Trans. Electronic Devices*, pp. 660–663, September 1968.

119. C. Kring and A. R. Newton,"A Cell-Replicating Approach to Mincut-based Circuit Partitioning.," *Proceedings of IEEE International Conference on Computer-Aided Design*, pp. 2–5, November 1991.

120. B. Krishnamurthy, "An Improved Mincut Algorithm for Partitioning VLSI Networks," *IEEE Transactions on Computers*, pp. 438–446, 1984.

121. J. P. Krusius, "System Interconnection of High Density Multi-Chip Modules," *International Conference on Advances in Interconnection and Packaging*, 1390, pp. 261–270, 1990.

122. J. P. Krusius and W. E. Pence, "Analysis of Materials and Structure Tradeoffs in Thin and Thick Film Multichip Packages," *Electronic Components Conference*, pp. 641–646, 1989.

123. K. Kucukcakar and A. C. Parker, "CHOP: A Constraint-Driven System-Level Partitioner," *Design Automation Conference*, pp. 514–519, June 1991.

124. J. Lam and J. Delosme, "Perfomance of a New Annealing Schedule," *Proceedings of the 25th Design Automation Conference*, pp. 306–311, 1988.

125. B. S. Landman and R. L. Russo, "On a Pin Versus Block Relationship for Partitions of Logic Graphs," *IEEE Transactions on Computers*, C-20, pp. 1469–1479, December 1971.

126. D. P. LaPotin, "Early Assessment of Design, Packaging and Technology Tradeoffs," *International Journal of High Speed Electronics*, 2(4), pp. 209–233, 1991.

127. D. P. LaPotin, T. R. Mazzawy, and M. L. White, "Early Package Analysis: Considerations and Case Studies," *Computer*, pp. 30–40, 1993.

128. C. Y. Lee, "An Algorithm for Path Connections and Its Applications," *IRE Transactions on Electronic Computers*, 1961.

129. Y. C. Lee, H. T. Ghaffari, and J. M. Segelken, "Internal Thermal Resistance Of a Multichip Packaging Design For VLSI-Based Systems," *Proc. 38th Electron. Components Conf. (ECC)*, pp. 293–301, 1988.

130. C. E. Leiserson and F. M. Maley, "Algorithms for Routing and Testing Routability of Planar VLSI Layouts," *Proceedings of the 17th Annual ACM Symposium on Theory of Computing*, pp. 69–78, 1985.

131. H. J. Levinstein, "Multichip Packaging Technology for VLSI-based Systems," *Proc. IEEE Int. Solid State Circuits Conf.*, pp. 224–225, February 1987.

132. K. F. Liao, M. Sarrafzadeh, and C. K. Wong, "Single-Layer Global Routing," *Proceedings of the Fourth Annual IEEE International ASIC Conference and Exhibit*, pp. 14-4.1–14-4.4, 1991.

133. M. Mahalingam, "Thermal Management in Semiconductor Device Packaging," *Proc. of the IEEE*, 73(9), pp. 1396–1404, September 1985.

134. L. Maliniak, "CAD Frameworks ride a Rough Road to Success," *Electronic Design*, pp. 36–42, August 6, 1992.

135. L. Maliniak, "ESDA Boosts CAE Technology to Higher Levels," *Electronic Design*, pp. 61–72, December 2, 1993.

136. L. Maliniak, "Constraint-Driven Tools Lay Out High-End Designs," *Electronic Design*, pp. 139–140, February 1992.

137. P. P. Marcoux, *Fine Pitch Surface Mount Technology*, Van Nostrand Reinhold, 1992.

138. J. C. Mather and J. K. Hagge, "Material Requirements for Packaging Multichip Modules," *Proc. NEPCON (Anaheim, CA)*, pp. 1135–1142, February 1990.

139. R. Maurer, "The AT&T MCM Packaging Program," *Proc. 1st Int. MCM Conf.*, p. 28, 1992.

140. T. Mazzullo, "Thin Film Hybrid Circuits: Requirements for Production Tooling," *Printed Circuit Design*, pp. 16–19, September 1989.

141. T. Mazzullo, "Introduction to MCM Design," *Proceedings of the Third Annual PCB Design Conference*, pp. 85–92, 1994.

142. J. F. McDonald, "The Trail of Wafer-Scale Integration," *IEEE Spectrum*, pp. 32–39, October 1984.

143. M. C. McFarland, A. C. Parker, and R. Camposano, "Tutorial on High-Level Synthesis," *25th Design Automation Conference*, pp. 330–336, 1988.

144. B. T. Merriman, "New Low Coefficient of Thermal Expansion Polyimide for Inorganic Substrates," *Proc. 39th ECC*, pp. 159–159, May 1989.

145. G. Messner, "Laminate Technology for Multichip Modules," *Electronic Packaging and Production*, pp. 32–40, October 1992.

146. N. Metropolis, A. Rosenbluth, and M. Rosenbluth, "Equation of State Calculations by Fast Computing Machines," *Journal of Chemistry and Physics*, pp. 1087–1092, 1953.

147. J. Nelson, "Implementing MCM-C Technology," *Electronic Packaging and Production*, pp. 26–30, October 1992.

148. C. A. Neugebauer, "Approaching Wafer Scale Integration from the Packaging Point of View," *Proc. IEEE Int. Conf. Computer Design*, pp. 115–120, October 1984.

149. C. A. Neugebauer, "Comparison of VLSI Packaging Approaches to Wafer Scale Integration," *Proc. IEEE Custom Integrated Circuits Conf.*, pp. 31–37, 1985.

150. C. A. Neugebauer and R. O. Carlson, "Comparison of Wafer Scale Integration with VLSI Packaging Approaches," *IEEE Transactions on Components, Hybrids and Manufacturing Technology*, Vol. CHMT-10, No. 2, pp. 184–189, June 1987.

151. A. R. Newton, "Design Technology Requirements for Multi-Chip Modules," *Wescon Conference Record*, pp. 534–538, 1991.

152. C. Niessen, "Hierarchical Design Methodologies and Tools for VLSI Chips," *Proceedings of the IEEE*, 71(1), pp. 66–75, 1983.

153. J. Novellino, "MCMs Demand a DFT Strategy," *Electronic Design*, pp. 50–60, October 14, 1993.

154. T. Ohsaki, "Electronic Packaging in the 1990s – A Perspective From Asia," *IEEE Trans. CHMT*, p. 254, 1991.

155. T. Ohtuski, *Partitioning, Assignment and Placement*, North-Holland, 1986.

156. K. Okutani, K. Otsuka, K. Sahara, and K. Satoh, "Packaging Design of a SiC Ceramic MultiChip RAM Module," *Proc. 1984 Int Electronic Packaging Society Conf.*, pp. 299–304, 1984.

157. G. W. Pan, F. A. Prentice, S. K. Zahn, A. J. Staniszewski et al. "The Simulation of High-Speed, High-Density Digital Inteconnections in Single Chip Packages and Multichip Modules," *IEEE Transactions on Components, Hybrids, and Manufacturing Technology*, 15(4), pp. 465–477, August 1992.

158. J. E. Pascente, "X-Ray Inspection and Testing of Multichip Modules," *ISHM Proc. (Chicago, IL)*, pp. 352–357, October 1990.

159. M. Pecht and B. T. Sawyer, "PCB Design for Thermal Reliability," *Printed Circuit Design*, pp. 12–16, May 1987.

160. W. Pence, and J. Krusius, "The Fundamental Limits for Electronic Packaging and System," *IEEE Trans CHMT*, p. 176, 1987.

161. W. E. Pence, *Electrical, Thermal, and Architecture Aspects of VLSI Packaging and Interconnects for High Speed Digital Computers*, PhD thesis, Cornell University, 1989.

162. W. E. Pence and J. P. Krusius, "Packaging Design and Analysis via System Analysis," *Proceedings of Electronic Materials and Processing Congress*, pp. 103–108, 1989.

163. L. Pillage, Z. Cendes, and J. White, "Interconnect and Packaging Analysis," *ICCAD-92 Tutorial* 2, 1992.

164. S. Prasitjutrakul and W. J. Kubits, "A Timing-Driven Global Router for Custom Chip Design," *Proceedings of the Design Automation Conference*, pp. 48–51, 1990.

165. B. Preas, M. Pedram, and D. Curry, "Automatic Layout of Silicon-on-Silicon Hybrid Packages," *Proceedings of IEEE Design Automation Conference*, pp. 394–399, 1989.

166. N. R. Quinn, "The Placement Problem as Viewed from the Physics of Classical Mechanics," *Proceedings of the 12th Design Automation Conference*, pp. 173–178, 1975.

167. N. R. Quinn, and M. A. Breuer, "A Force Directed Component Placement Procedure for Printed Circuit Boards," *IEEE Trans. Circuits and Syst.*, pp. 377–388, June 1979.

168. RACAL-REDAC, "MCM Design Suite Product Description," *RACAL-REDAC*, 1994.

169. J. Reche, "High Density Multichip Module Fabrication," *Int. J. Hybrid Microelectronics*, 13(4), pp. 91–99, December 1990.

170. M. L. Resnick, "SPARTA: A System Partitioning Aid," *IEEE on Computer-Aided Design*, pp. 490–498, October 1986.

171. M. L. Resnick, "SPARTA: A System Partitioning Aid," *IEEE Transactions on Computer-Aided Design*, pp. 490–498, 1986

172. F. Romeo and A. Sangiovanni-Vincentelli, "Convergence and Finite Time Behavior of Simulated Annealing," *Proceedings of The 24th Conference on Decision and Control*, pp. 761–767, 1985.
173. F. Romeo, A. S. Vincentelli, and C. Sechen, "Research on Simulated Annealing at Berkeley," *Proceedings of IEEE International Conference on Computer Design*, pp. 652–657, 1984.
174. R. Rossi and P. Machiesky, "Mechanical Properties of a New Polyimide Thin Film Dielectric Interlayer for Multichip Modules," *Solid State Technology*, pp. S1–S4, February 1990.
175. T. Sakurai, "Approximation of Wiring Delay in MOSFET LSI," *IEEE Journal of Solid-State Circuits*, pp. 418–426, 1983.
176. P. A. Sandborn, "A Software Tool for Technology Tradeoff Evaluation in Multichip Packaging," *Proceedings of the Eleventh International Electronics Manufacturing Technology Symposium (IEMT)*, pp. 337–341, 1991.
177. P. A. Sandborn, K. Drake, and R. Ghosh, "Computer Aided Conceptual Design of Multichip Systems," *Proceedings of the Custom Integrated Circuits Conference*, pp. 29.4.1–29.4.4, September 1993.
178. P. A. Sandborn and H. Moreno, *Conceptual Design of Multichip Modules and Systems*, Kluwer Academic Publishers, 1993.
179. G. Saucier, D. Brasen, and J. P. Hiol, "Partitioning With Cone Structures," *Proceedings of IEEE International Conference on Computer-Aided Design*, pp. 236–239, November 1993.
180. A. Schlitz, A. Vareille, and L. Trevenot, "Chip Insertion Techniques for Multichip Packaging," *Proc. 1st Int. MCM Conf.*, p. 9, 1992.
181. D. G. Schweikert and B. Kernighan, "A Proper Model for the Partitioning of Electrical Circuits," *Proceedings of the 9th Design Automation Workshop*, pp. 57–62, 1972.
182. C. Sechen and K. W. Lee, "An Improved Simulated Annealing Algorithm for Row-Based Placement," *Proceedings of the IEEE International Conference on Computer-Aided Design*, pp. 478–481, 1987.
183. C. Sechen and A. Sangiovanni-Vincentelli, "The Timber Wolf Placement and Routing Package," *IEEE Journal of Solid-State Circuits*, Sc-20, pp. 510–522, 1985.
184. N. A. Sherwani, *Algorithms for VLSI Design Automation*, Kluwer Academic Publishers, 1993.
185. M. Shih, E. S. Kuh, and R. Tsay, "Integer Programming Techniques for Multiway System Partitioning Under Timing and Capacity Constraints," *Proceedings European DAC*, pp. 294–298, February 1993.
186. M. Shin, E. S. Kuh, and R. Tsay, "High-Performance-Driven System Partitioning on MultiChip Modules," *Proceedings of 29th Design Automation Conference*, pp. 53–56, June 1992.
187. R. E. Sigliano, "Thin Film MCM-D: The Future of Performance? " *Inside ISHM*, 20(2), pp. 21–25, March/April 1993.
188. M. Sono, "Packaging Technology for ASICs," *Fujitsu Scientific Tech. J.*, 4, pp. 432–445, December 1988.
189. J. Soukup, "Fast Maze Router," *Proceedings of 15th Design Automation Conference*, pp. 100–102, 1978.

190. R. K. Spielberger, "Silicon-on-Silicon Packaging, " *IEEE Trans. Components, Hybrid, Manuf. Technol.*, 7, pp. 193–196, June 1984.

191. M. Sriram and S. M. Kang, "A New Layer Assignment Approach for MCMs," *Technical Report UIUC-BI-VLSI-92-01*, The Beckman Institute, University of Illinois at Urbana-Champaign, 1992.

192. W. Steigrandt and M. F. Ehman, "MCM's Impact on User-Supplier Relationships in the 1990's," *Hybrid Circuit Technology*, pp. 24–29, March 1991.

193. Cadence Design Systems, "Allegro Product Description," *Cadence Design Systems*, 1994.

194. T. G. Tessier, "Overview of MCM Technologies: MCM-D," *ISHM 92 Proc.*, pp. 235–247, 1992.

195. T. G. Tessier and P. Garrou, "Overview of MCM Technologies: MCM-D," *ISHM '92 Proceedings*, pp. 235–247, 1992.

196. S. K. Tewksbury, *Wafer-Level System Integration: Implementation Issues*, Kluwer Academic Press, Boston, 1989.

197. K. Tokouchi, N. Kamahara, and K. Niwa, "Packaging Technology for High Speed Computers-Multilayer Glass/Ceramic Circuit board," *Proc. Int. Symp. Microelectronics (ISHM)*, pp. 183–186, 1991.

198. P. A. Trask, "High Density Multichip Interconnect: Military Packaging System for the 1990s," *SPIE/OPTCON-90*, pp. 1390–1408, November 1990.

199. J. R. Trent, "Test Philosophy for Multichip Modules," *The International Journal of Microcircuits and Electronic Packaging*, 15(4), 1992.

200. D. B. Tuckerman, *Heat-Transfer Microstructures for Integrated Circuits*, PhD thesis, Standford University, 1984.

201. R. R. Tummala, "Multichip Packaging: A Tutorial, " *Proceedings of the IEEE*, 80(12), pp. 1924–1941, December 1992.

202. R. R. Tummala, H. R. Potts, and S. Ahmed, "Packaging Technology for IBM's Latest Mainframe Computers, " *Proc. 41th Electronic Components and Technology Conf.*, pp. 682–688, 1991.

203. R. R. Tummala and E. J. Rymaszewski, *Microelectronics Packaging Handbook*, Van Nostrand Reinhold, 1989.

204. R. A. Walker and R. Camposano, *Logic Design for Array-Based Circuits: A Structured Design Methodology*, Academic Press, 1992.

205. G. R. Weihe, "High Density Multichip Solutions, " *Proc. NEPCON (Anaheim, Ca)*, pp. 1121–1129, February 1991.

206. N. H. E. Weste and Kamran Eshraghian, *Principles of CMOS VLSI Designing*, Addison-Wesley Publishing Company, 1992.

207. D. E. White, *A Survey of High-Level Synthesis Systems*, Kluwer Academic Publishers, 1991.

208. T. W. Williams, and K. P. Parker, "Design for Testability: A Survey," *Proceedings of the IEEE*, 71(1), pp. 98–112, 1983.

209. D. K. Wilson, "Topological Aspects of Systems Partitioning, " *Design Policy Conference*, pp. 148–154, 1982.

210. Q. Yu, S. Badida, and N. Sherwani, *A Performance-Driven Three-Dimensional Approach for MCM Routing*, *Technical Report*, Department of Computer Science, Western Michigan University, September 1993.

211. G. W. Zobrist, *VLSI Fault Modeling and Testing Techniques*, Ablex Publishing Corporation, 1992.

INDEX